Spatio-Temporal Models for Ecologists

Ecological dynamics are tremendously complicated and are studied at a variety of spatial and temporal scales. Ecologists often simplify analysis by describing changes in density of individuals across a landscape, and statistical methods are advancing rapidly for studying spatio-temporal dynamics. However, spatio-temporal statistics is often presented using a set of principles that may seem very distant from ecological theory or practice. This book seeks to introduce a minimal set of principles and numerical techniques for spatio-temporal statistics that can be used to implement a wide range of real-world ecological analyses regarding animal movement, population dynamics, community composition, causal attribution, and spatial dynamics. We provide a step-by-step illustration of techniques that combine core spatial-analysis packages in R with low-level computation using Template Model Builder. Techniques are showcased using real-world data from varied ecological systems, providing a toolset for hierarchical modelling of spatio-temporal processes. *Spatio-Temporal Models for Ecologists* is meant for graduate level students, alongside applied and academic ecologists.

Key Features:

- Foundational ecological principles and analyses
- Thoughtful and thorough ecological examples
- Analyses conducted using a minimal toolbox and fast computation
- Code using R and TMB included in the book and available online

James T. Thorson is a statistical ecologist at the Alaska Fisheries Science Center within the National Marine Fisheries Service. His research interests include population dynamics, life-history theory, and methods for the sustainable management of natural resources. He has taught graduate-level courses in hierarchical modelling and spatio-temporal statistics at University of Washington.

Kasper Kristensen is a Senior Researcher at Danish Technical University. His research interests include spatio-temporal statistics and computational methods. He developed the R-package TMB, which is seeing increased use throughout ecology. For example, TMB is the computational backend for R-package glmmTMB, which has been cited over 3000 times from 2017-2022.

Chapman & Hall/CRC
Applied Environmental Series

Series Editors

Douglas Nychka, *Colorado School of Mines*
Alexandra Schmidt, *Universidade Federal do Rio de Janero*
Richard L. Smith, *University of North Carolina*
Lance A. Waller, *Emory University*

For more information about this series, please visit: https://www.crcpress.com/
Chapman--HallCRC-Applied-Environmental-Statistics/book-series/CRCAPPENVSTA

Spatio-Temporal Models for Ecologists

James T. Thorson and Kasper Kristensen

CRC Press

Taylor & Francis Group

Boca Raton London New York

CRC Press is an imprint of the
Taylor & Francis Group, an **informa** business

A CHAPMAN & HALL BOOK

Contents

Preface

Many of the largest problems facing society must be confronted using information developed by scientists. Major scientific problems often involve measurements and processes that vary over space and time, and research regarding methods for spatio-temporal analysis has been published across a wide range of disciplines. Perhaps for this reason, there is a proliferation of vocabulary, statistical methods, and computational machinery used for spatio-temporal analysis. This wide range of vocabulary and methods can be daunting for anyone who is interested in learning about these methods.

In this textbook, we aim to provide a gentle introduction to spatio-temporal analysis for advanced undergraduate and graduate-level students, as well as scientists conducting theoretical or applied research regarding ecology, earth sciences, and public health. The presentation involves code in the R statistical environment highlighted in green [184] as well as C++ code that is compiled by Template Model Builder highlighted in blue [122]. The presentation assumes conceptual familiarity with linear algebra and differential equations. However, the reader is not expected to solve equations analytically. The text instead presents a minimal toolbox for numerical analysis and computational statistics, where the analyst writes high-level code describing a given model and then flexible software automates the derivatives and integrals required for hierarchical modelling of spatio-temporal processes.

The book is intended to cover a logical sequence from foundational to advanced topics, with ecological and statistical concepts introduced in parallel (Table 0.1). In particular:

- *Introductory*: the first two chapters introduce the ecological and statistical foundation for subsequent chapters. Chapter 1 shows how to approximate individual dynamics (a Lagrangian viewpoint) using discretized statistical models (an Eulerian viewpoint), and Chapter 2 introduces how to efficiently estimate parameters for hierarchical models using the Laplace approximation;

- *Basic*: Chapters 3 then introduces univariate time-series models using population-dynamics examples, while Chapter 4 extends this to multivariate time-series models and analyzes individual movement from a Lagrangian viewpoint. Chapters 5 introduces computationally efficient methods for two-dimensional spatial correlations, and Chapter 6 discusses applications within the context of sampling theory and spatial integration to estimate population abundance. Finally, Chapter 7 introduces the use of covariates for ecological inference, as well as their role when combining data from multiple sampling protocols;

- *Advanced*: Chapter 8 is the first to introduce an interaction of spatial and temporal variation, and discusses the distinction between interannual and seasonal variation. Chapter 9 then introduces the importance of ecological and physical teleconnections, and uses both exploratory and confirmatory factor models for their analysis. Chapter 10 returns to the topic of movement from an Eulerian viewpoint, and presents efficient methods to estimate a habitat preference function using individual tracks and point-count data. Finally, Chapter 11 concludes by demonstrating how multivariate spatio-temporal models can be fitted using covariates, phylogenetic, and trait information, and how results can be used to analyze ecological communities.

TABLE 0.1: Book chapters, including their ecological focus and the statistical skills. Ecological topics are sequenced such that statistical skills build upon those from past chapters.

Chap.	Ecological focus	Statistical skill
1	Individual demography and habitat usage	Generalized linear models
2	Variable population density among habitat patches	Hierarchical models; Laplace approximation
3	Population dynamics	Univariate state-space models
4	Animal movement	Multivariate state-space models
5	Spatial variation unexplained by covariates	Two-dimensional spatial covariance
6	Spatial inference and sampling designs	Spatial integration and preferential sampling
7	Benefits of covariates in ecological analysis	Causal analysis using structural equation models; Integrated species distribution models
8	Spatio-temporal inference including seasonal and interannual variation	Cyclic splines and separable space-time covariance
9	Physical and ecological teleconnections	Exploratory factor analysis using empirical orthogonal functions; confirmatory factor analysis using spatially varying coefficients
10	Linking individual and population-scale movement	Specifying and solving a partial differential equation for movement using a continuous time Markov Chain; Fitting this model to individual tracks and point-count samples
11	Community assembly and biogeography	Including trait and phylogenetic information in joint species distribution models

We typically assume that readers are familiar with content from earlier chapters, but also include links to earlier content for those who read chapters out of sequence. We seek to demonstrate concepts using an interesting and varied set of real-world examples (Table 0.2) drawn from earth science (oceanography and atmospheric science), ecology (population dynamics, community ecology, and ecosystem modelling), and global change biology (climate and health science). All software, data, and resulting figures/tables are available online[1], and we encourage readers to modify and repurpose this code for other analyses. Despite our

---

[1]See https://github.com/james-thorson/Spatio-temporal-models-for-ecologists/ which is organized by chapter

TABLE 0.2: Case-study data used in this book, and the scientific discipline that is illustrated by their analysis.

Case study	Scientific focus and discipline
Barro Colorado vegetation census plots	Spatial dynamics
Bering Sea fish population dynamics	Spatial and population dynamics; Climate science
Northern fur seal track using satellite tag	Movement ecology
Bald eagle counts in Breeding Bird Survey	Spatial dynamics; Movement ecology
Ozone concentrations	Atmospheric science; Human health
Multiple survey methods for red snapper in the Gulf of Mexico	Spatial dynamics; Sampling methods
Arctic sea ice concentrations	Oceanography; Climate science
Pacific cod archival tag in the Aleutian Islands	Movement ecology; Oceanography
Habitat utilization for twenty birds in Western US states	Community ecology; Evolutionary analysis

best efforts, we also anticipate that the printed version will include bugs, and that changes in software dependencies will cause the code as printed to stop working. We intend to make corrections and updates to code available via GitHub, while using numbered releases to distinguish different versions.

We envision that many statistical ecologists are presented with an ever-increasing hodge-podge of statistical and computational techniques. However, limits to individual and collective memory suggest that only a small and general set of techniques will continue to be used over time. We therefore emphasize the importance of a *minimal toolbox*, which allows ecologists to fit a wide range of ecological analyses using a small set of computational and software skills. To see the importance of a minimal toolbox, consider the wide use of the Generalized Linear Model [162] in both introductory statistics courses and applied ecology. This simple model structure (introduced in Chapter 1) provides ecologists with a single procedure for model fitting, testing, and evaluation, which then includes analysis of variance, contingency tables, probit analysis, and linear regression as special cases. In this textbook, we propose a minimal toolbox for spatio-temporal analysis of ecological data, and in fact, proceed by extending the generalized linear model framework itself.

To complement this minimal toolbox, we also present a *comprehensive word bank* that is associated with spatio-temporal models. This includes a huge range of specialized terms, e.g., infill and sprawl asymptotics, separable models, Gaussian processes, and semivariograms. We feel that this large word bank is important, both as a starting point for readers to follow up on individual topics, and to emphasize foundational concepts and methodologies. Rather than defining these terms mathematically, we aim to introduce them via ecological examples so that readers can immediately see what insight each term can provide. We then encourage interested readers to use this word bank as a starting point for further specialized reading (which presumably will then provide further mathematical details).

Based on our experience, we have chosen to use Template Model Builder [122] and the R statistical platform [184] to develop the minimal toolbox. Specifically, we include TMB because:

- *Flexibility*: it facilitates fitting a wide range of linear and nonlinear models that give rise to correlations among variables across space and time while giving analysts detailed control over model specification;

- *Simplification*: it includes high-level functions that simplify common operations including the construction of covariance matrices, defining separable spatio-temporal processes, and computing the matrix exponential;

- *Automation*: it can apply the Laplace approximation for maximum-likelihood estimation or Markov chain Monte Carlo for Bayesian estimation, so that the analysts can focus on model specification rather than computational methods;

- *Computational efficiency*: it automatically computes gradients (using automatic differentiation) while detecting model sparsity (for efficient use of the Laplace approximation), and provides an interface to the Eigen library for advanced users.

To complement these benefits, we include R because:

- *Familiarity*: it is widely used by ecologists for importing, pre-processing, and visualizing data and model outputs, and therefore provides a general and familiar software interface;

- *Spatial data types*: it includes general functionality for spatial, temporal, and phylogenetic data and therefore simplifies the process of importing, exporting, and plotting these common types of ecological data. We specifically emphasize using the core R-packages sf [175] and terra [90] to read, process, project, and plot spatial data;

- *Extensibility*: it includes a wide range of packages that can be used to post-process the output from custom-built statistical models, and therefore extends the ecological inference that can be efficiently made.

We therefore believe that this combination of R and TMB provides a minimal toolbox for theoretical and applied ecologists, while still allowing for efficient and flexible development of statistical models. Ultimately, we hope that this combination of minimal toolbox and comprehensive word bank allows applied ecologists to engage with and contribute to the growing field of spatio-temporal modelling.

List of Figures

List of Tables

Part I

Introductory

1

Statistical Models for Individual-based Processes

1.1 Ecological Dynamics and Individual-based Models

Ecologists study a tremendous range of topics, including for example:

- How do different mechanisms of genetic mutation contribute to evolutionary dynamics [92]?

- Why do particular traits arise independently in unrelated taxa, and how does this depend upon the environment [183]?

- Under current global emissions of climate-altering gases and aerosols, how will the abundance, distribution, and function of ecosystems change [131]?

- What is the diversity of microbes in the human body, and how does this affect behavior and disease [51]?

These different questions have resulted in different research communities using different sampling and analytical methods, and it can be intimidating to look for any unifying framework that can unite theoretical and applied ecologists.

Despite this complexity, we seek a unified treatment of ecological dynamics. By "treatment" we mean a computational and analytical machinery that allows us to describe the ecological variables of interest, make accurate predictions while measuring our uncertainty, and simultaneously infer the impact of multiple mechanisms. To accomplish these various goals, we generally define two things in the following:

1. *Individuals*, which could be an individual organism, gene, or population;

2. *Marks*, which are properties of each individual that can change over time. These marks can include location, body size, age, maturity or energetic status, or other traits that are relevant to explain ecological patterns.

At its lowest level, we therefore describe dynamics from a *Lagrangian viewpoint*, where we seek to track individuals and how their characteristics (i.e., marks) change over time. However, the Lagrangian viewpoint often becomes cumbersome and computationally impractical when the number of individuals increases. In these cases, we then switch to describing dynamics using an *Eulerian viewpoint*. Instead of tracking individuals and marks, the Eulerian viewpoint tracks the number of individuals having a given mark within a given area. For example, we might use the Lagrangian viewpoint to track the location $S_j(t)$ of five individual $j \in \{1, 2, 3, 4, 5\}$ individuals as a mark over time t. However, with 1000 individuals we might switch to the Eulerian viewpoint to instead track the number $n_{s,t}$ of individuals within each area s at time t. In the following, we typically seek to introduce concepts using

DOI: 10.1201/9781003410294-1

a Lagrangian viewpoint, and then transition to the Eulerian viewpoint when addressing that same concept for a large number of individuals.

In particular, the Lagrangian viewpoint generally involves tracking the exact value of a given mark, e.g., location $S_j(t)$, for individual j at any continuous time t. Changes in the value of each mark (it's "dynamics") are then typically represented by some differential equation, e.g., where the location of animal $S_j(t)$ might change as it moves in the direction of preferred habitats, or the size $W_j(t)$ changes as it grows. In some special cases, it might be possible to calculate dynamics exactly, e.g., by solving the differential equation analytically. In practice, however, an ecologist will often specify dynamics for which there is no known exact solution, so dynamics must be approximated. In other cases, an ecologist might want to explore different specifications for dynamics that involve different types of analytical solution, and may not want to learn different strategies to solve each version. This again leads us to seek a general approach to approximate the outcome of a specified ecological process. Approximations can vary greatly from high to low quality, where poor-quality approximations will result in large errors relative to a hypothetical exact solution. In the following, we generally seek to approximate individual-based dynamics using *discretization*, which we define:

- As methods that divide continuous space and/or time into a set of discrete and non-overlapping intervals (for time) or areas (for space), thereby replacing continuous dynamics (using differential equations) with discrete dynamics (using difference equations and linear algebra);

- Where these methods are defined such that we can increase the resolution (i.e., the number of intervals and/or cells), and doing so will cause the approximation to be arbitrarily precise.

By emphasizing discretization, we hope to allow ecologists to balance their needs for accuracy (resulting from a high-resolution discretization), model flexibility (not being restricted to dynamics that can be solved analytically), and computational speed (resulting from initial exploration using low-resolution discretization).

Ecologists generally recognize that "nothing in biology makes sense except in the light of evolution" [45, 46]. From this we might take a few demographic principles that are true (with appropriate context) across all of ecology:

1. All taxa must be composed of individual organisms that start their life with some assortment of traits;

2. Similarly, all taxa are composed of individual organisms that produce new individuals, where these produced individuals ("children") have some similarity by descent to those individuals produced them (their "parents"), whether by sexual or asexual production;

3. Finally, all organisms will eventually die (such that the frequency of different traits can change without the total population size increasing infinitely).

Beyond this (nonexhaustive) list of consequences of evolution, ecology is generally constrained by physical and thermodynamic constraints, e.g.:

1. New individuals are typically smaller than the organisms that propagate them, and therefore will typically grow in size during life;

2. Organisms cannot teleport through space, but must instead be transported continuously to move;

3. Organisms must maintain some minimal metabolism to live, and death is irreversible (i.e., each organism will be born and die only once).

These demographic, evolutionary and thermodynamic principles imply that we can track the "timeline" for each individual organism from the time of its birth to death, and also track the value of different "marks", and thereby record the basic information about the individuals that make up a taxon. These principles also suggest a basic role for several marks, i.e., taxon c, location $s(t)$, time of birth t_{birth}, time of death t_{death}, age $a(t)$, size $w(t)$, and so on. Continuous marks (such as location) can typically take any value from a specified continuous domain, while categorical marks (such as the species for a given individual) will take one of a specified set of levels. Constructing machinery in this way is familiar to many ecologists as an *individual-based model* (IBM) or *agent-based model*, and the performance of an IBM can be evaluated by determining whether a small set of individual rules or behaviors can reproduce system-level properties [72].

We also note that ecologists will often have different types of data, which arise from different ways of measuring ecological processes. Here, we classify these based on how the data must be stored in a database, and arrive at five types of data: point, areal, individual measurements, flow, and trajectory data (Table 1.1). Different ecological systems will be relatively easy or hard to sample using these different categories. For example, it is now feasible to directly measure animal foraging and energy expenditures using animal's tags, such that data on individual habitat use and energy budgets is increasingly available for birds, mammals, and other animals [39]. By contrast, measuring the movement of individual plants during their dispersal stage (i.e., pollen) or small insects remains difficult, and therefore point and areal data might be easier to collect than trajectories for these stages and taxa. However, we seek a framework that can visualize and analyze all five categories of ecological data. Given this context, we proceed by introducing two examples of IBM, and use these examples to demonstrate that we can often simplify these individual dynamics to define a computationally efficient statistical model.

TABLE 1.1: Categorizing spatial data for ecological analyses by what type of spatial reference system is required to store it, based upon [197].

Data category	Ecological example	Spatial object
Point data	Location for an individual	Point referenced data
Areal data	Count arising from visual sampling in a defined area	Areal referenced data
Individual attributes	Size measurement for an individual	Attribute of an individual
Flow	Radar measurement of birds passing over a weather station	Directed graph data
Trajectory	A sequence of measurements of an animal's location over time	Sequence of points

1.2   Time-to-event Model for Individual Timelines

We first illustrate individual-based models by simulating the "timeline" for each individual, which we define as the time of birth B_i and death D_i for each individual i[1]. We specify a constant birth rate, with $r = 0.2$ individuals born per year. This constant birth rate implies that the time elapsed from one birth B_i to the next B_{i+1} follows an exponential distribution:

$$\Pr[B_{i+1} - B_i > \Delta_t] = e^{-r\Delta_t} \tag{1.1}$$

where we then record the birth time B_i for each individual occurring over 100 years.

Similarly, we specify a Gompertz function for survival [68]. This Gompertz function is commonly fitted to survival-at-age tables, which list the proportion of individuals that are still alive at a given age. If we define the probability $S(a)$ that an individual will survive to age a, i.e., $S(a) = \Pr[(D_i - B_i) > a]$ and define the derivative $\frac{d}{da}S(a) = -m(a)S(a)$ then $m(a)$ is the mortality rate. The Gompertz function specifies the mortality rate as:

$$m(a) = be^{ka} \tag{1.2}$$

where b is the initial mortality rate when $a = 0$, and k defines an exponential increase or decrease in mortality rate with age. The corresponding probability of surviving to a given age is then [263]:

$$S(a) = e^{\frac{b}{k}(1-e^{ka})} \tag{1.3}$$

For illustration purposes in the following, we specify a low initial mortality rate $b = 0.001\,\mathrm{yr}^{-1}$, where the mortality rate accelerates exponentially at a rate of $k = 0.1\,\mathrm{yr}^{-1}$. To simulate the time elapsing between birth and death, however, we must convert the Gompertz survival function to the distribution of ages upon death (termed the "lifetime density function"). In the terminology of a survival analysis, the Gompertz survival function is converted to a lifetime cumulative distribution function $F(a) = 1 - S(a)$, and the derivative $f(a)$ of the lifetime cumulative distribution $F(a)$ is then the lifetime density function (Eq. 1.4):

$$f(a) = \frac{d}{da}F(a) \tag{1.4}$$

In this example, we calculate the derivative symbolically using the function D in R, and the function lifetime_density outputs this derivative (Code 1.1). We then use *rejection sampling* as explained in Appendix B.3 to simulate values drawn from this lifetime density function $f(a)$ (Code 1.2).

CODE 1.1: R code applying symbolic derivative to calculate the lifetime density function.

```
1  # Calculate lifetime density function f = d/da( 1 - S(a) )
2  lifetime_density = function(a, b, k){
3    # Write expression for 1 - S(a)
4    gompertz_mortality = expression( 1-exp(b/k*(1-exp(k*a))) )
5    # Symbolic derivative with respect to age a
6    density_function = D(gompertz_mortality, "a")
7    # Evaluate derivative using specified values
8    eval(Expr)
9  }
```

---

[1]See https://github.com/james-thorson/Spatio-temporal-models-for-ecologists/Chap_1 for code associated with this chapter.

CODE 1.2: R code for rejection-sampling simulation of birth-death process.

```r
1  # Paraeters
2  b = 0.001 # Initial annual mortality rate
3  k = 0.1   # Acceleration in mortality rate
4  birth_rate = 0.2 # animals per year
5  n_t = 100 # Time domain  to simulate
6
7  # Calculate peak of lifetime density for use in rejection sampling
8  Opt = optimize(f=lifetime_density, interval=c(0,100), maximum=TRUE, b=b, k=k)
9
10 # Rejection sampling for age at death
11 simulate_lifespan = function(b,k){
12   while(TRUE){ # Repeat as long as necessary
13     A = runif(n=1,min=0,max=150)
14     M = lifetime_density(a=A, b=b, k=k)
15     rand = runif(n=1,min=0,max=Opt$maximum)
16     if(rand<M) return(A) # If accepted, save value
17   }
18 }
19
20 # Sample births using sequence of exponential distributions
21 birth = death = vector()
22 while( max(0,birth,na.rm=TRUE)<n_t ){
23   birth = c(birth, max(0,birth,na.rm=TRUE) + rexp(n=1, rate=birth_rate) )
24 }
25
26 # Sample deaths from lifespace after birth
27 for( i in seq_along(birth) ){
28   death[i] = birth[i] + simulate_lifespan( b=b, k=k )
29 }
```

One simulation from this individual-based model (Fig. 1.1) shows, e.g., that population size increased up until the maximum age of about 60 years, and then has a peak population size at around this time, arising due to stochastic variation in deaths occurring around that time. For now, we mainly note that this continuous process for births and death arising at one scale (i.e., for individuals) can then be coarsened to calculate stochastic variation in a function (i.e., population size), and that the function representing total population size $n(t)$ will change by at most one individual over a sufficiently small step size (i.e., $n(t) - 1 \leq n(t + \delta) \leq n(t) + 1$ as δ approaches zero).

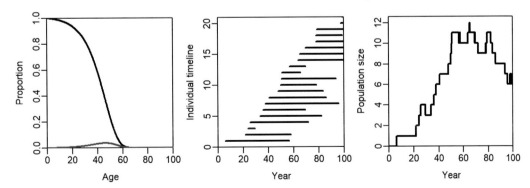

FIGURE 1.1: Gompertz survival function $\Pr[(D_i - B_i) > a]$ (left panel black line) and resulting lifetime density function $f(a)$ (left panel red line), along with one realization of individual timelines representing birth and death times over 100 simulated years (middle panel) and the total population size (right panel) resulting from this individual-based simulation.

TABLE 1.2: List of methods used throughout the textbook to calculate the slope of a line that is tangent to a specified function (called *differentiation*) [10].

Name	Description
Analytical	Memorizing several well-known transformations and applying them in sequence to a function $f(x)$ to define a derivative function $f'(x)$ that can later be evaluated to compute the derivative for any input x
Numerical (a.k.a. finite difference)	Evaluating a function (e.g., using a computer program) at several nearby locations, and approximating the derivative near those locations from the difference between these evaluated values;
Automatic	In simplest form (called "forward-mode AD"), writing a computer program that computes an output $f(x)$ from inputs x using on a series of elementary operations, where the program evaluates these operations in sequence using supplied input values to calculate $f(x)$ but also evaluates the derivative of these elementary operations (and automatically implements the chain rule) to compute $f'(x)$, without directly storing an expression for $f'(x)$ [71, 258];
Symbolic	Writing an expression to evaluate $f(x)$, and passing this to a computer program that calculates an expression for $f'(x)$. This expression might become very long as a result of the chain rule, but the expression can be written down and later evaluated to calculate $f'(x$ for alternative inputs x.

This example of sampling death times D_i from a negative derivative of the Gompertz survival function illustrates calculating a *symbolic derivative*. We will calculate derivates repeatedly throughout the book, alternatives to symbolic differentiation include numerical, analytic, and automatic differentiation (Table 1.2). Ecologists generally receive some background in analytical differentiation, but we suspect that many practicing ecologists find these methods intimidating such that methods requiring analytical differentiation will find limited use by applied ecologists. As a result, we will generally emphasize numerical and automatic differentiation, which can efficiently implemented using software. However, we will sometimes also present analytical derivatives if they are sufficiently simple to explain, and will use symbolic differentiation to check results from automatic differentiation.

We could instead approximate these individual dynamics using a model that tracks total population size n_t in a set of discrete times (e.g., using years $t \in \{1, 2, ..., t_{max}\}$ and ages $a \in \{01, 2, ..., a_{max}\}$), and tracking the count of births $b_{t,a}$ and deaths $d_{t,a}$ for each age a in each year t.

$$b_{t,a=0} \sim \text{Poisson}(\lambda = 0.2)$$
$$d_{t,a} \sim \text{Binomial}\left(n_{t,a}, \frac{F(a+1) - F(a)}{S(a)}\right) \qquad (1.5)$$
$$n_{t+1,a+1} = n_{t,a} + b_{t,a} - d_{t,a}$$

where total abundance is calculated by summing across ages:

$$n_t = \sum_{a=0}^{a_{max}} n_{t,a} \tag{1.6}$$

The Lagrangian viewpoint becomes infeasible when tracking millions of individuals, so it is convenient to understand both approaches to birth-death processes.

1.3   Spatial Point Process for Habitat Utilization

Alternatively, we can specify an individual-based model (the Lagrangian viewpoint) as a spatial point process, involving marks for location s for each individual, and sum across space to calculate the total abundance within a specified spatial domain. To illustrate, we envision an immobile organism (i.e., a germinated plant) distributed along a gradient in elevation along the side of a mountain.

CODE 1.3: R code simulating habitat utilization from an immobile organism arising from a point-process model.

```
1  library(mvtnorm)
2  library(stars)
3
4  # parameters
5  SD_km = 100
6  peak_km = 2 # km
7  best_elevation = 1
8  SD_logelev = 0.5
9  n_indiv = 500
10
11 # Simulate elevation and density
12 get_elevation = function(loc){
13    e0 = dmvnorm(c(0,0), mean=c(0,0), sigma=SD_km^2*diag(2))
14    peak_km * dmvnorm(loc, mean=c(0,0), sigma=SD_km^2*diag(2)) / e0
15 }
16 get_density = function(loc){
17    elev = get_elevation(loc)
18    exp(-1 * ( log(elev/best_elevation) )^2 / SD_logelev^2)
19 }
20
21 # Get values on grid for plotting
22 loc_gz = expand.grid( "x"=seq(-200,200,length=100),
23                       "y"=seq(-200,200,length=100) )
24 elev_g = get_elevation(loc_gz)
25 d_g = get_density(loc_gz)
26
27 # Function for rejection sampling for locations of individuals
28 max_density = optim( fn=get_density, par=c(0,100),
29                      control=list(fnscale=-1))$value
30 simulate_location = function( ... ){
31    loc = NULL
32    while(is.null(loc)){
33       samp = runif(n=2,min=-200,max=200)
34       D = get_density(samp)
35       rand = runif(n=1,min=0,max=max_density)
36       if(rand<D) loc = samp
37    }
38    return(loc)
39 }
40 loc_i = t(sapply( 1:n_indiv, FUN=simulate_location ))
```

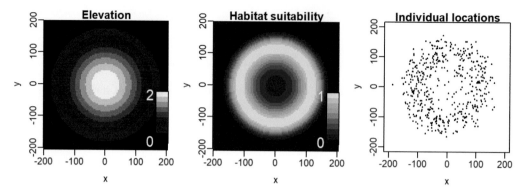

FIGURE 1.2: Elevation (left), habitat quality (middle), and realized distribution for 500 individuals.

We specifically simulate elevation as a normal distribution with a peak of 2 km and a standard deviation of 100 km. We further envision that habitat suitability is greatest at 1 km elevation, and declines as the squared log-ratio of elevation to this preferred value (Code 1.3). We first illustrate the spatial distribution resulting from these assumptions (Fig. 1.2). Organisms are distributed independently in space (Fig. 1.2), such that the count of individuals N_A within a given area A follows a Poisson distribution:

$$N_A \sim \text{Poisson}(\lambda_A)$$
$$\lambda_A = \int_{s \in A} \lambda(s) ds \tag{1.7}$$

hence the term *Poisson point process* for the resulting distribution of animal locations.

We next discretize this Poisson point process by dividing the spatial domain into square grid cells, and treating abundance in each grid cell as if it follows a Poisson distribution. This is an approximation, because we will use the elevation at the center of each grid cell as representing density for the entire cell. However, this approximation can become arbitrarily precise as grid cells become smaller, so it qualifies as a "discretization method" (recalling definition from Section 1.1). We then fit a generalized linear model to abundance N_i in each grid cell A_i for simulated organisms.

1.4 Generalized Linear Models

A *generalized linear model* (GLM) involves five components:

1. A response variable y_i that we seek to explain and/or predict for each sample i;

2. A set of covariates $\mathbf{x}_i$ that we use to explain the response variable, where covariates are specified as data and their values represent information associated with that sample;

3. A vector of coefficients β that we estimate from the model. The product of coefficients and covariates is called the "linear predictor" $p_i = \mathbf{x}_i \beta$;

4. A *link function* g, where the inverse-link function transforms the linear predictor into the mean of the responses such that $g(\mu) = p_i$;

5. A probability distribution f for responses given the inverse-link transformed linear predictor, $g^{-1}(\mathbf{x}_i\beta)$, where additional "measurement" parameters might be used to define this distribution.

In the preceding example, we specify a log-linked Poisson distribution, where the log-density depends on log-elevation. We specifically envision a circumstance where log-densities are highest at some intermediate elevation, and this is easy to specify using a quadratic function (i.e., including log-elevation and log-elevation squared as covariates):

$$N_i \sim \text{Poisson}(\lambda_i)$$
$$\log(\lambda_i) = \beta_0 + \beta_1\log(E_i) + \beta_2\log(E_i)^2 \tag{1.8}$$

To demonstrate this point, we bin the area into 50 km by 50 km grid cells (Fig. 1.3), calculate the total abundance in each cell, and fit that with a log-linked Poisson GLM. To do so, we use the R function glm (Code 1.4); while fast, this approach does not give any particular insight into the mechanics of how the estimated value of parameters is identified, or how standard errors are calculated.

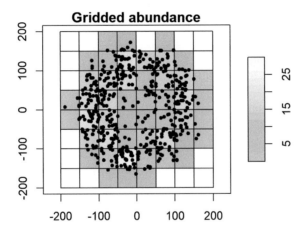

FIGURE 1.3: A grid overlayed on the spatial domain, showing the location of individuals as well as the gridded value of above-ground-biomass, generated by counting the number of individuals in a given grid cell.

CODE 1.4: R code fitting GLM to point process.

```
1  # make sample-level data frame
2  samples = data.frame("x"=loc_i[,1], "y"=loc_i[,2] )
3  samples = st_as_sf( samples, coords=c("x","y") )
4
5  # Get count in each grid cell
6  grid_size = 10
7  grid = st_make_grid( st_bbox(c(xmin=-200, xmax=200, ymin=-200, ymax=200)),
8                       cellsize=grid_size )
9  grid_i = st_intersects( samples, grid )
10 N_i = tapply( rep(1,nrow(samples)),
11               INDEX = factor(unlist(grid_i),levels=1:length(grid)),
12               FUN = sum )
13
14 # Convert to a data frame for 'glm'
15 Data = data.frame( st_coordinates(st_centroid(grid)),
16                    "N"=ifelse(is.na(N_i),0,N_i) )
17 Data$elev = get_elevation( Data[,c("X","Y")] )
```

```
18
19 # Fit with canned GLM software
20 fit = glm( N ~ log(elev) + I(log(elev)^2), data=Data, family=poisson )
```

1.5   Likelihood Estimation of Generalized Linear Models

We next seek to replicate the GLM using a lower-level analysis that allows us to better understand the process by which parameters are estimated. To do so, we take a detour into *maximum likelihood theory* [57].

To illustrate maximum likelihood theory in a simple context, we envision that an analyst has some data y and can also specify a function $y \sim f(\theta)$ that is assumed to generate the data given unknown parameters θ. We seek an estimator for θ based on data y, and one intuitive estimator is to pick the value of θ that would then result in a high probability that the data take the values we observed y. This *maximum likelihood estimator* is (in a very loose sense) taking the hypothesized probability of data and using it in reverse to identify plausible values for parameters. The likelihood of data is often written as $\mathcal{L}(\theta; y)$ and is then called the *likelihood function.* Maximum likelihood estimation involves identifying the value of parameters θ that maximizes the likelihood function. However, multiplying numbers often results in a computed likelihood that is larger than the largest digit that can be stored in computer software, or smaller than the smallest digit that can be stored (called *numerical overflow and underflow,* respectively). We therefore instead compute the natural logarithm of the likelihood (the "log-likelihood") such that numerical over and underflow is less likely to occur:

$$\hat{\theta} = \text{argmax}_\theta \log\mathcal{L}(\theta; y) \tag{1.9}$$

To understand the properties of this estimator $\hat{\theta}$, we first remember that if two events A and B are independent, then the probability that these two events both occur is equal to the product of their individual probabilities, $\Pr(A, B) = \Pr(A)\Pr(B)$. The likelihood is itself a reinterpretation of the probability density $p(Y|\theta)$, so it also shares this property:

$$p(Y|\theta) = \prod_{i=1}^{n_i} p(Y_i|\theta) \tag{1.10}$$

and replacing this with the log-likelihood:

$$\log\mathcal{L}(\theta; y) = \sum_{i=1}^{n_i} \log\mathcal{L}(\theta; y_i) \tag{1.11}$$

This then shows that we can maximize the sum of log-likelihoods for each datum to calculate the total log-likelihood, as long as residuals are independent.

We note that using $\hat{\theta}$ as an estimator for unknown parameters θ has several useful properties:

1. *Asymptotic consistency:* if there's some value of parameters θ such that function $f(\theta)$ is a perfect description of the true data-generating process, then as data are added the maximum likelihood estimation $\hat{\theta}$ will asymptotically approach the value θ.

2. *Estimating uncertainty:* we can estimate the uncertainty in parameters by calculating how quickly the log-likelihood declines as we change parameter values away

from the maximum likelihood estimator $\hat{\theta}$. This involves measuring the curvature of the log-likelihood around estimator $\hat{\theta}$, and one useful measurement is the matrix of 2nd derivatives for $\log \mathcal{L}(\theta; y)$ with respect to θ evaluated at its maximum value $\hat{\theta}$, called the *Hessian matrix* $\mathbf{H}$. The matrix inverse of the Hessian, $\mathbf{H}^{-1}$ approximates the variance of $\hat{\theta}$ asymptotically (i.e., as the number of samples becomes large).

In summary, maximum likelihood theory suggests that we can fit our Generalized Linear Model by calculating the loglikelihood for each datum, summing these, and then identifying the value of parameters $\hat{\theta}$ that maximizes this loglikelihood, and this estimator will have good performance in terms of asymptotic consistency and uncertainty estimates.

1.6   Estimation using Template Model Builder

We next introduce *Template Model Builder* (TMB) [122]. TMB is an R-package that allows an analyst to write a "template file" that calculates the loglikelihood; this template is written in C++ and therefore requires some basic familiarity with writing loops, indexing, etc. Having written this template, however, TMB can then either calculate the likelihood, or use automatic differentiation to calculate a variety of gradients of the loglikelihood with respect to parameters. These values are passed back to R, and nonlinear minimizers in R can use the loglikelihood and its gradients to efficiently identify the value of θ that maximizes the likelihood. Calculating a "cheap gradient" is key to using TMB to generalize almost all statistical operations that might interest us.

CODE 1.5: TMB code defining the joint negative log-likelihood `jnll` for a generalized linear model using a Poisson distribution and log-link function.

```
1  #include <TMB.hpp>
2  template<class Type>
3  Type objective_function<Type>::operator() ()
4  {
5    // Data
6    DATA_VECTOR( y_i )
7    DATA_MATRIX( X_ij );
8
9    // Parameters
10   PARAMETER_VECTOR( beta_j );
11
12   // Global variables
13   Type jnll = 0;
14   int n_i = X_ij.rows();
15   int n_j = X_ij.cols();
16   vector<Type> log_mu(n_i);
17   log_mu.setZero();
18
19   // Probability of data conditional on fixed and random effect values
20   for( int i=0; i<n_i; i++){
21     for( int j=0; j<n_j; j++){
22       log_mu(i) += beta_j(j) * X_ij(i,j);
23     }
24     jnll -= dpois( y_i(i), exp(log_mu(i)), true );
25     // Option to simulate new data given parameters
26     SIMULATE{
27       y_i(i) = rpois( exp(log_mu(i)) );
28     }
29   }
```

```
30
31   // Return values to R
32   SIMULATE{REPORT(y_i);} // If simulating new data, hand the value back to R
33   REPORT( log_mu );
34   return jnll;
35 }
```

In the case of a GLM, we first write the template file that is interpreted by TMB (Code 1.5). We see that TMB requires several blocks of code:

1. A header that defines the language, which will be identical in most files (lines 1–4 in Code 1.5);

2. Functions (starting with syntax `DATA_`) that define data passed from R (lines 5–7);

3. Functions (starting with syntax `PARAMETER_`) that define parameters passed from R (lines 9–12);

4. Code defining variables that are used internally during TMB calculations (lines 14–17), where these are typically declared as scalars, matrices, arrays, etc., using a C++ type called `Type`. Declaring all variables as type `Type` then allows TMB to track either the value of variables, or their gradient with respect to parameters, depending upon which of these is useful at a given stage of parameter estimation;

5. Code calculating the joint negative log-likelihood `jnll` based on data and parameters (line 21–31);

6. Code that passes calculated values back to R, with the `jnll` in particular defining the joint negative log-likelihood.

We will use TMB repeatedly, and will use a similar formatting of code blocks every time.

CODE 1.6: R code to compile and fit a GLM built in TMB.

```
1  # Compile and load TMB
2  library(TMB)
3  compile( "poisson_glm.cpp" )
4  dyn.load("poisson_glm")
5
6  # Make covariate data
7  formula = ~ log(elev) + I(log(elev)^2)
8  X_ij = model.matrix( formula, data=Data )
9
10 # Built inputs for TMB object
11 data = list( "y_i"=na.omit(Data)$N, "X_ij"=X_ij )
12 params = list( "beta_j"=rep(0,ncol(X_ij)) )
13
14 # Build object
15 Obj = MakeADFun( data=data, parameters=params )
16
17 # Optimize and get standard errors
18 Opt = nlminb( objective=Obj$fn, grad=Obj$gr, start=Obj$par )
19 Opt$SD = sdreport( Obj )
```

Finally, this code is compiled and linked as a *dynamically linked library* in R (Code 1.6). This then allows for a cheap evaluation of the log-likelihood or its gradients. We can compare the output from this lower-level calculation with estimates using function `glm` to fit a generalized linear model (Table 1.4). We see here that the parameters estimates and standard errors for `beta_j` are identical to those using `glm`. We can also plot the estimate of density, and compare it with the known true value (Fig. 1.4).

TABLE 1.3: Estimated quadratic effect of elevation on simulated density of above-ground-biomass including estimated values, standard errors, T-value, and associated probability that the estimate differs from zero.

	Estimate	Std. Error	T-value	Prob(>0)
Intercept	2.957	0.054	54.507	0
log(elevation)	-0.097	0.13	-0.743	0.457
log(elevation)^2	-2.653	0.218	-12.143	0

TABLE 1.4: Estimated quadratic effect of elevation on simulated density (see Table 1.3 for details).

	Estimate	Std Error
Intercept	2.957	0.054
log(elevation)	-0.097	0.13
log(elevation)^2	-2.653	0.218

1.7 Evaluating Model Fit

Finally, ecologists will often want to compare model predictions with available data to determine whether there's evidence that the model is inconsistent with data. Conceptually, one way to do this is to simulate data from the fitted model and then compare these simulated data with the real data to see if they are substantially different. More formally, we can calculate *simulation residuals* to calculate the quantile for each datum. This involves the following steps:

1. For each observation, simulate new data conditional upon the maximum likelihood estimates of parameters $\hat{\theta}$;

2. Calculate the proportion of simulated values that are less than the real sample, and also the proportion that are less than or equal.

CODE 1.7: R code for calculating and plotting the quantile residual diagnostic.

```
1  # Simulate 100 new data sets
2  Sim = sapply(1:100, FUN=function(i) Obj$simulate()$y_i )
3
4  # Make function for plots
5  plot_ecdf = function( y, ysim, ... ){
6    plot(ecdf(ysim), xlim=max(ysim)*c(0,1), ... )
7    x_intercept = y
8    y_min = mean(ysim<y)
9    y_max = mean(ysim<=y)
10   y_intercept = runif(1,y_min,y_max)
11   arrows( x0=x_intercept, x1=x_intercept, y0=0, y1=y_intercept )
12   arrows( x0=x_intercept, x1=0, y0=y_intercept, y1=y_intercept )
13 }
14
15 # Make plot showing two observations
16 par( mar=c(3,3,1,1), mgp=c(2,0.5,0), mfrow=c(1,2) )
17 plot_ecdf( y=Data$N[4], ysim=Sim[4,], main="ECDF: sample 4" )
18 plot_ecdf( y=Data$N[6], ysim=Sim[6,], main="ECDF: sample 6" )
```

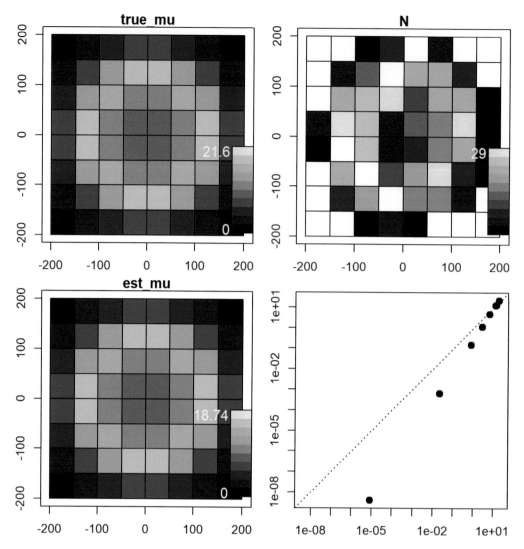

FIGURE 1.4: True density (top-left), sampled total size (top-right), estimated density (bottom-left), and a comparison of estimated vs. true density (bottom-right) for the GLM fitted using TMB.

For continuous-valued data, there is zero probability that any simulated sample will exactly match the observation, so the two values in Step-2 (i.e., the proportion less and the proportion less than or equal) will be identical, and the p-value can be calculated from either one. Using the Poisson distribution (or other distributions for count data), however, the cumulative distribution for each sample follows a stair-step pattern because every simulated or real-world sample is a positive integer. As a result, many samples might match a given observation and the two values in Step-2 would give different p-values. For discrete-valued data, we therefore add another step:

3. Randomly sample a p-value from a uniform distribution between these two values calculated in Step-2.

This extra step is called the Probability Integral Transform (PIT) [50], and it allows us to calculate a p-value for any mixture of discrete and continous data (see Fig. 1.5). Step 1 involves simulating new data from the fitted model, which we enabled already using

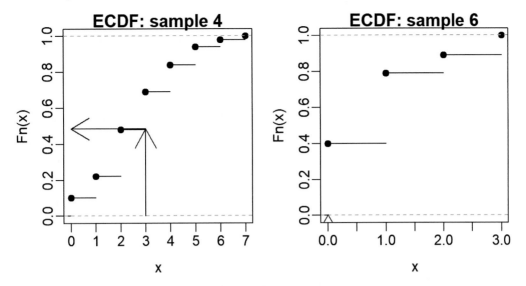

FIGURE 1.5: An illustration of how to calculate PIT residuals from a fitted model, which involves simulating from the fitted distribution for each datum and then calculating the quantile associated with each observation. In the case of multiple simulations having the same value as an observation, the quantile is drawn from a uniform distribution within this range (see right-hand panel for example).

the function **SIMULATE** in the TMB code (line 28 and 34 of 1.5). We can then trigger this simulation by calling the function obj\$simulate() in R. Steps 2–3 are then done in R (Code 1.7).

We can evaluate model fit by evaluating the quantile Q_i associated with each fitted sample y_i. The quantiles for all samples are then compared with a uniform distribution; if the model is well specified, these quantiles will tend to follow a uniform distribution. We can therefore make a *quantile-quantile plot* and in some cases calculate a set of statistics that measure how different the simulation residuals are from a uniform distribution. In general, we will calculate these PIT residuals and summary statistics using the DHARMa package [84] in R, where DHARMa then automatically takes care of calculations and plotting. In this case, the quantile-quantile plot shows no evidence of model misspecification (Fig. 1.6).

CODE 1.8: R code plotting quantile residuals using DHARMa.

```
library(DHARMa)

# Create DHARMa object with PIT residuals
dharmaRes = createDHARMa(simulatedResponse = Sim,
  observedResponse = data$y_i,
  fittedPredictedResponse = exp(Obj$report()$log_mu),
  integer = FALSE)

# Plot
plot( dharmaRes )
```

We acknowledge that many ecologists are more familiar with Pearson residuals, which are calculated as the difference between a predicted and observed sample, divided by the estimated standard deviation for that sample, where these Pearson residuals are then

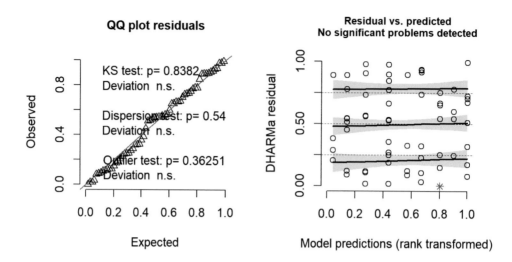

FIGURE 1.6: Standard output from DHARMa, showing the quantile-quantile plot (left panel) with several goodness-of-fit tests listed (where a value >0.05 is non-significant ("n.s."), and also showing the rank-transformed predictions vs. residuals to see if there's a trend in either mean or 25/75th percentiles.

compared with a normal distribution. However, Pearson residuals have two major problems:

1. *Skewness*: pearson residuals are typically compared with a normal distribution. However, a model might specify a distribution (e.g., a lognormal) that has a higher probability of extreme events than the normal distribution predicts. In these cases, the Pearson residuals will seem to indicate a high number of outliers, even when those might be consistent with the distribution that was specified in the model;

2. *Misleading patterns*: pearson residuals involve calculating the difference between predicted and observed values. However, for count data, this difference is constrained to be an integer, such that Pearson residuals are sometimes constrained to some particular values and not others. This can result in visual artefacts when Parson residuals are plotted, and these then distract attention from other patterns that might actually identify model lack-of-fit.

Importantly, both of these problems are addressed by using PIT residuals.

CODE 1.9: R code visualizing an estimated covariate-response function and associated standard errors.

```
1 # Sample new parameters from estimation covariance
2 beta_rj = mvtnorm::rmvnorm( n=500, mean=Opt$par, sigma=Opt$SD$cov.fixed )
3 x_i = seq( 0.05, 2, length=1000 )
4 X_ij = model.matrix( formula, data=data.frame("elev"=x_i) )
5
6 # Calculate response curve for each simulated parameter-vector
7 yhat_ir = scale(X_ij %*% t(beta_rj), scale=FALSE)
8 ybar_i = scale(X_ij %*% Opt$par, scale=FALSE)
```

```
 9  ytrue_i = scale(-1 * (log(x_i) - log(best_elevation))^2 / SD_logelev^2, scale
        =FALSE)
10  yci_zi = apply( yhat_ir, MARGIN=1, FUN=quantile, prob=c(0.05,0.95) )
11
12  # Plot marginal effect
13  plot( x=x_i, y=ybar_i, log="x", type="l", lwd=2, col='blue', xlab="elevation"
        , ylab="log-relative density", ylim=c(-30,2) )
14  lines( x=x_i, y=ytrue_i, lwd=2, col="black")
15  polygon( x=c(x_i,rev(x_i)), y=c(yci_zi[1,],rev(yci_zi[2,])), col=rgb
        (0,0,1,0.2) )
```

In additional to plotting residuals to identify any evidence of poor model fit, ecologists often use model selection to identify a parsimonious model from a set of candidate models. In general, ecologists can explore parsimony by conducting a crossvalidation experiment [190], i.e.:

1. Randomly divide the data into two bins, which are either fitted (*in bag*) or withheld (*out-of-bag*) from the model;

2. Fit the model to the in-bag data;

3. Use the fitted model to predict the out-of-bag data, either by computing their predictive log-likelihood or some other performance metric;

4. Repeat these steps with different samples for in-bag and out-of-bag data, calculate average predictive performance across these samples, and select a model with best average predictive performance.

However, conducting a crossvalidation experiment can be time-consuming for large models. As a result, statisticians have developed approximations to the expected predictive performance that can be calculated from a fitted model without a crossvalidation experiment. For example, ecologists often compute the Akaike Information Criterion (AIC) [1] and have been trained to report estimates arising from the model with the lowest AIC score [18]. This model selection criterion favors models with good fit (i.e., a high log-likelihood) as well as fewer parameters, and can be viewed as one way to penalize covariate-response parameters towards zero [96]. Penalizing the number of model parameters arises because increasing the number of parameters also increases the expected difference in performance between fitted and new data (i.e., the log-likelihood for in-bag vs. out-of-bag data). Alternative model-selection criteria can be adapted for the hierarchical models that we introduce in later chapters [269, 248], although we generally do not emphasize model-selection in the following.

Another important way to explore model fit is to visualize how predictions change for different covariate values. Given that we are simulating data, we know the true covariate-response curve, and can evaluate model performance by comparing the true and estimated response curve relating log-density to elevation. To do so, we sample from the estimated covariance of fixed effects using a multivariate normal approximation. We plot this estimated response against the known effect of elevation to evaluate how well the model did at reconstructing habitat preferences (Code 1.9). We observe, however, that the fitted response curve using this coarse-scale discretization (left-hand panel of Fig. 1.7) does not perfectly match the known (simulated) effect of elevation. We therefore refit using a finer discretization involving 10 km grid cells (instead of 50 km cells). This closer match (right-hand panel of Fig. 1.7) provides visual confirmation that representing a Poisson point process as a Poisson GLM can in fact converge on the true data-generating process. As noted in Section 1.1, choosing a spatial scale for discretization requires balancing computational cost against other study goals. We therefore recommend that analysts start with a coarse scale during

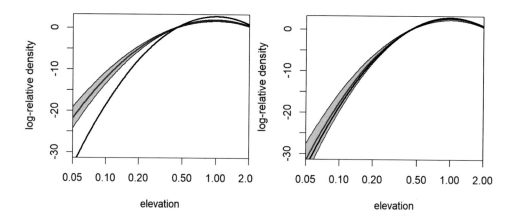

FIGURE 1.7: A comparison of the estimated response curve (blue line) and its 95% confidence interval (blue shaded area) against the true response curve (black line) using coarse 50 km × 50 km grid cells (left-hand panel), or a finer-scale discretization, i.e., 10 km × 10 km grid cells.

early model development, increasing the resolution as the model becomes more refined, and both test and communicate the sensitivity of results to this choice.

In later chapters, we will also automate the computation of these response curves using the marginaleffects package [2]. To do this, we must define four functions for the model, and the marginaleffects package subsequently calculates effects and standard errors (Code 1.10). This allows us to easily visualize the estimated responses to covariates (Fig. 1.8) using high-level plotting code (Code 1.11).

CODE 1.10: R code defining functions used by the marginaleffects package to visualize covariate-response curves.

```
1  # Function to get coefficients for TMB model
2  get_coef.custom_tmb = function(model, param, ...){
3    out = model$parhat[[param]]
4    names(out) = rep(param, length(out))
5    return(out)
6  }
7
8  # Function to get variance-covariance for TMB model
9  get_vcov.custom_tmb = function(model, param, ...){
10   rows = which( names(model$opt$par) == param )
11   array( model$opt$SD$cov.fixed[rows,rows],
12     dim = rep(length(rows),2),
13     dimnames = list(rep(param,length(rows)),rep(param,length(rows))) )
14 }
15
16 # Function to change coefficients for TMB model
17 set_coef.custom_tmb = function(model, newpar, param, ...){
18   model$parhat[[param]] <- newpar
19   return(model)
20 }
21
22 # Function to get predictions when changing coefficients
```

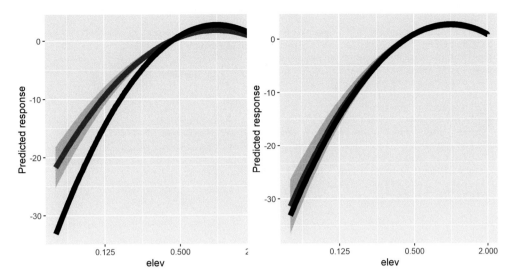

FIGURE 1.8: Same as Fig. 1.7 but using the `marginaleffects` package to generate response curves, and `ggplot2` [260] to plot the response functions.

```
23  get_predict.custom_tmb = function(model, newdata, param, center=FALSE, ...){
24    # build original model.frame
25    frame0 = model.frame( formula=model$formula, data=model$data )
26    terms0 = terms( frame0 )
27    xlevels = .getXlevels( terms0, frame0 )
28    # get new design matrix
29    terms1 = delete.response( terms0 )
30    frame1 = model.frame( terms1, newdata, xlev=xlevels )
31    X_ik = model.matrix( terms1, frame1 )
32    gamma_k = get_coef.custom_tmb(model, param)
33    # Calculate linear predictor and format output
34    yhat_i = X_ik %*% gamma_k
35    if(center==TRUE) yhat_i = yhat_i - mean(yhat_i)
36    out = data.frame( rowid=seq_along(yhat_i[,1]), estimate=yhat_i )
37    return(out)
38  }
39
40  # Change marginaleffects options to define 'custom_tmb' class
41  options("marginaleffects_model_classes" = "custom_tmb")
```

CODE 1.11: R code for using the `marginaleffects` package to produce covariate-response plots.

```
1   # Source marginaleffects functions
2   library(marginaleffects)
3   library(ggplot2)
4   source( "../Shared_functions/marginaleffects.R" )
5
6   # Define input expected by marginaleffects functions
7   fit = list( obj = Obj,
8               opt = Opt,
9               data = Data,
10              formula = formula,
11              parhat = Obj$env$parList() )
12  class(fit) = "custom_tmb"
13
```

```
14  # Get prediction
15  pred = predictions( fit, param="beta_j", newdata=data.frame("elev"=x_i) )
16  pred$true = -1 * (log(x_i) - log(best_elevation))^2 / SD_logelev^2
17  pred$true = pred$true - mean(pred$true)
18  pred[,c("estimate","conf.low","conf.high")] = pred[,c("estimate","conf.low","
        conf.high")] - mean(pred[,"estimate"])
19
20  # Plot using ggplot
21  ggplot( as.data.frame(pred), aes(elev, estimate)) +
22    geom_line( aes(y=estimate), color="blue", size=3 ) +
23    geom_ribbon( aes( x=elev, ymin=conf.low, ymax=conf.high), fill=rgb
        (0,0,1,0.2) ) +
24    geom_line( aes(y=true), color="black", size=3 ) +
25    scale_x_continuous(trans='log2') +
26    labs(y="Predicted response")
```

1.8   Chapter Summary

In summary, we have showed that:

1. We can specify most ecological dynamics as an individual-based model by en-
   visioning a set of "individuals" where each individual has one or more "marks".
   These marks then represent an individual timeline (birth, death) and characteristics (size, location). These marks may change over time, where, e.g., size changes
   due to growth and location changes due to movement. In these cases, we are tracking changes using a "Lagrangian" viewpoint. However, this Lagrangian viewpoint
   becomes computationally infeasible as the number of individuals increase;

2. We can generally approximate an individual-based viewpoint using an "Eulerian"
   viewpoint, which tracks the number of individuals while dividing marks into bins.
   For example, location (a continuous variable) can be binned into a series of grid-
   cells, or size (a continuous variable) can be divided into size bins [152]. In some
   cases, this approximation can become arbitrarily precise when binning data at a
   finer resolution, and we call this "discretization" to distinguish it from the broader
   category of approximation methods;

3. An Eulerian viewpoint can in some cases be fitted as a Generalized Linear Model
   (GLM). The GLM includes a linear predictor (calculated from covariates and
   estimated coefficients), a link function, and a specified distribution;

4. A GLM can be fitted using maximum likelihood methods, which involves specifying a probability distribution for data, using this to calculate the log-likelihood
   of parameters given data, and maximizing this with respect to parameters;

5. The asymptotic performance of a maximum likelihood model can be understood
   using the central limit theory, and the model can be explored using PIT simulation
   residuals and by plotting marginal response curves using partial dependence plots.

We will use this statistical machinery throughout the remainder of the book. We next explore
cases when model residuals are correlated. The GLM assumed that residuals are independent
(such that the log-likelihood can be summed across data), so correlated residuals then violate
the assumptions involved in our analysis here.

1.9 Exercise

Revisiting the time-to-event model for individual timelines (Section 1.2), what is the equilibrium abundance given the birth-rate and Gompertz survival parameters presented? Please determine this in multiple ways and confirm the answer by comparing these. Consider either simulating the sampling-based solution over a long time period and taking the average after equilibrium has been reached, or solving analytically (in the absence of stochasticity) by noting that population size is the product of birth rate and life expectancy, where the latter can be solved by integrating the survival function either analytically or using the `integrate` function in R.

2

Hierarchical Models and Laplace Approximation

2.1 Why Hierarchical Models are Necessary

In Chapter 1, we saw that most ecological processses can be represented as an individual-based model (IBM), and that often we can approximate these IBMs using a statistical model fitted using maximum likelihood. We also saw that maximum likelihood estimation involves specifying a model that results in a probability distribution for data given some set of parameters θ, and then maximizing this log-likelihood with respect to parameters to get the estimator $\hat{\theta}$. However, there are at least three different reasons to extend maximum likelihood estimation to include some variables that are marginalized across (called *random effects*).

1. *Distribution for data has no simple expression*: In many cases, we cannot write a probability density for data directly as a function of parameters. For example, in Jolly-Seber capture-recapture models [109, 203], we observe a sequence of times when a tagged organism is encountered, and want to infer birth and death times while estimating survival and detection probabilities. It is not immediately clear how to write out the probability density for when the animal was encountered given survival and detection probabilities. However, it is possible to write down this probability when introducing additional latent (unobserved) processes (i.e., the latent time of birth and death for each animal), and then defining the data conditional upon these;

2. *Data are not independent*: As we saw, it is often convenient to write the full log-likelihood by calculating the log-likelihood for each datum individually and then summing across these. However, these two are equivalent only when residuals are independent for each sample. In the following, we are concerned with situations when residuals are correlated across space or time, and these correlated residuals will make it impossible to calculate the log-likelihood of data by simply summing the log-likelihood of each datum individually. For example, ecologists are trained to avoid *pseudoreplication* that arises when experimental replicates are not independent, e.g., because experimental sites are located near one-another and hence have similar ecological characteristics that are not under experimental control [100]. However, we might still gain useful information while analyzing data from a pseudoreplicated design by modelling correlations that result from these uncontrolled factors;

3. *Improved performance using shrinkage estimators*: Statisticians described the Stein paradox decades ago [213], wherein an analyst can average measurements for different things (termed "sampling units") to decrease the expected error for

DOI: 10.1201/9781003410294-2

each. This method is paradoxical because it could involve averaging measurements for sampling units that are completely unrelated, like batting averages for baseball players and the proportion of rainy days in different cities. Statisticians later resolved this paradox by defining the concept of *shrinkage*, where we estimate the variance in some measurement among sampling units, and the ratio of measurement and among-unit variance defines how much we "shrink" estimates towards one-another [53]. Random effects introduce a general method for defining *shrinkage estimators* that can often improve statistical efficiency.

There is a large vocabulary of different ways of describing models that include random effects, including *hierarchical models, mixed-effect models, multi-level models, data augmentation, empirical Bayes*, and many others. These different vocabularies arise from different ways that random effects can be derived, and have subtle differences in how models are interpreted. However, these differences are not relevant in the following, so we will proceed by referring to *mixed-effects models* as those that have both random and fixed effects.

Before we continue further, however, we will illustrate the first reason using a common ecological example, wherein an analyst samples the total biomass in a given area. This arises frequently in fisheries surveys [146], but also for insect light traps [272], plant pollen traps [64], and other settings.

2.1.1   Example: Compound distributions

In Section 1.3, we introduced a Poisson point process, where organisms are located randomly and independently such that the number of individuals in a specified area follows a Poisson distribution[1]. However, ecologists often study not just the density of individuals but also their sizes. To explore this, we might assume that individuals have two marks, representing location S_i and size measured as organism biomass W_i. We specifically seek to identify a distribution for the total biomass B in a given area A, $B = \sum_{i=1}^{N} W_i$.

To envision this process, we specify a distribution for the number of animals N in a defined area and the size of each such animal W_i. For simplicity, we will assume that animals are distributed independently, such that abundance N follows a Poisson distribution, and that organism sizes follow a Gamma distribution:

$$N \sim \text{Poisson}(\lambda_A)$$
$$W_i \sim \text{Gamma}(k, \theta) \tag{2.1}$$

where $\lambda_A = \int_{s \in A} \lambda(s) ds$ and $\lambda(s)$ is the expected density of individuals at any location s.

The resulting distribution for B is called a *compound probability distribution*. It is straightforward to simulate from this distribution by:

1. simulating the number of organisms N from a Poisson distribution;

2. for each organism, simulate it's biomass from a Gamma distribution;

3. add together the biomass from each organism

where these steps correspond to the components in Eq. 2.1. This corresponds to taking a draw from the probability distribution from the compound distribution, which we call a Poisson-gamma distribution because it is formed from these two component processes. We can therefore simulate from this compound distribution, even though we can't write down the probability distribution $f(B|\lambda_A, k, \theta)$ in a simple expression. Unfortunately, we

---

[1]See https://github.com/james-thorson/Spatio-temporal-models-for-ecologists/Chap_2 for code associated with this chapter.

need this probability distribution to then calculate the likelihood and use this to estimate parameters.

To calculate this probability density (or the density of any other compound distribution), we could first try calculating the density given a hypothetical value for N, corresponding to some perfect knowledge about how many organisms are present in that area:

$$f_{N=0}(B|N=0,k,\theta) = \begin{cases} 1 & \text{if } B = 0 \\ 0 & \text{if } B > 1 \end{cases}$$

$$f_{N=1}(B|N=1,k,\theta) = \text{dGamma}(B|k,\theta)$$
$$f_{N=2}(B|N=2,k,\theta) = \text{dGamma}(B|2k,\theta) \tag{2.2}$$

$$...$$

$$f_{N=n}(B|N=n,k,\theta) = \text{dGamma}(B|nk,\theta)$$

where $\text{dGamma}(B|k,\theta)$ evaluates a gamma probability density function for response B given shape k and scale θ. We additionally know the probability that $N = n$ individuals are present follows a Poisson distribution. Writing this more succinctly, we see:

$$f(B|\lambda_A,k,\theta) = \sum_{N=0}^{\infty} \text{dGamma}(B|Nk,\theta)\text{dPoisson}(N|\lambda_A) \tag{2.3}$$

where dPoisson evaluates the Poisson probability mass function. We will use specialized terms for the different components of this expression. We call:

- N a *latent variable*, because it affects the likelihood of our data B without itself being directly observable;

- $\text{dPoisson}(N|\lambda_A)$ the marginal distribution for the latent variable;

- $\text{dGamma}(B|Nk,\theta)$ the conditional distribution for the data given a value of the latent variable;

- $\text{dGamma}(B|nk,\theta)\text{dPoisson}(n|\lambda_A)$ the "joint distribution" because it includes the distribution of both our observable data B and our unobservable random effect;

- $f(B|\lambda_A,k,\theta)$ the marginal distribution for the data because it is calculated by *marginalizing* across the latent variable, N.

This example therefore takes the marginal distribution for a latent variable, the conditional distribution for the data given this latent variable, and uses these to calculate the marginal distribution for the data. We can later do the same process in reverse to calculate the likelihood of parameters k, θ, and λ.

This compound Poisson-gamma distribution arises often in ecological studies, and statisticians have developed efficient techniques to evaluate it as a Tweedie distribution [58]:

$$W_A \sim \text{Tweedie}(\mu_A, \phi_A, p) \tag{2.4}$$

where $\mu_A = \lambda_A k\theta$, $\phi_A = \theta(k+1)\mu_A^{\frac{-1}{k+1}}$, and $p = \frac{k+2}{k+1}$ [233]. Importantly, the Tweedie distribution includes a non-zero probability of observing mass $B = 0$, and this arises when the Poisson distribution yields a count of zero animals. However, the compound Poisson-gamma distribution then also admits continuous values for total mass $B > 0$.

Other compound distributions that arise often for ecologists include:

- *Negative binomial distribution*: the negative binomial arises as a Poisson distribution, where the Poisson intensity λ^* itself follows a Gamma distribution.

$$\lambda^* \sim \mathrm{Gamma}(k, \theta)$$
$$N \sim \mathrm{Poisson}(\lambda^*) \tag{2.5}$$

where shape $k = \frac{1}{\sigma^2}$ and scale $\theta = \lambda \sigma^2$, and σ is the coefficient of variation for the Gamma distribution while λ is again the average density of individuals. This compound distribution is useful to represent a point process where animals are clustered in space or time, where the gamma distribution represents fine-scale variation in density [133];

- *Lognormal-Poisson distribution*: as an alternative to the negative binomial, we could instead assume that spatial clustering arises from a lognormal process:

$$\log(\lambda^*) \sim \mathrm{Normal}(\mu, \sigma^2)$$
$$N \sim \mathrm{Poisson}(\lambda^*) \tag{2.6}$$

where σ is the log-standard deviation of overdispersion, and μ is the log-density. Although the lognormal-Poisson distribution is less widely used than the negative binomial, normally distributed variation in the log-intensity is the simplest form of what's called a *log-Gaussian Cox process*;

- *Dirichlet-multinomial distribution*: in a point process involving C categorical marks (i.e., taxa, sizes, ages, etc.), the multinomial distribution is useful to represent the count N_c for each category $c \in \{1, 2, ..., C\}$ given a known total count $n = \sum_{c=1}^{C} N_c$ across categories. The Dirichlet-multinomial distribution then generalizes this by approximating a Poisson point process where each species is spatially clustered [230]. For example, we might specify an average proportion $\mathbf{p}$ for each category, but where a given site has proportion $\mathbf{p}^*$ that differs due to small-scale variation in habitat:

$$\mathbf{p}^* \sim \mathrm{Dirichlet}(\alpha \mathbf{p})$$
$$\mathbf{N} \sim \mathrm{Multinomial}(\mathbf{p}^*, n) \tag{2.7}$$

where the variance of the Dirichlet distribution decreases as α increases, such that the compound Dirichlet-multinomial distribution reverts to a standard multinomial distribution as $\alpha \to \infty$.

In each of these compound distributions, the marginal distribution is then obtained by marginalizing across the outcome of some underlying latent variable. However, it is helpful to be familiar with these distributions because:

1. *Closed-form calculation*: the negative binomial and Dirichlet-multinomial distributions also have a closed form expression for evaluating the likelihood, and therefore can be fitted with little computational cost;

2. *Computational efficiency*: the Tweedie distribution is implemented in many packages, using several computational tricks to promote computational efficiency;

3. *Simplified explanation*: each of these distributions is commonly encountered across ecological models, and therefore can be described without needing extensive background for readers.

2.1.2 A Quick Derivation of Random Effects

We next seek to generalize the process that was presented in Section 2.1.1 when calculating the compound Poisson-gamma distribution, so that we can use it during maximum likelihood estimation. To do so, we first introduce two concepts:

- *Axiom of total probability*: say we have two events X and Y. We can re-write a joint probability distribution $\Pr(X, Y)$ as the product of a conditional and marginal distribution:

$$\Pr(X, Y) = \Pr(Y|X)\Pr(X) \tag{2.8}$$

  This is how we've formulated each compound distribution (e.g., 2.7, 2.6, and 2.5), where it is easier to specify a conditional probability $\Pr(Y|X)$ and the marginal probability of a latent variable $\Pr(X)$ than specify the joint probability $\Pr(X, Y)$;

- *Law of total probability*: similarly, we can marginalize a joint probability to get a marginal probability. If the latent variable X is continuous-valued, this involves an integral:

$$\Pr(Y) = \int \Pr(X, Y)\mathrm{d}X \tag{2.9}$$

  Alternatively, if latent variable X is discrete-valued with values Ω, this involves a summation as we already saw in Eq. 2.3:

$$\Pr(Y) = \sum_{x \in \Omega} \Pr(Y, X = x) \tag{2.10}$$

  In either case, it tends to require some careful thought to find a computationally efficient way to marginalize across latent variable X.

In essence, the axiom of total probability allows us to construct a joint probability from a marginal and conditional probability, while the law of total probability allows us to recover a marginal from a joint probability.

Next, we introduce a vector of latent variables ϵ, where we can define a distribution for data given these variables, $\Pr(Y|\theta, \epsilon)$, and can also specify a distribution for them $\Pr(\epsilon|\theta)$. By the Axiom of Total Probability, the product of these two is the joint probability of data and random effects. Then, by the Law of Total Probability, we can integrate this to get the marginal likelihood of parameters:

$$\mathcal{L}(\theta; Y) = \int \Pr(Y|\theta, \epsilon)\Pr(\epsilon|\theta)\mathrm{d}\epsilon \tag{2.11}$$

This derivation shows that we can still calculate $\hat{\theta}$ by maximizing $\log \mathcal{L}(\theta; y)$, but in some cases it requires introducing random effects ϵ which we then marginalize across (either using summation or integration).

After identifying $\hat{\theta}$, we also define the "empirical Bayes" predictions for random effects:

$$\hat{\epsilon} = \mathrm{argmax}_\epsilon \Pr(Y|\hat{\theta}, \epsilon)\Pr(\epsilon|\hat{\theta}) \tag{2.12}$$

These empirical Bayes predictions $\hat{\epsilon}$ are affected both by data $\Pr(Y|\hat{\theta}, \epsilon)$ but are also shrunk towards a distribution $\Pr(\epsilon|\hat{\theta})$ that is affected by estimated parameters $\hat{\theta}$, such that the empirical Bayes estimator generalizes the properties of James-Stein shrinkage. However, calculating $\mathcal{L}(\theta; Y)$ requires an integral across ϵ, which we call "random effects".

TABLE 2.1: List of methods used throughout the textbook to calculate the area or volume under a function (called *integration*), where these methods provide an inverse for differentiation (Table 1.2).

Name	Description
Analytical	Memorizing many well-known transformations and applying them in clever ways (e.g., using u-substitution) to function $f(x)$ to define the antiderivative $F(x)$ that can later be evaluated to calculate a definite or indefinite integral given some bounds
Monte Carlo	Evaluating function $f(x)$ at some randomized set of points, and computing the integral via the evaluated values at those randomized points
Rejection sampling	Randomly sampling points x from the domain of $f(x)$ and either rejecting or accepting them, where the accepted points then approximates the target function $f(x)$ and can be used to integrate some transformation of that function (Section B.3)
Markov chain Monte Carlo (MCMC) sampling	Randomly sampling a sequence of points x and either accepting or rejecting changes that occur during that sequence, such that the sequence of values then approximate the target function $f(x)$, with use similar to rejection sampling
Laplace approximation	Approximating the joint likelihood $f(x)$ as if it follows a normal distribution with the same peak value, location, and curvature (see Section 2.2)

2.2 Introducing the Laplace Approximation

We next introduce the *Laplace approximation*, which can be a computationally efficient way of computing the integral (called *integration*) across random effects. Ecologists generally receive some background in integration and may recall that it can be conducted analytically by identifying an antiderivative for a given function. However, analytical integration is generally harder than analytical differentiation, and we again suspect that most ecologists are unlikely to investigate a function by integrating it analytically. We therefore introduce several techniques to compute an integral (Table 2.1) but generally emphasize the Laplace approximation due to its computationally efficiency when combined with automatic differentiation (Table 1.2) [208].

At it's core, the Laplace approximation replaces any integral with a multivariate normal distribution that has similar properties, and calculates the area under that multivariate normal distribution [208]. To approximate the integral for a function $g(\theta)$ it requires calculating:

1. *Location of peak*: the value $\hat{\theta} = \text{argmax}_\theta(\log(g(\theta)))$ that maximizes the function;

2. *Value at peak*: the function value $\log(g(\hat{\theta}))$ at this peak;

3. *Curvature at peak*: the curvature of the function at this peak, measured as the second derivative of $\log(g)$ around $\hat{\theta}$, $h = \frac{d^2}{d\theta^2} \log(g(\hat{\theta}))$.

The Laplace approximation then substitutes g with a normal distribution having those same three values. The normal distribution has density:

$$\text{dNormal}(x|\mu, \sigma) = \frac{1}{\sigma\sqrt{2\pi}} e^{-0.5(x-\mu)^2 \sigma^{-2}} \tag{2.13}$$

where σ is the standard deviation and μ is the mean of the distribution, and (like any probability) the area under this distribution is equal to 1. Therefore, if a distribution g has the same shape as the normal distribution, but has peak with value $g(\hat{x})$ and $\log(g)$ has curvature h, then the integral for g will be approximately:

$$\log\left(\int g(\theta)d\theta\right) \approx \sqrt{2\pi} \log\left(g(\hat{\theta})\right) 0.5h \tag{2.14}$$

Similarly, the Laplace approximation can be applied to approximate a function $g(\theta)$ where θ has two or more dimensions. In this case, we approximate the function using a multivariate normal distribution, which has density:

$$\text{dMVN}(\mathbf{x}|\mu, \Sigma) = \frac{\sqrt{|\mathbf{Q}|}}{(2\pi)^{k/2}} e^{-0.5(\mathbf{x}-\mu)^T \mathbf{Q}(\mathbf{x}-\mu)} \tag{2.15}$$

where $\mathbf{Q} = \Sigma^{-1}$ is the inverse of the covariance matrix and $|\mathbf{Q}|$ is the determinant of this matrix. $\mathbf{Q}$ will come up often in subsequent chapters, and is often specifically called the *precision matrix*. Importantly, the precision matrix $\mathbf{Q}$ and not the covariance matrix Σ appears in the multivariate normal distribution, so we will often look for ways to compute the precision matrix $\mathbf{Q}$, rather than constructing the covariance matrix Σ and then inverting it (because this matrix inverse step is computationally expensive). As we saw in Section 1.5, we can measure the curvature using precision $\mathbf{Q}$ by calculating a matrix of second derivatives of $\log(g)$, called the Hessian matrix $\mathbf{H}$.

Extending Eq. 2.14, if a function has peak with value $g(\hat{\theta})$ with $\log(g)$ has curvature $\mathbf{H}$, we can approximate the integral using a multivariate normal distribution with those same characteristics:

$$\log\left(\int g(\theta)d\theta\right) \approx (2\pi)^{k/2} \log\left(g(\hat{\theta})\right) 0.5|\mathbf{H}| \tag{2.16}$$

where $|\mathbf{H}|$ is the determinant of this Hessian matrix. This then reverts to Eq. 2.14 when g has only a single dimension but generalizes it for two or more dimensions, and we present both because the univariate version might be more familiar to some readers.

We first show how the Laplace approximation is used to approximate a univariate integral (Eq. 2.14) using a simple example with a known "Chi-squared" distribution, so we can compare the Laplace approximation with the true value. The chi-squared distribution has a known density function:

$$\Pr(X = x|k) = \frac{1}{2^{k/2}\Gamma(k/2)} x^{k/2-1} e^{-x/2} \tag{2.17}$$

Taking the log and eliminating the component that does not depend on x:

$$\log\left(\Pr(X = x|k)\right) \propto f(x|k) = \left(\frac{k}{2} - 1\right)\log(x) - \frac{x}{2} \tag{2.18}$$

We then calculate the first derivative:

$$f'(x|k) = \left(\frac{k}{2} - 1\right)x^{-1} - \frac{1}{2} \tag{2.19}$$

The peak of this function then occurs where $f'(x|k) = 0$, and solving Eq. 2.19 for x shows that this occurs when $\hat{x} = k - 2$. Substituting this into Eq. 2.19 and then taking the second derivative yields:

$$f''(\hat{x}|k) = \frac{1}{2(k-2)} \tag{2.20}$$

We can therefore approximate the chi-squared distribution with a normal distribution that has the same peak and a variance equal to $1/f''(\hat{x}|k)$:

$$\log\left(\Pr(X = x|k)\right) \propto \log\left(\text{dNormal}(x|\mu = k - 2, \sigma^2 = 2(k-2))\right) \tag{2.21}$$

Alternatively, we can do this without calculating derivatives algebraically, and instead use the function **D** in R to do "symbolic differentiation" (see Code 2.1). We used symbolic differentiation previously in Section 1.2 to simplify calculating the distribution for age-at-death, and here show that the same can be done when calculating the Laplace approximation. As this code demonstrates, the area under this normal approximation can be calculated as:

CODE 2.1: R code visualizing Laplace approximation for the chi-squared distribution using symbolic differentiation.

```
1  # Calculate derivatives for log-density of chi-square distribution
2  log_chi2 = function(x, k, order=0 ){
3    # Define log of chi-squared distribution
4    Expr = expression( log(1/(2^(k/2)*gamma(k/2)) * x^(k/2-1) * exp(-x/2)) )
5    # Calculate symbolic derivatives
6    if(order>=1) Expr = D(Expr,"x")
7    if(order>=2) Expr = D(Expr,"x")
8    out = eval(Expr)
9  }
10
11 # Define function for plotting
12 plot_laplace = function( k, xlim=c(0,k*3), ... ){
13   Opt = optimize( log_chi2, interval=xlim, k=k, order=0, maximum=TRUE )
14   H = -1 * log_chi2( x=Opt$maximum, k=k, order=2 )
15   x = seq(xlim[1],xlim[2],length=1000)
16   ytrue = exp(log_chi2(x,k=k))
17   plot( x=x, y=ytrue, type="l", main=paste0("k = ", k), lwd=2, ... )
18   ynorm = dnorm( x, mean=Opt$maximum, sd=sqrt(1/H) )
19     ynorm = ynorm / max(ynorm) * exp(Opt$objective)
20   #Equal to:  ynorm = exp(Opt$objective) * exp( -0.5*(x-Opt$maximum)^2*H )
21   lines( x=x, y=ynorm, col="blue", lwd=2 )
22   dens_laplace = exp( log(sqrt(2*pi)) + Opt$objective - 0.5*log(H) )
23   legend("topright", bty="n", legend=signif(dens_laplace,3), text.col="blue")
24 }
25
26 # Plot for three chi-squared distributions
27 par(mfrow=c(1,3), mar=c(3,3,3,1), mgp=c(2,0.5,0), xaxs="i")
28 plot_laplace(k = 5)
29 plot_laplace(k = 25)
30 plot_laplace(k = 100)
```

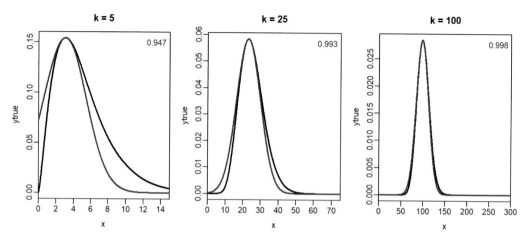

FIGURE 2.1: A chi-square distribution with $k = 5, 25, 100$ degrees of freedom (black lines) and the normal approximation (blue lines) that has the same mode, peak, and curvature (2nd derivative evaluated at the peak), and listing the area under the curve for the normal approximation calculated using the Laplace method (blue legend).

$$\int e^{f(\theta)} d\theta = \sqrt{2\pi} e^{f(\hat{\theta}) - 0.5 f''(\hat{\theta})} \tag{2.22}$$

As we showed in Section 1.2, we could instead approximate the Chi-squared density function by directly sampling from it. This could be useful, e.g., if we knew the density function but did not already know its mean or standard deviation. In this example, we again use Rejection Sampling, and display the mean of the samples in each panel.

Comparing the Laplace approximation (Fig. 2.1) and Rejection Sampling (Fig. 2.2), we see that both approaches can be used to approximate the shape and resulting properties of a

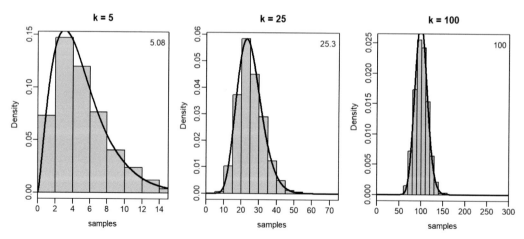

FIGURE 2.2: A chi-square distribution with $k = 5, 25, 100$ degrees of freedom (black lines) and the approximation obtained using Rejection Sampling for 1000 samples, listing the sample-based approximation to the mean in the top-right for each panel (which should be equal to k, and differs due to the finite sample size and sampling bounds that must be specified during Rejection Sampling).

function. However, the Laplace approximation is used in particular to approximate the area under a given curve (i.e., integrate a monotonic function), while sampling is typically used to calculate, e.g., the mean and standard deviation of a function when those are not already known. We next demonstrate how these two approximations can be used for maximum likelihood estimation.

2.3 Estimating Heterogeneity using a Generalized Linear Mixed Model

To summarize, recall that maximum likelihood involves identifying a vector of parameters that best explains the data:

$$\hat{\theta} = \text{argmax}_\theta (\log \mathcal{L}(\theta; y)) \tag{2.23}$$

and we have now defined a marginal likelihood that involves integrating across random effects by approximating them as a normal distribution:

$$\log \mathcal{L}(\theta; y) = \log \left(\int e^{f(\theta, \epsilon; y)} d\epsilon \right) \tag{2.24}$$

where $f(\theta, \epsilon; y) = \log \left(\Pr(y|\theta, \epsilon) \Pr(\epsilon|\theta) \right)$ and this approximation in turn involves calculating $\hat{\epsilon}$:

$$\hat{\epsilon} = \text{argmax}_\epsilon f(\theta, \epsilon; y) \tag{2.25}$$

So in terms of algorithm, using the Laplace approximation to fit a model with fixed and random effects involves:

1. Defining the joint log-likelihood function $f(\theta, \epsilon; y)$;

2. Proposing a value of fixed effects θ;

3. Optimize the joint log-likelihood with respect to random effects to identify $\hat{\epsilon}$;

4. Calculate the Hessian of $f(\theta, \hat{\epsilon})$ with respect to ϵ (termed the *inner Hessian matrix*);

5. Combine these two to approximate the log of the marginal likelihood;

$$\log \mathcal{L}(\theta; y) = f(\theta, \hat{\epsilon}; y) - 0.5 \log(|\mathbf{H}|) + \frac{N}{2 \log(2\pi)} \tag{2.26}$$

where N is the number of random effects, and noting that the term $\frac{N}{2 \log(2\pi)}$ does not depend upon parameters and can therefore be dropped without affecting the estimated parameters or standard errors;

6. Propose a new value of fixed effects θ, repeat steps 3–5, compare the new and old log-marginal likelihoods, and accept whichever one is higher;

7. Repeat step 6 until no further improvements are found;

8. Calculate the Hessian of the log-marginal likelihood $\log \mathcal{L}(\theta; y)$ with respect to fixed effects (termed the *outer Hessian matrix*). The inverse of this outer Hessian matrix is then an estimate of the covariance in estimated fixed effects (recall Section 1.5).

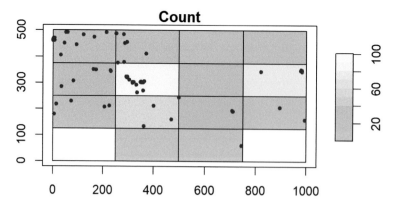

FIGURE 2.3: Gridded numerical abundance of *Vismia baccifera* for 16 cells in the Barro Colorado census in 1985.

To demonstrate this algorithm, we use real-world data measuring densities of different tree species in *Barro Colorado*, a long-term ecological experiment in Panama that records the exact location and species for every woody tree and shrub at least 1 cm in diameter within a 1 km by 500 m area over a sequence of years, 1982, 1985, 1990, 1995, 2000, 2005, and 2010 [33][2]. This field experiment has yielded a wealth of insights about species turnover and community stability, and we here use it to illustrate that real-world data can be analyzed using these statistical principles. We specifically illustrate data for *Vismia baccifera* in 1985, and grid the data into 16 square cells each 250 m by 125 m (Fig. 2.3).

Given that we have counts in each grid cell, we might start by specifying a Poisson GLM. However, it is apparent that this would not capture the variation among sites. Additionally, we might specifically wonder what is the variance of log-density among these sites. To do this, we instead specify a generalized linear mixed model (GLMM).

$$C_i \sim \text{Poisson}(e^{\epsilon_i})$$
$$\epsilon_i \sim \text{Normal}(\mu, \sigma^2) \tag{2.27}$$

where μ is the average log-density, ϵ_i is the log-density for site i, and σ^2 is the variance in log-density among sites.

CODE 2.2: R code for fitting a log-linked Poisson distribution in a generalized linear mixed model using package lme4 .

```
1  library(lme4)
2  Data$Site = factor(1:nrow(Data))
3  Lme = glmer( Count ~ 1 + (1|Site), data=Data, family=poisson(link = "log") )
```

This model is easily fitted using many packages in R, and we use the lme4 package for simplicity [9] (Code 2.2). In this example, family = poisson(link = "log") specifies a log-linked linear predictor, and formula Count ~ 1 + (1 | Site) indicates that we want to estimate the intercept (representing log-density μ) as well an additional intercept as

---

[2]We acknowledge R. Foster as plot founder and the first botanist able to identify so many trees in a diverse forest; R. Pérez and S. Aguilar for species identification; S. Lao for data management; S. Dolins for database design; plus hundreds of field workers for the census work, now over 2 million tree measurements; the National Science Foundation, Smithsonian Tropical Research Institute, and MacArthur Foundation for the bulk of the financial support. Data downloaded from https://repository.si.edu/handle/10088/20925.

random effect for each site (representing the centered deviation $\epsilon_i - \mu$ for each i). This function then generates a wide range of output, including the intercept and standard deviation estimates, as well as indicating that lme4 is already using the Laplace approximation:

```
1  Generalized linear mixed model fit by maximum likelihood (Laplace
       Approximation) ['glmerMod']
2   Family: poisson  ( log )
3  Formula: Count ~ 1 + (1 | Site)
4     Data: Data
5        AIC        BIC    logLik deviance df.resid
6  130.4195  131.9647  -63.2098 126.4195       14
7  Random effects:
8   Groups Name        Std.Dev.
9   Site   (Intercept) 1.739
10 Number of obs: 16, groups:  Site, 16
11 Fixed Effects:
12 (Intercept)
13       1.912
```

CODE 2.3: R code for joint likelihood of a log-linked Poisson GLMM.

```
1  # Joint negative-log-likelihood
2  joint_nll = function( prandom, pfixed, plist, Data, random ){
3    # Read in values
4    phat_fixed = relist(pfixed, plist[setdiff(names(plist),random)])
5    phat_random = relist(prandom, plist[random])
6    phat = c(phat_fixed,phat_random)
7    # Data likelihood
8    jnll = 0
9    for( i in 1:nrow(Data) ){
10     jnll = jnll - dpois(Data$Count[i],lambda=exp(phat$eps[i]),log=TRUE)
11   }
12   # Random effect distribution
13   for( i in 1:nrow(Data) ){
14     jnll = jnll - dnorm(phat$eps[i],mean=phat$logmu,sd=exp(phat$logsd),log=
         TRUE)
15   }
16   return(jnll)
17 }
```

We again want to replicate this calculation using lower-level methods that can subsequently be extended. To do so, we implement the inner and outer optimizers in R. We first define a function that calculates the joint likelihood (Code 2.3). We then use this to define a function that maximizes the joint likelihood with respect to random effects, calculates the inner Hessian matrix, and assembles the Laplace approximation to the marginal likelihood (Code 2.4). We later call this the *inner optimizer*. Finally, we use an optimizer to identify the maximum likelihood estimates for fixed effects (Code 2.5), which we later call the *outer optimizer*. Inspecting estimates (Table 2.2) shows that they agree with those from lme4 .

CODE 2.4: R code calculating the marginal likelihood by implementing an inner optimizer and calculating the Laplace approximation to the marginal likelihood.

```
1  # Outer optimization function
2  marg_nll = function( pfixed, plist, Data, random, jnll, what="laplace" ){
3    # Inner optimizer
4    prandom = unlist(plist[random])
5    inner = nlminb(start=prandom, objective=jnll, pfixed=pfixed, plist=plist,
         Data=Data, random=random)
6
7    # Calculate the Laplace approximation
8    inner$hessian = optimHess(par=inner$par, fn=jnll, pfixed=pfixed, plist=
         plist, Data=Data, random=random)
```

TABLE 2.2: Estimated parameters and standard errors using the Laplace approximation implemented using custom code in R.

	Estimate	Std. Error
logmu	1.44	NA
logsd	1.08	NA

```
9    inner$laplace = inner$objective + 0.5*log(det(inner$hessian)) - length(
        prandom)/2*log(2*pi)
10   if(what=="laplace") return(inner$laplace)
11   if(what=="full") return(inner)
12 }
```

CODE 2.5: R code for maximizing the marginal likelihood using an outer optimizer.

```
1  # Define inputs
2  plist = list("logmu"=0, "logsd"=0, "eps"=rep(0,nrow(Data)))
3  random = "eps"
4
5  # Identify MLE using outer optimizer
6  opt1 = nlminb( objective = marg_nll,
7                 start = unlist(plist[setdiff(names(plist),random)]),
8                 jnll = joint_nll,
9                 plist = plist,
10                Data = Data,
11                random = random,
12                control = list(trace=1) )
13
14 # Compute finite-difference approx. to outer Hessian
15 Hess = optimHess( par = opt1$par,
16                   fn = marg_nll,
17                   plist = plist,
18                   Data = Data,
19                   random = random,
20                   jnll = joint_nll )
```

For illustration, we also extend this demonstration to include additional code that applies a finite-difference approximation to obtain the gradient of the marginal likelihood function with respect to fixed effects (see Table 1.2). We can then use this gradient to aid the process of identifying the maximum-likelihood estimates for fixed effects in the outer optimizer (Code 2.6). Inspecting estimates now shows that estimates more closely agree with lme4 (comparing Tables 2.4 and 2.3), and that the final gradient of the log-marginal likelihood is suitably low (as required for a maximum likelihood model to be converged).

CODE 2.6: R code for adding finite-difference gradient for outer optimizer using Laplace approximation for mixed-effect estimation.

```
1  # Function to calculate finite difference gradient
2  grad = function( pfixed, ... ){
3    delta = 0.0001
4    gr = rep(0,length(pfixed))
5    for(i in seq_along(gr)){
6      dvec = rep(0,length(pfixed))
7      dvec[i] = delta
8      # Calculate central finite difference
9      val1 = marg_nll(pfixed=pfixed+dvec, ... )
10     val0 = marg_nll(pfixed=pfixed-dvec, ...)
11     gr[i] = (val1 - val0) / (2*delta)
12   }
13   return(gr)
```

TABLE 2.3: Estimated parameters, standard errors, and final gradients using the Laplace approximation implemented using custom code in R that utilizes finite-difference gradients.

	Estimate	Std. Error	Final gradient
logmu	1.91	0.476	-0.00720905362073
logsd	0.555	0.238	-0.000743

```
14  }
15
16  # Re-estimate parameters
17  opt2 = nlminb( objective = marg_nll,
18                 gradient = grad,
19                 start = unlist(plist[setdiff(names(plist),random)]),
20                 jnll = joint_nll,
21                 plist = plist,
22                 Data = Data,
23                 random = random,
24                 control = list(trace=1) )
25  Hess = optimHess( par = opt2$par,
26                    fn = marg_nll,
27                    plist = plist,
28                    Data = Data,
29                    random = random,
30                    jnll = joint_nll )
```

This process of writing an inner and outer optimizer and then the gradient function is cumbersome. We therefore turn to TMB to automate the entire process, including calculating the gradients of the joint likelihood, computing the Hessian and Laplace approximation, and computing the gradient of that Laplace approximation to the log-marginal likelihood with respect to fixed effects. As before, this involves writing the joint negative log-likelihood $f(\theta, \epsilon | y)$ in a template file in C++ (Code 2.7). This template file is then compiled and linked, and then the TMB object is built and optimized within R (Code 2.8). By doing this, TMB automatically implements the inner optimizer while providing both the marginal log-likelihood and its gradients for use in an outer optimizer within R. Inspecting estimates, it agrees more closely with lme4 than either implementation using base R (while also being much faster to write and run).

CODE 2.7: TMB code for a log-linked Poisson generalized linear mixed model.

```
1  #include <TMB.hpp>
2  template<class Type>
3  Type objective_function<Type>::operator() ()
4  {
5    // Data
6    DATA_VECTOR( y_i )
7
8    // Parameters
9    PARAMETER_VECTOR( eps_i );
10   PARAMETER_VECTOR( ln_mu );
11   PARAMETER( ln_sd );
```

TABLE 2.4: Estimated parameters, standard errors, and final gradients using the Laplace approximation implemented using TMB.

	Estimate	Std. Error	Final gradient
ln_mu	1.91	0.469	$-2.49292412824\text{E-}07$
ln_sd	0.555	0.218	$-7.46\text{E-}07$

```
12
13    // Global variables
14    Type jnll = 0;
15
16    // Probability of data conditional on fixed and random effect values
17    vector<Type> yhat_i(y_i.size());
18    for( int i=0; i<y_i.size(); i++){
19      yhat_i(i) = exp( eps_i(i) );
20      jnll -= dpois( y_i(i), yhat_i(i), true );
21    }
22
23    // Probability of random effects
24    for( int i=0; i<y_i.size(); i++){
25      jnll -= dnorm( eps_i(i), ln_mu(0), exp(ln_sd), true );
26    }
27    Type yhat_sum = yhat_i.sum();
28
29    // Return values to R
30    REPORT( jnll );   // Used in shrinkage demo
31    REPORT( yhat_i );
32    REPORT( yhat_sum );
33    ADREPORT( yhat_i );
34    ADREPORT( yhat_sum );
35    return jnll;
36  }
```

CODE 2.8: R code for running the generalized linear mixed model implemented using TMB.

```
1  # Compile and load TMB model
2  library(TMB)
3  compile("poisson_glmm.cpp")
4  dyn.load("poisson_glmm")
5
6  # Define inputs
7  data = list( "y_i"=Data$Count )
8  params = list( "eps_i"=rep(0,nrow(Data)),
9                 "ln_mu"=0,
10                "ln_sd"=0 )
11
12 # Compile TMB object while declaring random effects for inner loop
13 Obj = MakeADFun( data=data, parameters=params, random="eps_i" )
14
15 # Outer optimizer using gradients
16 Opt = nlminb( start=Obj$par, obj=Obj$fn, grad=Obj$gr )
17
18 # Compute standard errors including joint precision
19 Opt$SD = sdreport( Obj, getJointPrecision=TRUE )
```

We also approximate the uncertainty in estimated fixed effects θ or predicted random effects ϵ by constructing the *joint precision matrix* $\mathbf{Q}_{joint}$:

$$\mathbf{Q}_{joint} = \begin{pmatrix} \mathbf{H}_1 & -\mathbf{H}_1\mathbf{G} \\ -\mathbf{G}^t\mathbf{H}_1 & \mathbf{G}^t\mathbf{H}_1\mathbf{G} + \mathbf{H}_2 \end{pmatrix} \qquad (2.28)$$

where $\mathbf{H}_2$ is outer Hessian matrix, $\mathbf{H}_1$ is the inner-Hessian matrix, and $\mathbf{G}$ is the matrix of gradients of predicted random effects with respect to fixed effects (the *outer Jacobian matrix*). This construction of $\mathbf{Q}$ represents the precision (inverse-variance) arising from uncertainty about the value of fixed effects $\mathbf{H}_1$, uncertainty about the random effects given estimated fixed effects $\mathbf{H}_2$, how uncertainty about fixed effects affects uncertainty about random effects $\mathbf{H}_1\mathbf{G}$, and uncertainty about random effects accumulated from fixed effects $\mathbf{G}^t\mathbf{H}_2\mathbf{G}$. Inverting this joint precision then estimates the uncertainty in random effects

[113]. We can calculate this using our custom version of the Laplace approximation (Code 2.9), but TMB calculates it much faster automatically.

CODE 2.9: R code for calculating the joint precision matrix from the inner and outer Hessian matrices.

```
1  # Extract predicted random effects
2  fn = function( pfixed ){
3    inner = marg_nll( pfixed=pfixed, plist=plist, Data=Data, random=random,
       jnll=joint_nll, what="full" )
4    return(inner$par)
5  }
6  # calculate outer Jacobian (gradient) matrix
7  G = numDeriv::jacobian( func=fn, x=opt2$par )
8  # Inner and outer Hessian matrices
9  H1 = marg_nll( pfixed=opt2$par, plist=plist, Data=Data, random=random, jnll=
       joint_nll, what="full" )$hessian
10 H2 = Hess
11 # Assemble joint precision
12 Q_joint <- rbind(
13   cbind2( H1, -H1 %*% G ),
14   cbind2( -t(G) %*% H1, as.matrix(t(G) %*% H1 %*% G) + H2 )
15 )
```

Similar to Section 2.2, we can also compare the Laplace approximation with a sample-based approximation. In this case, we use *Markov Chain Monte Carlo* (MCMC) and specifically the No-U-Turn-Sampler NUTS [91] as implemented in Stan (see [156] for an accessible introduction to NUTS). MCMC generally involves generating a chain of samples from the *posterior distribution* (the product of the the joint likelihood function and Bayesian priors for fixed effects). In this case, we do not explicitly specify any Bayesian priors, so implicitly we use an uniform and unbounded prior for all fixed effects.

Conveniently, users can run Stan algorithms on a TMB object using R-package tmbstan [155] with no changes needed when defining the joint negative log-likelihood (Code 2.10). We compare these approaches for illustration purposes, and note that a real-world application would require consideration of Bayesian priors and post-hoc exploration of their impact on results. We also note that NUTS could instead be run on the log-marginal likelihood (calculated using the Laplace approximation) or on the joint negative log-likelihood (as used during the inner optimization step). Any difference between these two then indicates either error in the MCMC sampling algorithm or the Laplace approximation, and therefore any difference can be used to indicate that further investigation is needed. We then compare the estimated densities obtained from both the Laplace approximation and NUTS sampling (Fig. 2.4), and see that these are indistinguishable by eye (despite NUTS including implicit uniform priors).

CODE 2.10: R code for running STAN to sample from the generalized linear mixed model implemented using TMB.

```
1  library(tmbstan)
2  fit <- tmbstan(Obj, chains=1)
```

As noted in 2.1, hierarchical models are useful in part because they provide an objective way to implement *shrinkage*. To explain this in detail, we introduce the notation for a *generalized additive model* (GAM) [266, 86]. Similar to a GLM, a GAM identifies parameters by maximizing an objective function $\mathcal{P}$. It specifically calculates this objective function as the probability of the data given parameters but also includes an additional penalty on estimated coefficients:

$$\log \mathcal{P}(\theta; Y) = \log \Pr(Y|\theta) - \lambda \theta^{\mathbf{T}} \mathbf{S} \theta \qquad (2.29)$$

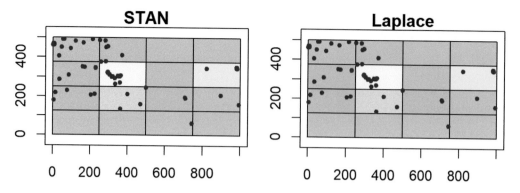

FIGURE 2.4: Predictions of gridded numerical abundance for 16 cells using the Laplace approximation or No-U-Turn-Sampling, both implemented using TMB and using the same color scale as Fig. 2.3.

where $\mathbf{S}$ is a matrix that represents penalties on GAM coefficients. Comparing this with maximum likelihood estimation (Eq. 2.26), we see that the GAM requires specifying a penalty matrix $\mathbf{S}$ and identifying the penalty λ, where the latter can be tuned based on fit to the data using a metric like *generalized cross-validation*. By contrast, maximum likelihood automatically estimates the penalty $0.5 \log(|\mathbf{H}|)$ during parameter estimation (Fig. 2.5). In GLMMs such as this, the penalty term $0.5 \log(|\mathbf{H}|)$ decreases as the variance of random effects increases, such that the marginal likelihood is maximized at a higher estimate of variance than the joint likelihood. In fact, the joint likelihood typically increases infinitely as the variance of random effects approaches zero, and the log-determinant of the Hessian is needed to pull the maximum likelihood estimate away from this degenerate solution.

Despite these differences between GLMMs and GAMs, both models include a component that *shrinks* the coefficients towards zero. Both are therefore ways of implementing a shrinkage estimator. Shrinkage estimation can then bridge continuously between two different generalized linear models:

- *Intercept-only GLM*: as the shrinkage becomes large (i.e., σ becomes small or λ becomes large), then $\epsilon_i = \mu$ for all sites in the GLMM or GAM and these approach a GLM with only a single intercept;

- *Site-level GLM*: as the shrinkage becomes large (i.e., variance σ^2 becomes large or penalty λ becomes small), then $\epsilon_i + \mu$ will approach the estimates that arise fitting a GLM with an intercept for each site.

The site-level GLM has more parameters (one intercept for each site) than the intercept-only GLM (one intercept for all sites). Intuitively then, using a shinkage estimator (whether GLMM or GAM) to estimate parameters results in an *effective degrees of freedom* that is intermediate between these two extremes.

We have illustrated that including random effects has many practical benefits (listed in Section 2.1). However, they also complicate the evaluation of model fit and interpretation of residuals. To see this, recall from Section 1.7 that we can calculate residuals by comparing model predictions $\hat{Y}$ with observations Y, and can use residuals to evaluate fit by constructing a quantile-quantile plot or applying a statistical test. However, the mixed-effects model now has two apparent ways to calculate residuals:

- *Conditional residuals* involve calculating model predictions conditional on maximum likelihood estimates of fixed effects $\hat{\theta}$ and empirical Bayes predictions of random effects $\hat{\epsilon}$;

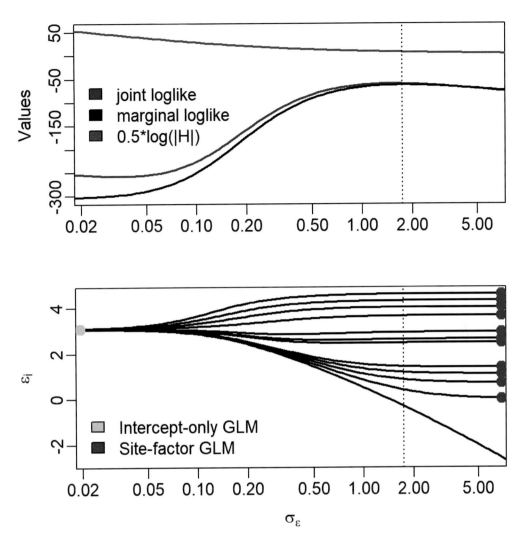

FIGURE 2.5: Depiction of shrinkage in the generalized linear mixed model (see Fig. 2.3), showing (top panel) the joint log-likelihood, marginal log-likelihood, and the penalty term $0.5 \log(|\mathbf{H}|)$ calculated from the inner Hessian matrix $\mathbf{H}$ (with maximum likelihood estimated shown as dotted vertical line), where the marginal loglikehood (black) is the joint loglikelihood (blue line) minus the penalty (red line) plus a constant. Also showing (bottom panel) the estimates of each site-level log-density for each site (y-axis) for different values of the random effect standard deviation (x-axis, using a log-scale), with the intercept-only GLM and site-level GLM estimates shown as green bullets.

- *Simulation residuals* involve acknowledging that random effects are estimated from a distribution $\hat{p}(\epsilon) \propto \Pr(Y|\hat{\theta}, \epsilon)\Pr(\epsilon|\hat{\theta})$ (recall Eq. 2.12). We could then simulate the value of random effects from this distribution, calculate model predictions given those simulated values, and then calculate residuals.

From this description, it is perhaps apparent that conditional residuals are easier to compute but do not properly acknowledge that random effects are estimated from a distribution. As a result, we cannot compute the expected distribution for model residuals under a properly specified model, and cannot use residuals to determine whether the model departs from this expectation. However, simulation residuals also pose a subtle problem, because the predictive distribution $\Pr(Y_i|\hat{\theta}, \hat{p}(\epsilon))$ for datum Y_i is informed by its observed value via the estimated distribution of random effects $\hat{p}(\epsilon)$. Calculating residuals in this way is circular (i.e., residuals for each datum are calculated from a predicted distribution that is informed by that datum) and this circularity then results in residuals that are not statistically independent. This lack of independence again makes it difficult to compute a "null distribution" for model residuals, or interpret whether any patterns are a substantial departure from this.

To address these projects arising from conditional and simulated residuals, TMB includes a generic implementation for *one-step-ahead* ("OSA") residuals [242]. Calculating OSA residuals involves ordering the data and constructing the predictive distribution for each datum conditional only on preceding data. These OSA residuals are then statistically independent, and simulations confirm that OSA residuals have better performance for identifying model mis-specification than either conditional or marginal residuals [246].

2.4   Inner Hessian Sparsity and Conditional Independence

We have seen that we can implement the Laplace approximation both in custom R code, and via TMB, and that this approximation gives very similar estimates of density to a sampling-based approximation or high-level functions in R. We next discuss the computational aspects of the Laplace approximation in more detail.

In our example, we have 16 coefficients in random effect ϵ, such that the *inner Hessian matrix* is 16 x 16. The Laplace approximation involves calculating the log-determinant of this inner Hessian matrix (Eq. 2.26), which becomes more computationally expensive as the number of random effects increases. To compensate, TMB tracks the inner Hessian as a *sparse matrix* using a triplet format, i.e., where it records the row index, column index, and value of the inner hessian matrix only for a pair of random effects that are correlated conditional upon a fixed value of parameters. This inner Hessian is a diagonal matrix in our example (i.e., the second derivative of any two sites is zero) and is therefore as sparse as possible for the GLMM we are using. We will often want to plot this inner Hessian to check its sparsity pattern (Fig. 2.6), where this image shows a nonzero value on the diagonal and is exactly zero along the off-diagonal.

There are several ways to conceptualize the sparsity of the inner Hessian:

1. *Conditional independence*: if a random effect ϵ_1 is independent of another ϵ_2 conditional upon a fixed value of for other fixed and random effects, then the inner Hessian matrix will be zero for that pair of random effects.

2. *Separability*: se have seen that the marginal likelihood is calculated by computing a multidimensional integral across random effects:

$$\mathcal{L}(\theta; y) = \int \Pr(y|\theta, \epsilon) \Pr(\epsilon|\theta)\mathrm{d}\epsilon \qquad (2.30)$$

However, this multidimensional integral can in some times be replaced by a series of lower-dimensional integrals. For example, in our GLMM example, we can replace a 16-dimensional integral with 16 1-dimensional integrals.

$$\mathcal{L}(\theta; y) = \prod_{i=1}^{n_i} \int \Pr(Y_i|\theta, \epsilon_i) \Pr(\epsilon_i|\theta)\mathrm{d}\epsilon_i \qquad (2.31)$$

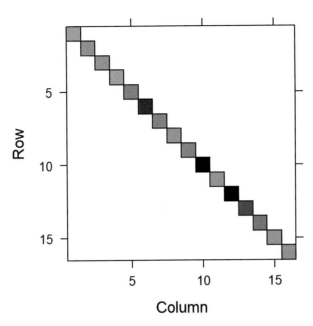

FIGURE 2.6: Hessian matrix for the 16 values of ϵ in our GLMM example, where white cells are known to be exactly zero and hence not tracked in the sparse representation of the matrix.

In other cases, the integral can be broken into multiple blocks. When using the Laplace approximation, breaking the integral into blocks is identical to specifying that the inner Hessian matrix has that same block structure.

3. *Graphical blocking*: a statistical model can generally be written as a graph where fixed effects, random effects, and data are vertices and any probability distribution that involves any pair of parameters is represented by an edge (Fig. 2.7). In a graphical model, two random effects will be correlated if either (1) they both point at a single datum, or (2) they have an arrow pointing from one to the other. In this case, however, there are no arrows between random effects (diamond), and similarly, there are no data that are connected by multiple random effects. Therefore, each random effect is independent conditional upon the value of fixed effects.

Clearly, we are using similar terms across these three explanations, so stating them as separate interpretations is somewhat artificial. However, each might be a convenient approach to think about sparsity in some cases. These same principles apply in later models, where the inner Hessian has a more complicated structure. Finally, we note that different parameterizations can result in an identical maximum likelihood estimate but different sparsity, and typically the more sparse version will run faster and with less computer memory required.

We also emphasize that the Laplace approximation involves alternating between an inner and an outer optimizer (see Code 2.4 and 2.5). This process will converge very slowly on

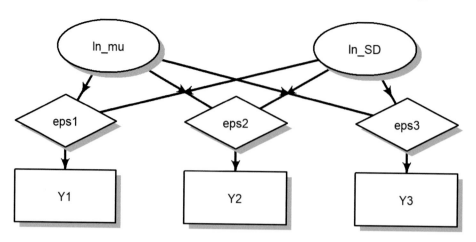

FIGURE 2.7: A graphical model for three of the 16 sites in our GLMM example (Eq. 2.27), showing data as boxes, random effects as diamonds, and fixed effects as circles, and showing the conditional dependencies as arrows.

the maximum likelihood estimate whenever fixed and random effect estimates are highly correlated (i.e., the Hessian of the joint likelihood has higher values between fixed and random effects). For example, our GLMM could instead have been written as:

$$C_i \sim \text{Poisson}(e^{\mu + \epsilon_i})$$
$$\epsilon_i \sim \text{Normal}(0, \sigma^2) \tag{2.32}$$

and this would result in identical estimates of fixed effects. However, it would also cause fixed effect μ and random effect ϵ to be highly correlated. We encourage readers to change the R code to use this parameterization and see how it affects estimation performance in particular when not using gradient information.

2.5   Addressing Common Problems During Inner or Outer Optimization

Finally, every analyst using TMB will encounter bugs and errors with some frequency:

- *TMB syntax errors*: some of these are evident when running the `compile` function to compile a CPP script using TMB and the file fails to compile. Typically, some informative (although difficult to read) messages are displayed in the terminal that can give clues to where the syntax error arises;

- *Out-of-bounds errors*: other bugs occur when running the `MakeADFun` function to build a TMB object. These typically occur when the CPP code is attempting to access a vector, matrix, or array value that is "out of bounds". This occurs, e.g., when defining a vector of length 10 but trying to access the 11th element. This type of error may crash the R session.

These compiler and out-of-bounds errors are programming bugs, and addressing them in detail is outside the scope of this book. However, we do specifically note two more types of error that have a statistical interpretation, where explaining their cause and potential solution can provide insight into model structure. These errors occur either in the outer or inner optimizer, and we further address them in detail below.

2.5.1 Inner Hessian failure

The first type of error occurs during the inner optimizer, where users might see a message like below:

```
1  iter: 1   value: 862.7342 mgc: 100 ustep: 0.0692153
2  iter: 2   value: 266.728 mgc: 904.1507 ustep: 0.2631617
3  iter: 3   value: 105.5603 mgc: 308.9029 ustep: 0.5130416
4  iter: 4   value: 78.98625 mgc: 92.27675 ustep: 0.7162976
5  iter: 5   value: 77.18475 mgc: 19.02569 ustep: 0.846359
6  iter: 6   value: 77.08321 mgc: 1.456402 ustep: 0.9199857
7  iter: 7   value: 77.05425 mgc: 0.05938868 ustep: 0.959163
8  iter: 8   value: 77.04899 mgc: 0.02162014 ustep: 0.9793707
9  iter: 9   value: 77.04861 mgc: 0.005434197 ustep: 0.9896326
10 iter: 10  value: 77.0486 mgc: 0.0007698619 ustep: 0.9948033
11 iter: 11  value: 77.0486 mgc: 5.440074e-05 ustep: 0.9973985
12 iter: 12  value: 77.0486 mgc: 1.917284e-06 ustep: 0.9986986
13 iter: 13  value: 77.0486 mgc: 3.414297e-08 ustep: 0.9993491
14 iter: 14  [1] NaN
```

where the final line is the terminal error message. For demonstration purposes, we have intentionally caused this error to occur by specifying an additional random effect that has no effect on the joint likelihood, i.e., by specifying $params\$eps\_i = rep(0, nrow(Data)+1)$, such that the final element of vector eps_i is not used in the calculation of either the probability of random effects or the likelihood of data. The terminal message indicates that the maximum gradient component ("mgc") of the inner optimizer has become close to zero (abbreviated in the R-terminal messages as mgc: 1.350213e-08). At this point, the inner optimizer stops because the low gradient suggests that the inner optimizer is converged. TMB then calculates the Hessian matrix, and attempts to calculate its determinant for use in the Laplace approximation (Eq. 2.26). However, the final element of eps_i has no effect on the joint likelihood, such that the gradient of the joint likelihood with respect to its value is zero. Similarly, the matrix of second derivatives is zero for any row or column involving this value. This then causes the Hessian matrix to not be invertible; technically it is only positive semi-definite, and has an eigenvalue that is identical to zero, which when inverted would involve dividing by zero (see Section B.4.1).

In this case, the error can be detected by extracting the inner Hessian using Obj\$env\$spHess(random=TRUE) . Plotting this immediately shows that the final random effect has no rows or columns that are non-zero (Fig. 2.8). In more complicated cases, we can instead build the model with all fixed effects "mapped off", and instead optimize the joint log-likelihood with respect to random effects (Code 2.11). This allows us to explore in detail what is happening during the inner optimizer, which is otherwise hidden from the user. This approach is helpful when the inner Hessian fails to converge for some values of fixed effects and not others.

CODE 2.11: R code for manually running the inner optimizer to diagnose lack-of-convergence.

```
1  # Map off fixed effects and treat random as fixed
2  FE = setdiff( names(params), "eps_i" )
3  map <- lapply( FE, function(x) factor(params[[x]]*NA) )
4  names(map) = FE
5  # Rebuild and run optimizer
6  Test = MakeADFun( data=data, parameters=buggy_params, map=map )
7  Opt_random = nlminb( start=Test$par, obj=Test$fn, grad=Test$gr )
```

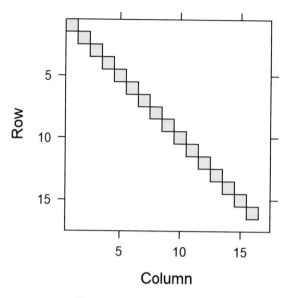

**Inner Hessian in TMB
with non-invertible inner Hessian**

Dimensions: 17 x 17

FIGURE 2.8: Hessian matrix for the 17 values of ϵ in our GLMM example when intentionally including a random effect that has no effect on the joint likelihood and therefore causes an error when calculating the inner Hessian used in the Laplace approximation.

In more complicated cases, it might not be apparent what combination of random effects is causing the inner Hessian to not be invertible. In this case, the analyst can extract the inner Hessian, calculate its eigendecomposition, and identify those eigenvalues that are zero (or very close to zero) (see Section B.4.1). The corresponding eigenvectors can then be inspected, to find what parameters are associated with non-invertible dimensions (Code 2.12). This approach can then identify a linear combination of parameters that is not invertible, and can prompt further thought about model restrictions.

CODE 2.12: R code for identifying non-invertible axes of the inner Hessian matrix.

```
1 # Explore inner Hessian
2 inner_hessian = optimHess( par=Opt_random$par, fn=Test$fn, gr=Test$gr )
3 # Check eigendecomposition
4 Eigen = eigen(inner_hessian)
5 bad_vectors = Eigen$vectors[,which(Eigen$values <= 0)]
6 data.frame( "Par"=names(Test$par), "MLE"=Test$par, "Status"=ifelse(bad_
     vectors>0.1,"Bad","Good") )
```

2.5.2 Outer Hessian failure

The second type of error occurs during the outer optimizer, where users might run sdreport and then see that standard errors are missing:

```
1 sdreport(.) result
2        Estimate Std. Error
```

```
3 ln_mu 1.9089677        NaN
4 ln_mu 0.0000000        NaN
5 ln_sd 0.5551555        NaN
6 Warning:
7 Hessian of fixed effects was not positive definite.
8 Maximum gradient component: 6.30028e-08
```

As indicated by the output, the Hessian of the outer optimizer (which optimizes fixed effects) is not positive definite. We can again use optimHess to construct and visualize the outer Hessian matrix, and inspect the eigendecomposition to determine which combination of parameters is responsible.

Addressing these topics from a more theoretical perspective, we note that a model is *estimable* if and only if:

1. we can identify a set of fixed effects that minimizes the negative marginal log-likelihood; and

2. the Hessian matrix is also positive definite.

Therefore, optimizing and confirming that the outer Hessian is positive definite is sufficient to confirm that the model is estimable for given a particular data set [99]. Alternatively, a model may not be *identifiable*. Identifiability refers to whether a model has parameters that could ever be estimable, given some ideal combination of data. If the model is not identifiability then it can never be estimable for any data set [106]. From the law of contraposition, we can also see that, if a model is estimable for any single data set, then the model is identifiable. On a practical level, it is often possible to specify a non-identifiable model in TMB; in fact, that is what we did by intentionally introducing a fixed effect that had no role in calculating the joint likelihood. In this case, we showed that we could detect that the model was not estimable for a simulated data set. Alternatively, we fitted the log-linked Poisson GLMM in Section 2.3 and calculated its standard errors. By observing that the model was estimable using that single data set, we also confirmed that the log-linked Poisson GLMM was identifiable.

2.6 Chapter Summary

In summary we have showed that:

1. Many real-world circumstances will result in a likelihood that cannot be easily written, either because model residuals are correlated and/or ecological processes involve unobserved ("latent") variables. In these cases, it is helpful to formulate the model while including latent variables, and then obtain the likelihood by marginalizing across these variables. This results in a marginal likelihood that can be optimized to identify maximum likelihood estimators for parameters;

2. The marginal likelihood can be used to identify maximum-likelihood estimates for variance parameters. The estimated variance then controls the magnitude of "shrinkage," wherein random effects are pulled towards some specified average value. This performs a similar role to the penalty term in a generalized additive model, and both provide a continuous bridge between two extremes along a continuum: either fitting a model without random effects, or fitting a model that treats each random effect as if it were estimated separately as a fixed effect;

3. Marginalizing across latent variables requires a multidimensional integral if random effects are continuous, or summation if random effects are discrete-valued. The Laplace approximation is a convenient approximation to identify the area under a smooth function, and therefore can be used to integrate across random effects. The Laplace approximation involves approximating a smooth function as if it where actually a multivariate normal distribution with the same peak and curvature. This approximation requires identifying the value of random effects that maximizes the joint likelihood, and then calculating the curvature of the joint likelihood around those values.

4. Using the Laplace approximation to fit a likelihood model with fixed and random effects involves an inner optimizer, which optimizes random effects conditional upon fixed effects and then calculates the Laplace approximation to the marginal likelihood. This inner optimizer is then embedded in an outer optimizer, which optimizes the fixed effects using the marginal likelihood.

5. The work of coding the inner and outer optimizer can be automated using Template Model Builder, and the performance of the Laplace approximation can be explored by comparison with sample-based methods like MCMC or Rejection Sampling. Errors arising during the inner or outer optimizers can often be diagnosed by extracting the inner or outer Hessian matrix, respectively, and then either visualizing this (in simple cases) or inspecting the eigendecomposition (in more complicated cases);

6. The Laplace approximation is computationally efficient when the Hessian matrix used in the inner optimizer (called the "inner Hessian") is sparse (i.e., contains many zeros). Sparseness can be automatically detected and plotted, but can also be explored by thinking about conditional independence, writing out a separable integral for random effects, or constructing a graph of fixed effects, random effects, and data, and exploring how these are linked by components of the joint likelihood.

2.7   Exercises

1. *Heterogeneity in proportions*: we introduced the concept of compound distributions in Section 2.1.1, and here expand this concept to represent heterogeneity in proportions. To do so, imagine a survival experiment including 10 microcosms, each containing 20 individuals that are subject to some experimental treatment. Imagine that some fraction $p = 0.5$ is expected to survive over the course of the experiment, such that the number of survivors C_i in each microcosm i is drawn from a Binomial distribution $C_i \sim \text{Binomial}(N = 20, p = p_i)$ and where $p_i = 0.5$ for each microcosm. First, simulate data from one replicate of this experiment. Then, fit a logit-linked generalized linear model using a Binomial distribution for survival in each microcosm (see Appendix B.2 for details), confirming identical results when using glm or when coding the generalized linear model in TMB by modifying Code 1.5. Next, repeat the experiment, but simulating data where each microcosm has a different survival rate drawn from a uniform distribution, $p_i \sim \text{Uniform}(0.2, 0.8)$ using p = runif(n=20,min=0.2,max=0.8) . How might you represent this heterogeneity in survival proportions within a generalized linear mixed model? Consider

modifying Code 2.7 to represent microcosm-specific survival rates using a Beta distribution (i.e., calculating `jnll -= dbeta(p_i,alpha,beta,true)` for the distribution of survival rates in each microcosm) while estimating two parameters `alpha` and `beta`, where the mean of the Beta distribution is then $\frac{\alpha}{\alpha+\beta}$. Please confirm that the estimated mean matches the true mean $\mathbb{E}(p_i) = 0.5$ of the simulated Uniform distribution.

2. *Comparison of residual estimators*: in Section 2.3, we fit a log-linked Poisson GLMM for gridded plant densities, and also introduce the distinction between conditional, marginal, and OSA residuals. Please compute conditional, marginal and OSA residuals for that data set. To do so, first review Section 1.7 to calculate conditional residuals, i.e., by computing PIT residuals conditional on empirical Bayes predictions of random effects. Then consider calculating marginal residuals by using `tmbstan` to sample from a single vector of random effects conditional on the maximum likelihood estimate of fixed effects, and computing PIT residuals for that vector of random effects. Finally, modify Code 2.7 to compute OSA residuals, consulting `?oneStepPredict` to identify syntax for code changes and using method=“cdf” to apply `oneStepPredict` to these the discrete-valued data. Next, simulate new data sets from the fitted model (i.e., new random effects conditional on estimated fixed effects, and data conditional on both), refit the model, and recompute these residuals. How do the residuals using real-world data compare with residuals computed from a correctly specified model, and what does this suggest about inference from that original model?

Part II

Basic

3

Population Dynamics and State-space Models

3.1 Population Models and Applied Ecology

Population theory (and associated models) are used by applied ecologists working in many different management contexts and for a wide range of taxa. For example, fisheries managers worldwide set catch quotas based on information from population models [151]. Similarly, assessments of sustainability inform whether further regulations are needed for many protected species of birds and mammals, and these assessments are based on fitting population models to available data [116]. Furthermore, the dynamics of human or animal disease are often explored by fitting population models to available incidence data [186].

To match this range of taxa and management contexts, population ecologists use a wide range of field, experimental, and analytical methods. However, these generally involve some combination of:

1. *Repeated monitoring*, i.e., field sampling that can record the abundance (number and/or biomass) of an entire population, or some portion that is either representative of the total or of interest by itself. For example, ecologists count the number of breeding pairs of Adélie penguins in the vicinity of the Palmer Long-Term Ecological Research Station in the Southern Ocean [26];

2. *Information regarding demographic rates and responses*, i.e., using targetted experimental or observation studies to learn about individual demographic processes, such as juvenile survival, growth, maturity, and mortality. For example, a researcher might conduct a reciprocal transplant experiment to measure how ants adapted to rural or urban environments fair when reared in each of those two habitats, and this might be used to predict changes in demographic rates during continued urbanization [145];

3. *Time-series analysis*, i.e., statistical analysis of abundance from field-sampling, combined with information about demographic rates from targetted studies, which is then fitted as a time-series model. The resulting analysis could then be used to predict the probability of abundance falling below a minimum threshold ("pseudo-extinction") or, e.g., predict the likely impact of future climate conditions on recovery for southern blue whales given the current moratorium on whaling [247].

In the following, we use population models to illustrate basic concepts about time-series analysis. We aim to develop familiarity with processes operating over a single dimension (i.e., time), which we can then build upon when introducing two dimensions (e.g., spatial coordinates) or more in later chapters.

DOI: 10.1201/9781003410294-3

3.2 Gompertz Model for Population Dynamics

To introduce population-dynamics models, let us assume that we can measure population abundance n_t in multiple years t and then calculate the annual population growth rate $\lambda_t = \frac{n_{t+1}}{n_t}$. Ecologists then seek to estimate some function that explains population growth rate, perhaps as a function $f(.)$ of current population size:

$$\lambda_t = f(n_t) \tag{3.1}$$

Ecologists often monitor this population growth rate to determine whether trends are sufficient to warrant management changes. For illustration, we present the simple Gompertz model for population abundance [42]:

$$\lambda_t = e^{\alpha - \beta \log(n_t)} \tag{3.2}$$

This model has several properties that are shared by most population-dynamics models:

1. There is some level of population abundance K (called "carrying capacity") such that increases and decreases cancel out and the population tends to remain at this level. Solving $\lambda_t = 1$ for n_t, we see that $K = e^{\alpha/\beta}$.

2. The population growth rate can be "density-dependent", such that an increase in log-abundance decreases population growth by β.

The Gompertz model can be reparameterized such that log-abundance $\log(n_t)$ in time t follows a linear model based on log-abundance $\log(n_{t-1})$ the year before:

$$\log(n_t) = \alpha + \rho \log(n_{t-1}) \tag{3.3}$$

where ρ is the magnitude of autoregression (Fig. 3.1), and this version is easily converted to the original parameterization in Eq. 3.2 by setting $\rho = 1 - \beta$. Equation 3.3 shows that the Gompertz model is a first-order *autoregressive process* for log-abundance $\log(n_t)$. The model has different implications for population dynamics based on the value of ρ:

- The population immediately returns to carrying capacity every year when $\rho = 0$;

- The population tends to return to its carrying capacity when $|\rho| < 1$;

- The population follows a random walk when $\rho = 1$; and

- The population has oscillatory dynamics when $\rho < 0$.

This ability to mimic a variety of density-dependent and -independent behaviors has therefore made the Gompertz model a common approach for detecting density dependence across a range of time-series [119].

Additionally, we note that the Gompertz model is an approximation to any population model of the form $x_t = f(x_{t-1})$ that results in a carrying capacity K. Specifically, the Gompertz model results from taking the first-order Taylor series expansion of f with respect to $\log(n)$ when abundance equals its carrying capacity K. As expected for a Taylor series expansion, the Gompertz model is ecologically implausible as abundance moves far from where the approximation is calculated (in this case, from carrying capacity). For example, intrinsic growth rate is often used as a "common currency" to compare productivity across

---

[1]See https://github.com/james-thorson/Spatio-temporal-models-for-ecologists/Chap_3 for code associated with this chapter.

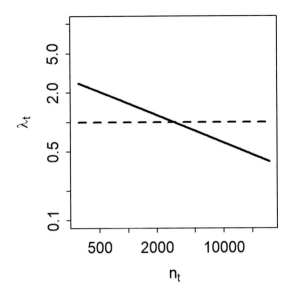

FIGURE 3.1: Visualizing the growth rate λ_t (y-axis) as a function of abundance n_t for the Gompertz model of population dynamics, showing that the log-population growth rate is linear with respect to log-abundance (i.e., plotting λ_t and n_t using log-scale axes as we do here). Carrying capacity occurs when the growth rate is 1.0 (which occurs here at 3000 animals) and the density dependence is 0.2, suggesting that abundance at half of carrying capacity will have production about 0.2 above replacement.

species, and is typically defined as $r_0 = \log(\lambda)$ as abundance goes to zero [25]. However, the Gompertz model has an infinite intrinsic growth rate, such that extinction can never occur. Despite being a poor approximation in these cases, the Gompertz model remains a useful starting point to illustrate time-series analysis of population models.

Given this simple model, we now proceed to demonstrate several techniques that are useful for population models generally. In particular, we will:

1. fit the model using standard linear estimation methods;

2. add process errors, and refit the model as a linear state-space model;

3. project the model beyond the range of measurements to illustrate how variance accumulates in time-series dynamics;

4. generalize the Gompertz model to continuous time by briefly introducing a complex and irregular autoregressive process [41, 55].

To start, we fit this model to a multi-decadal time-series of population abundance. We specifically use data from the Alaska Fisheries Science Center, which has conducted a bottom trawl survey annually following a fixed station design in the eastern Bering Sea using the same survey protocols and gear since 1982 [126][2]. We specifically access data for flathead sole (*Hippoglossoides elassodon*), and this stock supports a substantial commercial fishery with annual landings of approximately 10,000 metric tons [154]. To fit this model, we introduce the concept of *process errors* ϵ_t. There is a different estimated value for process errors ϵ_t in each year t, and this value represents the net effect of all mechanisms that

---

[2]Data are publicly available and were obtained from https://apps-afsc.fisheries.noaa.gov/RACE/groundfish/survey_data/data.htm.

affect changes in biomass beyond what is predicted by the Gompertz population model. These mechanisms might include age-structured effects, changes in fishery catches, or a nonlinear shape in the function relating biomass changes to current biomass [229]. We start by assuming that the survey data are a perfect measurement of population biomass, i.e., that there is no measurement error:

$$\log(b_{t+1}) = \alpha + \rho \log(b_t) + \epsilon_t$$
$$\epsilon_t \sim \text{Normal}(0, \sigma_\epsilon^2) \tag{3.4}$$

where this linear model is then simple to fit in R (Code 3.1). The model estimates a carrying capacity of approximately 500,000 tons, i.e., in the right-hand panel of Fig. 3.2 as the value of b_t when the production curve intersects $\lambda_t = 1$. Similarly, density dependence is significantly different from zero (i.e., the population does not appear to follow a random walk), as shown by the confidence interval for the production curve not including the dotted line showing $\lambda_t = 1$.

CODE 3.1: R code to fit Gompertz dynamics as a linear model.

```
1  # Get abundance in KG
2  Index_t = read.csv( file="Biomass_index.csv" )
3  Index_t = array(Index_t[,2], dimnames=list(Index_t[,1]))
4  Brange = range(Index_t) * c(0.8,1.2)
5  Prange = c(0.5,2)
6
7  # Fit Gompertz model with process errors
8  y = log(Index_t[-1])
9  x = log(Index_t[-length(Index_t)])
10 Lm = lm( y ~ 1 + x )
```

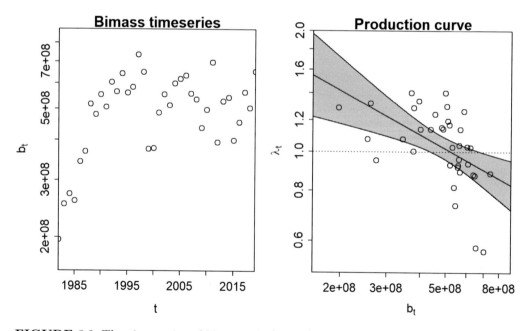

FIGURE 3.2: The time-series of biomass b_t (y-axis) over time t (x-axis) for flathead sole (left panel), and the resulting production λ_t (y-axis) vs. biomass (x-axis) (right panel, where each dot corresponds to b_t and $\lambda_t = b_{t+1}/b_t$ in a single year) along with the fitted production curve from a Gompertz model (where the blue shaded area shows the 95% confidence interval).

3.3 Semivariance and Correlation Functions

Before proceeding further with estimates of population dynamics, we use the Gompertz model to illustrate tools that are useful for understanding or specifying a wide range of spatio-temporal models. Specifically, we might wonder how much log-abundance $\log(b)$ is expected to change between two times t and $t + \Delta t$. This can be answered in several ways [37].

1. *Semivariance*: we can compute the variance of the difference between $\log(b_t)$ and $\log(b_{t+\Delta t})$ for different values Δt, and this is called the "variogram". However, this difference captures the variance arising at both times, so it is helpful to divide it by two to calculate the *semivariogram* $\gamma(\Delta t)$:

$$\gamma(\Delta t) = \frac{1}{2}\text{Var}(\log(b_t) - \log(b_{t+\Delta t})) \tag{3.5}$$

    Many population models are "stationary" in the sense that the semivariance approaches a plateau (the *sill* or "stationary variance") as Δ_t increases;

2. *Correlation function*: in models with a stationary distribution (i.e., finite sill), we can instead express the similarity between two values as a correlation function $C(\Delta t)$. In these cases, it is easy to convert between semi-variance and correlation functions:

$$C(\Delta_t) = 1 - \frac{\gamma(\Delta t)}{\sigma^2_{stationary}} \tag{3.6}$$

    where $\sigma^2_{stationary}$ is the stationary variance for a given process.

In later sections, we will often specify a model via the correlation function, because we find it is often more intuitive and simple to explain. However, the semi-variance remains a more general measurement for autocorrelation, because it applies to both stationary and nonstationary processes (i.e., a random-walk process), while a correlation function can only be defined for stationary processes (i.e., when $\sigma^2_{stationary}$ can be defined).

To illustrate these two functions, we simulate the Gompertz model assuming a carrying capacity K of one million metric tons, calculate the semivariogram and convert it to the correlation function. We also repeat this with a noisy measurement of abundance I_t, obtained as:

$$\log(I_t) \sim \text{Normal}(\log(b_t), \sigma^2_I) \tag{3.7}$$

where σ^2_I is the variance of measurement errors. Results from this simulation experiment (Figure 3.3) then illustrate a few principles:

- Different combinations of autocorrelation $\rho = \{0.5, 0.75\}$ and the magnitude of process errors $\sigma_\epsilon = \{0, 0.4\}$ result in a semivariogram that plateaus, such that the Gompertz model is stationary for these tested values. In fact, the Gompertz model accumulates variance as a power series:

$$\text{Var}(\log(b_t)) = \sum_{\Delta_t=0}^{\infty} \rho^{2\Delta_t} \sigma^2_\epsilon \tag{3.8}$$

and this power-series is converges if $|\rho| < 1$, such that $\sigma^2_{stationary} = \sigma^2_\epsilon/(1-\rho^2)$;

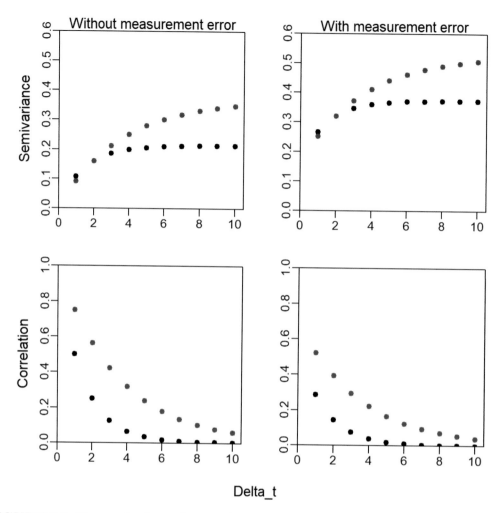

FIGURE 3.3: The semivariance (top row) and correlation function (bottom row) for a Gompertz population model, calculated either for log-abundance without measurement error (left column, $\sigma_I^2 = 0^2$) or a noisy measurement of log-abundance (right column, $\sigma_I^2 = 0.4^2$), given a constant process errors variance $\sigma_\epsilon^2 = 0.4^2$ and either moderate ($\rho = 0.5$ in black points) or strong ($\rho = 0.75$ in red points) autocorrelation.

- The time-series that includes measurement errors, $\log(I_t)$ has higher semivariance for all lags Δ_t than a model without measurement errors. Given the shape of the semivariogram, we could extrapolate its intersection with the y-axis when $\Delta_t = 0$. This value is sometimes called the *nugget*, and its value is an estimator for the variance of *measurement errors* (i.e., the magnitude of variance that would occur if we had replicated but independent samples in a single time);

- The time-series with higher autocorrelation $\rho = 0.75$, corresponding to weaker density dependence (red points), takes longer to plateau than the time-series with lower autocorrelation $\rho = 0.5$ (black points);

- The correlation function for the time-series that does not include measurement errors, $\log(b_t)$, appears to follow an exponential function, $C(\Delta_t) = e^{-\rho\Delta_t}$. The correlation function is in fact easy to calculate for the Gompertz model, but it is sometimes helpful to simulate values from a population model and then calculate the variogram and resulting correlation function for models where an analytical solution is not known.

3.4 State-Space Population Model

Returning to the example for flathead sole in the eastern Bering Sea, we know that the bottom trawl survey has some degree of error in its measurement of total population abundance, both due to limited sample sizes and interannual variation in the vertical and horizontal distribution of fish and resulting availability to the summer survey [165]. We therefore next fit this model including two sources of variation:

1. *Process errors*: as before, process errors ϵ represent a change in biomass beyond what is explained by the Gompertz model, with process-error variance σ_ϵ^2;

2. *Measurement errors*: however, we also now acknowledge that the survey does not perfectly measure biomass, but instead has some measurement error with variance σ_I^2.

Including two sources of variance in a time-series model is conventionally called a *state-space model*, and estimating parameters requires integrating across one or the other form of variance (either explicitly or implicitly). Given that this state-space model involves integrating across random effects, we can interpret it as another form of mixed-effect model (see Section 2.1).

We can implement this state-space Gompertz population model by specifying:

$$\log(b_t) \sim \begin{cases} \text{Normal}(\log(b_0), \sigma_\epsilon^2) & \text{if } t = 1 \\ \text{Normal}(\alpha + \rho\log(b_{t-1}), \sigma_\epsilon^2) & \text{if } t > 1 \end{cases} \tag{3.9}$$

$$\log(I_t) \sim \text{Normal}(\log(b_t), \sigma_I^2)$$

This model reverts to the previous process-error model (Eq. 3.4) when the variance of measurement errors $\sigma_I^2 = 0$ is estimated at zero, and reverts to a measurement-error model when the variance of process errors $\sigma_\epsilon^2 = 0$ is estimated at zero. We could instead parameterize this model by defining process errors ϵ_t as a random effect. As noted in Section 2.4, however, this alternative parameterization results in a less sparse model (i.e., more non-zero elements of the inner Hessian matrix) so we choose instead to treat biomass $\log(b_t)$ as a random effect.

Identifying maximum likelihood estimates for parameters $\theta = \{\alpha, \rho, \log(b_0), \sigma_\epsilon^2, \sigma_I^2\}$ while treating b_t as a random effect in each year involves calculating:

$$\mathcal{L}(\theta; y, x) = \int e^{f(x, \theta, y)} \mathrm{d}x \tag{3.10}$$

where $f(x, \theta, y) = \log(\Pr(y|\theta, x)) + \log(\Pr(x|\theta))$ and we change variables $x_t = \log(b_t)$ and $y_t = \log(I_t)$ to allow compact notation. We then factor the joint probability of random effects $\Pr(x|\theta)$ into a series of separable conditional distributions:

$$\log(\Pr(x|\theta)) = \log(\Pr(x_1)) + \log(\Pr(x_2|x_1)) + ... + \log(\Pr(x_T|x_{T-1})) \tag{3.11}$$

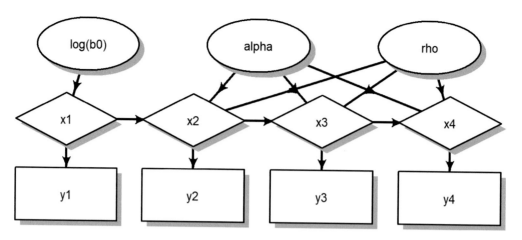

FIGURE 3.4: A graphical model for the first four years in our Gompertz conditional autoregressive model, showing data $y_t = \log(I_t)$ as boxes, random effects $x_t = \log(b_t)$ as diamonds, and fixed effects $\{\alpha, \rho, \log(b_0)\}$ as circles (while dropping variance parameters $\{\sigma_\epsilon^2, \sigma_I^2\}$ for visual clarity), and showing the conditional dependencies as arrows.

where we drop the dependency upon θ on the right-hand-side for simplicity of notation. The joint distribution (Eq. 3.10) can be factored into these conditional probabilities (Eq. 3.11) because the Gompertz model has a 1st-order *Markov structure*, wherein the probability for state x_t depends only on its value in the immediately preceding time x_{t-1}.

From this factored version, we can see that x_{t+1} is statistically independent of x_{t-1} conditional upon the value of x_t. In the graphical representation (Fig. 3.4), this is seen where random effects x_1 and x_3 do not both connect to any single datum (box), and are also not connected directly by any arrow, and hence expect that the inner Hessian should be sparse (recall rules from Section 2.4). Alternatively, we can express this as a *simultaneous autoregressive process* (SAR), and see Appendix B.5 for more details.

CODE 3.2: TMB code to specify a state-space Gompertz model.

```
1  #include <TMB.hpp>
2  template<class Type>
3  Type objective_function<Type>::operator() ()
4  {
5    // Data
6    DATA_VECTOR( log_b_t );
7    DATA_VECTOR( log_bnew_z );
8    DATA_VECTOR( simulate_t );
9
10   // Parameters
11   PARAMETER( log_d0 );
12   PARAMETER( log_sigmaP );
13   PARAMETER( log_sigmaM );
14   PARAMETER( alpha );
15   PARAMETER( rho );
16   PARAMETER_VECTOR( log_d_t );
17
18   // Objective funcction
19   Type jnll = 0;
20
21   // Probability of random coefficients
22   jnll -= dnorm( log_d_t(0), log_d0, exp(log_sigmaP), true );
23   for( int t=1; t<log_b_t.size(); t++){
```

```
24    if( simulate_t(t) == 1 ){
25      SIMULATE{
26        log_d_t(t) = rnorm( alpha + rho*log_d_t(t-1), exp(log_sigmaP) );
27      }
28    }
29    jnll -= dnorm( log_d_t(t), alpha + rho*log_d_t(t-1), exp(log_sigmaP),
        true );
30  }
31
32  // Probability of data conditional on fixed and random effect values
33  for( int t=0; t<log_b_t.size(); t++){
34    if( !R_IsNA(asDouble(log_b_t(t))) ){
35      jnll -= dnorm( log_b_t(t), log_d_t(t), exp(log_sigmaM), true );
36    }
37  }
38
39  // Predicted production function
40  vector<Type> log_out_z( log_bnew_z.size() );
41  for( int t=0; t<log_bnew_z.size(); t++){
42    log_out_z(t) = alpha + rho * log_bnew_z(t);
43  }
44  ADREPORT( log_out_z );
45  SIMULATE{ REPORT( log_d_t ); }
46
47  // Reporting
48  return jnll;
49 }
```

Coding this joint negative log-likelihood in TMB involves specifying fixed and random effects, and we follow Eq. 3.9 in defining log-biomass log_d_t as a coefficient that we will then specify as a random effect (Code 3.2). We also include additional code (i.e., lines using the SIMULATE function) that we will discuss in detail in a later section. Fitting this model in R is similarly straightforward (Code 3.3), given that TMB will calculate the marginal likelihood (i.e., apply the inner optimizer and calculate the Laplace approximation) automatically.

CODE 3.3: R code to run the state-space Gompertz model implemented in TMB.

```
1  # Compile
2  library(TMB)
3  compile( "gompertz.cpp" )
4  dyn.load( dynlib("gompertz") )
5
6  # Build inputs
7  Data = list( "log_b_t"=log(Index_t), "log_bnew_z"=xpred, "simulate_t"=rep(0,
       length(Index_t)) )
8  Parameters = list( "log_d0"=0, "log_sigmaP"=1, "log_sigmaM"=1, "alpha"=0, "
       rho"=0, "log_d_t"=rep(0,length(Index_t)) )
9  Random = "log_d_t"
10
11 # Build and fit object
12 Obj = MakeADFun(data=Data, parameters=Parameters, random=Random )
13 opt_CAR = nlminb( start=Obj$par, obj=Obj$fn, grad=Obj$gr )
14 opt_CAR$SD = sdreport( Obj )
```

Before running the model, however, we check the sparsity of the inner Hessian to confirm that we have used an efficient parameterization (Fig. 3.5). As expected for a first-order Markov process, this shows that we have used a parameterization whereby the distribution for $\log(b_{t+1})$ is independent of $\log(b_{t-1})$ conditional upon the value of $\log(b_t)$. Finally, we run the model and estimate uncertainty in both biomass and the production function:

Comparing the state-space (Fig. 3.6) and process-error models (Fig. 3.2) shows that the state-space model estimates a true but unobserved time-series of biomass that is more

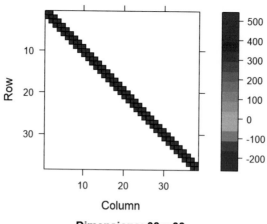

Dimensions: 38 x 38

FIGURE 3.5: Sparsity pattern of the inner Hessian for the state-space Gompertz model, showing that random effects are conditionally independent for any two years that are not adjacent.

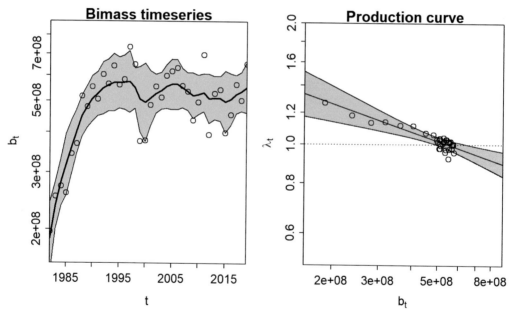

FIGURE 3.6: The time-series of the biomass index for flathead sole in the eastern Bering Sea (left panel) along with estimated biomass (red line, with red shading showing the 95% confidence interval), and the resulting production (where each dot corresponds to b_t and $\lambda_t = b_{t+1}/b_t$ in a single year) along with the fitted production curve from a Gompertz model (blue line, with blue shading showing the 95% confidence interval), which can be compared with Fig. 3.2.

smooth than the survey observations, and also estimates population growth rates that are shrunk toward the predictions from the Gompertz population model. As introduced in Section 2.1, we expect that this shrinkage estimator for biomass will have lower expected error than using the raw survey data as an estimate of population biomass. Similarly, the estimated magnitude of process and measurement errors determines how much biomass is shrunk (left-hand side of Fig. 3.6) relative to how much production is shrunk toward the Gompertz model (right-hand side of Fig. 3.6).

3.5 Conditional vs. Joint Covariance Modeling

So far in this chapter, we have developed the state-space Gompertz model for population dynamics by factoring the joint log-likelihood into a series of conditional log-likelihood components (Eq 3.11). We then code that expression in TMB, and TMB assembles a graph representing the dependency among variables and data, which is conceptually similar to the graphical model in Fig. 2.7. TMB uses this graph and applies a feature called *automatic sparseness detection* to identify all pairs of random effects that are conditionally independent of one-another, and hence detect the sparsity of the resulting Hessian matrix (e.g., in Fig. 3.5). As we will see, automatic sparseness detection is extremely handy for most users, who then do not have to manually specify which random effects are conditionally independent.

In some cases, however, it is helpful to understand alternative approaches to implementing this same model, which provide different statistical insights or options for writing code. Here, we next reparameterize this model by simultaneously calculating the probability of all random effects, i.e., calculating $\Pr(\mathbf{b})$ directly. We seek to show that both versions give nearly identical parameter estimates. In later chapters, we will often introduce a model by expressing the probability of random effects using a series of conditional distributions, because we find that this derivation is often more intuitive. We will then sometimes replace this specification with a joint distribution, which can often be written using fewer lines of code and hence is easier to read and maintain as software. We therefore think that both versions can be useful in different circumstances.

The joint version of the Gompertz model [240] is written as:

$$\log(\mathbf{b}) = \log(b_0)\rho^{\mathbf{t}} + \frac{\alpha}{1-\rho} + \epsilon$$
$$\epsilon \sim \mathrm{MVN}(\mathbf{0}, \sigma_\epsilon^2 \mathbf{R}_\epsilon) \tag{3.12}$$
$$\log(I_t) \sim \mathrm{Normal}(\log(b_t), \sigma_I^2)$$

where $\log(b_0)$ is the initial log-biomass in time t_0 relative to car, which follows an exponential decay toward carrying capacity as time increases, $\mathbf{t} = \{0, 1, ..., T-1\}$ is the vector of times after the initial modeled interval, such that $\rho^{\mathbf{t}}$ is the amount that initial biomass has decayed by each time $\mathbf{t}$. Additionally, $\epsilon \sim \mathrm{MVN}$ indicates that ϵ follows a multivariate normal distribution, $\mathbf{0}$ is a vector of T zeros, and $\mathbf{R}_\epsilon$ is a T by T matrix containing the correlation in ϵ_t between any two times t_1 and t_2:

$$r_{t_1,t_2} = \rho^{|t_2 - t_1|} \tag{3.13}$$

such that:

$$\mathbf{R}_\epsilon = \begin{bmatrix} 1 & \rho^1 & \rho^2 & \rho^3 \\ \rho^1 & 1 & \rho^1 & \rho^2 \\ \rho^2 & \rho^1 & 1 & \rho^1 \\ \rho^3 & \rho^2 & \rho^1 & 1 \end{bmatrix} \tag{3.14}$$

This correlation matrix is dense, i.e., has a nonzero value everywhere. However, recall from Section 2.2 that calculating the multivariate normal distribution (Eq. 2.15) only requires knowing the inverse of the covariance $\mathbf{Q}_\epsilon = (\sigma_\epsilon^2 \mathbf{R})^{-1}$, called the "precision matrix" [199]. Conveniently, this precision matrix is sparse and can be approximated (while ignoring small differences at the beginning of the time-series which we call *boundary conditions*) as:

$$\mathbf{Q}_\epsilon = (\sigma_\epsilon^2 \mathbf{R}_\epsilon)^{-1} = \sigma_\epsilon^{-2} \begin{bmatrix} 1+\rho^2 & -\rho & 0 & 0 \\ -\rho & 1+\rho^2 & -\rho & 0 \\ 0 & -\rho & 1+\rho^2 & -\rho \\ 0 & 0 & -\rho & 1+\rho^2 \end{bmatrix} \tag{3.15}$$

In fact, this sparse form matches the "tri-diagonal" form of the inner Hessian matrix that was seen earlier for the factored version of the Gompertz model (Figure 3.5). Interested readers can see Appendix B.5 for more information on deriving 3.15 for the Gompertz model.

CODE 3.4: TMB code to fit a state-space Gompertz model by constructing an AR1 precision matrix and then using the density namespace to evaluate the multivariate normal density function.

```
1  #include <TMB.hpp>
2  template<class Type>
3  Eigen::SparseMatrix<Type> Q_ar1( Type rho, int n_t ){
4    // Compute SparseMatrix precision
5    Eigen::SparseMatrix<Type> Q( n_t, n_t );
6    for(int t=0; t<n_t; t++){
7      Q.coeffRef( t, t ) = 1 + pow(rho,2);
8      if(t>=1) Q.coeffRef( t-1, t ) = -rho;
9      if(t>=1) Q.coeffRef( t, t-1 ) = -rho;
10   }
11   Q /= (1 - pow(rho,2));
12   return Q;
13 }
14
15 template<class Type>
16 Type objective_function<Type>::operator() ()
17 {
18   // Data
19   DATA_VECTOR( log_b_t );
20   DATA_VECTOR( log_bnew_z );
21
22   // Parameters
23   PARAMETER( log_delta );
24   PARAMETER( log_sigmaP );
25   PARAMETER( log_sigmaM );
26   PARAMETER( alpha );
27   PARAMETER( rho );
28   PARAMETER_VECTOR( eps_t );
29
30   // Global variables
31   Type jnll = 0;
32   int n_t = log_b_t.size();
33   Eigen::SparseMatrix<Type> Q( n_t, n_t );
34   Q = Q_ar1( rho, n_t );
```

TABLE 3.1: Estimated parameters and standard errors for the Gompertz population model using a conditional or joint specification of the covariance.

CAR param	Estimate	Std. Error	SAR param	Estimate	Std. Error
log_d0	19	0.143	log_delta	−1.06	0.147
log_sigmaP	−2.82	0.807	log_sigmaP	−2.29	0.739
log_sigmaM	−1.91	0.223	log_sigmaM	−1.93	0.216
alpha	4.66	1.12	alpha	4.57	1.01
rho	0.768	0.056	rho	0.773	0.0504

```
35
36   // Probability of random coefficients
37   jnll += density::GMRF(Q)( eps_t );
38
39   // Probability of data conditional on fixed and random effect values
40   vector<Type> log_d_t( n_t );
41   for( int t=0; t<n_t; t++){
42     log_d_t(t) = log_delta*pow(rho, t) + alpha/(1-rho) + exp(log_sigmaP)*
       eps_t(t);
43     jnll -= dnorm( log_b_t(t), log_d_t(t), exp(log_sigmaM), true );
44   }
45   ADREPORT(log_d_t)
46
47   // Predicted production function
48   vector<Type> log_out_z( log_bnew_z.size() );
49   for( int t=0; t<log_bnew_z.size(); t++){
50     log_out_z(t) = alpha + rho * log_bnew_z(t);
51   }
52   ADREPORT( log_out_z );
53
54   // Reporting
55   return jnll;
56 }
```

We can therefore implement this joint version of a Gompertz population model in TMB by calculating the precision matrix as a sparse matrix using the C++ package Eigen . The Eigen package requires some special syntax, e.g., using Q.coefRef(a,b)=c to assemble the triplet-format precision matrix where row a and column b of matrix Q has value c . We then use a built-in package density in TMB to calculate the multivariate normal distribution using this precision matrix (Code 3.4). Comparing the maximum likelihood estimates of parameters for the conditional and joint versions shows some small differences, resulting from a different treatment of boundary conditions for random effects (Table 3.1). However, these differences have little effect on resulting estimates of biomass or the production function (Figure 3.7).

3.6 Seasonal/missing Data and Forecasting

So far, we have shown how to implement a state-space population-dynamics model using either conditional or joint specification of the covariance among states $\log(b_t)$. We next summarize how this is generalized to deal with several practical issues that arise in ecological time series, namely:

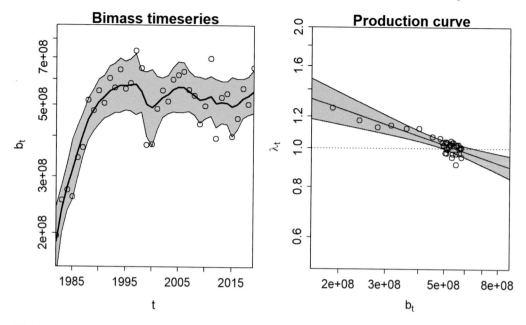

FIGURE 3.7: Output from the joint implementation of the Gompertz population model (see Figure 3.6 for details).

1. *Seasonal data*: how can we deal with data that arise irregularly, i.e., by defining a process in continuous time?

2. *Missing data*: how can be fit models when data are missing for some times?

3. *Forecasting*: how can we forecast dynamics into the future?

First, we might have irregularly spaced samples of population dynamics. For example, the bottom trawl survey in the eastern Bering Sea typically occurs with midpoint in July but ranges from June to August depending upon logistical constraints in each year. Therefore, instead of indexing time as an integer $t \in \{1, 2, ..., n_t\}$, we might define time as a continuous variable $t_{min} \leq t \leq t_{max}$. In this case, we replace the autoregressive model (Eq. 3.4) with a generalization called an *irregular autoregressive (IAR) model*. This IAR model specifies $\log(b_{t+\Delta_t})$ based on the continuous-valued time Δ_t elapsed since the previous observation:

$$\log(b_{t+\Delta_t}) = (1 - \rho^{\Delta_t})\frac{\alpha}{1 - \rho} + \rho^{\Delta_t} \log(b_t) + \epsilon_t$$
$$\epsilon_t \sim \text{Normal}(0, \sigma_\epsilon^2 \sqrt{1 - \rho^{2\Delta_t}})$$

(3.16)

where the autocorrelation, intercept and standard deviation variance of ϵ all depend upon the time elapsed. Notably, the autocorrelation parameter ρ is now defined as the correlation for a reference time-interval. This continuous-time model could then be fitted to samples $I(t)$ in each of those times. This IAR can instead be parameterized as an *Ornstein-Uhlenbeck process* by replacing $\rho = e^{-\theta}$ [41], although we present the IAR to highlight the similarity to Eq. 3.9.

In Section 3.5, we discussed how the Gompertz model with equal time-intervals can be parameterized by either specifying (1) the conditional distribution for biomass b_t at each time given the previous b_{t-1}, or (2) the distribution for biomass **b** for all times jointly. In Eq. 3.16, we use the former *conditional specification* but we could instead use the latter *joint specification* for the IAR model (which we later recommend as an Exercise for the reader).

We can therefore interpret this IAR as a *Gaussian process*, which defines a probability distribution for the biomass function $b(t)$ at any continuous time t, such that biomass for any finite set of times will follow a multivariate normal distribution [187]. In later chapters, we will repeatedly define a multivariate normal distribution for process errors across space and/or time. In many cases, a set of spatially or temporally varying random effects that follows a joint multivariate normal distribution can be viewed as a draw from a Gaussian process, and the model can be viewed as a distribution for a function-valued variable [226].

The autocorrelation for the IAR in Eq. 3.16 is constrained to be positive, but otherwise the model reduces to the equal-interval Gompertz model (Eq. 3.4) when time intervals Δ_t are constant, and parameters are identical with appropriate rescaling of $\Delta_t = 1$. As a further extension, we could instead use a *complex irregular autoregressive model* (CIAR). This CIAR [55] represents oscillatory time-series dynamics by introducing a latent variable representing the momentum in the rate of population change, analogous to the momentum of a pendulum that results in its oscillatory movement. Viewed more formally, we can model the state-variable as a *complex number* $x = y + zi$, where real component y and imaginary component z both follow an IAR, but we only observe the real component of this variable which is then interpreted as log-biomass $y = \log(b)$. Treating the state-variable as a complex number allows an autoregressive model for $\log(b)$ to include oscillations, where the expected frequency arises from estimated parameters, similar to the original oscillatory dynamics that arise in the equal-interval Gompertz model whenever $\rho < 0$.

Similarly, it is easy to calculate the Gompertz model with missing data: this simply involves excluding any data that are missing from the calculation of the joint loglikelihood function (which we accomplished using `R_IsNA` in Code 3.2). However, the performance of the state-space model will likely change as data are progressively removed. To illustrate this, let's start by supposing we have a model where the state-space Gompertz model can be fitted using maximum likelihood, where both variance parameters $\hat{\sigma}_P$ and $\hat{\sigma}_M$ are estimated to be greater than zero. As data are removed, one or the other of these parameters will often be estimated at zero, corresponding to a model that includes only measurement or process variability. In this case, the state-space model is not "estimable" (see Section 2.5.2), and the model instead collapses to a simpler model without measurement errors ($\sigma_I = 0$) or process errors ($\sigma_\epsilon = 0$).

Finally, ecologists often seek to project likely population dynamics into the future. However, population projections are uncertain for several important reasons:

1. *Parametric uncertainty*: typically, we are uncertain about the parameters representing population dynamics, and forecasts using different values will result in different dynamics. We can measure this uncertainty by the standard errors of fixed effects;

2. *Uncertainty about historical states*: similarly, forecasts are initialized based on the population abundance in the final year of data. This population abundance is typically uncertain, and uncertainty about initial conditions will then propagate forward during forecasts. We can measure this uncertainty using the predictive variance of random effects representing historical biomass;

3. *Future variability*: finally, the state-space Gompertz model identifies process errors, representing the cumulative impact of unknown factors that affect population dynamics. We can measure this uncertainty using the predictive variance of random effects that occur during the forecast period.

Different combinations of these are more or less easy to extract using previous TMB code (Code 3.2) and compact code in R (Code 3.5). For example:

- *Total uncertainty*: We can calculate the net effect of all forms of variability by including future years but treating them as missing data (i.e., using `R_IsNA` to skip data that are inputted as NAs). We then use `TMB::sdreport` to calculate standard errors for biomass in those forecasted years;

- *Future variability*: We can isolate the effect of uncertainty about future process errors by including the `SIMULATE` function in the TMB file to simulate new values for random effects that occur during the forecast period (where we identify which years to resimulate using the vector `simulate_t`), but otherwise leaving the fixed and other random effects at their estimated values;

- *Uncertainty about historical states*: We can approximate the effect of uncertainty about historical states by taking random draws from the precision of random effects. We specifically use a function `obj$env$MC` that is available in any TMB object, which takes Monte Carlo samples from a fitted model.

CODE 3.5: R code to extract different types of forecasting uncertainty.

```
1  # Full uncertainty
2  out_full = data.frame( "year" = years,
3                         "pred" = as.list(opt_CAR$SD,"Est")$log_d_t,
4                         "SE" = as.list(opt_CAR$SD,"Std")$log_d_t )
5
6  # Samples from forecast variability
7  log_d_tz = sapply( 1:1000,
8                     FUN=\(i) Obj$env$simulate(Obj$env$last.par.best)$log_d_t )
9  out_1 = data.frame( "year" = years,
10                     "pred" = apply(log_d_tz,MARGIN=1,FUN=mean),
11                     "SE" = apply(log_d_tz,MARGIN=1,FUN=sd) )
12
13 # Samples from forecast uncertainty + state uncertainty
14 u_zr = Obj$env$last.par.best %o% rep(1, 1000)
15 MC = Obj$env$MC( keep=TRUE, n=1000, antithetic=FALSE )
16 u_zr[Obj$env$random,] = attr(MC, "samples")
17 log_d_tz = sapply( 1:1000,
18                    FUN=\(i) Obj$env$simulate(par=u_zr[,i])$log_d_t )
19 out_2 = data.frame( "year" = years,
20                     "pred" = apply(log_d_tz,MARGIN=1,FUN=mean),
21                     "SE" = apply(log_d_tz,MARGIN=1,FUN=sd) )
```

By combining these different options, we can therefore identify the relative importance of different sources of ecological uncertainty. We illustrate this by including (1) future variability, (2) future variability combined with historical uncertainty, or (3) all three sources of variation (Fig. 3.8) in our fit to survey data for flathead sole in the Bering Sea. In this case, the future process variation (left panel) is almost as uncertain as the cumulative uncertainty arising from all three sources. We therefore conclude that the precision of forecasts would be improved more by an improved model of population dynamics that would then reduce the estimated magnitude of process errors, rather than trying to improve our knowledge of parameters or historical states.

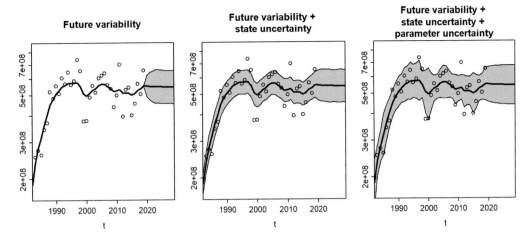

FIGURE 3.8: Forecasts of flathead sole biomass from the state-space Gompertz model, including only future variability (left panel), future variability and historical uncertainty (middle panel), or also including parametric uncertainty (right panel).

3.7 Chapter Summary

In summary, we have showed that:

1. Population-dynamics models generally define a probability distribution for population growth rates based on the abundance in preceding times. A simple approximation to population dynamics results in the Gompertz population model, and this defines a first-order autoregressive process;

2. If a stochastic population dynamics model is combined with a probability distribution for measurements of abundance, then the resulting "state-space" model can be estimated by specifying process errors (or the state-variable) as a random effect;

3. In an autoregressive model like Gompertz population dynamics, the degree of autoregression defines a "semivariogram". If dynamics are stationary, this semivariance function can instead be represented as a covariance function. Similarly, an autoregressive model can be specified either by factoring the covariance of states into a series of conditional distributions or by constructing this joint covariance directly. Furthermore, it is more efficient to construct the joint precision directly than constructing the joint covariance and then have to invert it. Joint and conditional covariance models can be specified to give identical results, or will sometimes differ slightly given small differences in implied boundary conditions;

4. An autoregressive model generally has a sparse precision matrix for random effects, as can be shown by using automatic sparseness detection in TMB, using a graphical model to illustrate dependencies, or by deriving it from a simultaneous autoregressive process;

5. It is straightforward to fit a state-space model given missing or unevenly spaced data. Similarly, forecasts involving a state-space model can decompose uncertainty into components representing future variability, uncertainty about historical states that define the initial conditions for a forecast, and parametric uncertainty, and all versions are straightforward to obtain using TMB.

3.8   Exercise

In Eq. 3.16, we introduce an IAR model by specifying the conditional distribution for each time t given the time $t - 1$ preceding it. We also claimed that this IAR could instead be described by specifying the joint distribution for biomass at all modeled times, similar to what's done in Section 3.5. Please write down this joint specification of the IAR model, implement both in TMB, and confirm that both given similar results when applied to the flathead sole data set.

4

Individual Movement

4.1 Movement Ecology

All organisms must move at some point in their life. Examples include an Arctic tern (*Sterna paradisaea*) migrating 40,000 km between the poles every year, an apple falling off the tree and rolling a meter before taking root, or freshwater mussel larvae that parasitize fish gills to hitch a ride upriver. Given that movement is ubiquitous, the movement of organisms is important in (nearly) all ecological theories and is analyzed using a wide range of techniques. For example:

- The Island Theory of Biogeography derives area-diversity relationships from postulating some ongoing process by which species are transported from a mainland habitat (species pool) to an island [140];

- Optimal Foraging Theory seeks to identify those behaviors (e.g., foraging location and time spent feeding) that are expected to maximize individual fitness [120], and from there predict how behaviors would change under alternative forage densities and predatory controls;

- Modern Coexistence Theory seeks to classify the mechanisms by which a population might persist over time, defining the correlation between productivity and density arising from a combination of habitat differences and dispersal abilities [25];

- The Movement Ecology Paradigm [160] seeks to unify the analysis of movement for any organism by describing the organism's internal state, locomotive, sensory, cognitive, and navigational capabilities, and the external factors affecting movement.

Clearly, any investigation of spatio-temporal processes in ecology must include some general approaches to study the movement of organisms that includes these various mechanisms.

Despite the ubiquity of movement in ecological theory, some theories and analytical methods include movement *implicitly* while others do so *explicitly*. For example, models for population growth are often derived from implicit assumptions about how individuals are distributed spatially and how increases in local density inhibit population growth rates [196]. By contrast, other ecological models explicitly represent the drivers and rates of individual movement, e.g., when representing population rescue in a meta-population model of checkerspot butterflies occurring in multiple habitat patches [81]. Explicit movement models often proceed by tracking the movement of individuals between habitat patches or across continuous space, while implicit models often derive some steady-state properties of aggregate dynamics from proposed movement rules [130]. Both approaches are useful, but aggregate ("implicit") properties can generally be obtained from an explicit movement model, while it is harder to go the other way (i.e., to identify an explicit movement process from a resulting spatial pattern). We therefore focus the following on how to estimate movement explicitly, so that readers can then use estimates to simulate animal movement

DOI: 10.1201/9781003410294-4

to generate aggregate (implicit) patterns. In some cases, the aggregate patterns resulting from explicit movement assumptions can then be compared with population-scale data to refute or support hypothesized movement models.

In Chapter 1, we introduced Individual-Based Models (IBM) by demonstrating how population dynamics arise from birth and death timing of individuals, or habitat utilization arises from the location and size of individuals. We similarly claimed that IBMs can be efficiently fitted to data by converting from this Lagrangian viewpoint (tracking changes in attributes for individuals over time) to an Eulerian viewpoint (tracking densities of animals based on their attributes). However, a full description of individual dynamics must also include how individual location changes over time. Most ecological movements can be accounted for with three basic processes, each with its own simple mathematical expression:

1. *Taxis*, where animals move intentionally and predictably toward their preferred habitats, and where the emphasis is upon identifying the habitat preference function that governs taxis;

2. *Drift*, where animals move along a known path, typically drifting due to their physical environment. Passive drift occurs, e.g., for dispersal of pollen due to prevailing winds, or dispersal of aquatic larvae due to oceanographic currents. Drift is also a generic term for directional movement where the habitat preference is not known or explicitly represented;

3. *Diffusion*, where animals move in a way that is not otherwise explained as taxis or passive drift, and therefore appears to be "random" relative to known or hypothesized movement drivers. In the following, we use the term "diffusion" both in its strict sense (i.e., using the Eulerian viewpoint, where densities tend to revert to the average of nearby locations) and a looser sense (i.e., using the Lagrangian viewpoint to represent diffusive movement for an individual following a random-walk process).

Both taxis and drift are sometimes classified as *advection*. However, we distinguish between these two components of advection for technical reasons that we will discuss later. Similarly, diffusion is a statistical residual that remains unexplained after predicting movement based on advection. It is therefore useful to start with a model of diffusion, and subsequently describe how we can attribute movement using other mechanisms. For some ecological purposes, we might seek to describe taxis and drift so accurately that there is no remaining movement to describe as diffusion. For other ecological analyses, however, it might not be necessary to account for taxis and drift explicitly, and we might be content by estimating a diffusion rate with no other mechanism involved.

4.2   Defining Diffusion and Taxis

We start by defining location S_t along one dimension, $S \in \mathbb{R}$ for an individual at any time t[1]. We start by explaining all movement as "diffusion", i.e., as a random-walk process that results in diffusive movement:

$$S(t + \Delta_t) = S(t) + \delta \tag{4.1}$$

---

[1]See https://github.com/james-thorson/Spatio-temporal-models-for-ecologists/Chap_4 for code associated with this chapter.

where:

$$\delta \sim \text{Normal}(0, \Delta_t \sigma^2) \tag{4.2}$$

and where σ^2 is the rate of random movement such that the variance in displacement over interval interval Δ_t is $\Delta_t \sigma^2$. We can instead write this using a diffusion coefficient $D = \frac{\sigma^2}{2}$, although the random movement rate σ^2 likely provides more intuition regarding the process involved.

If we allow the length of time-intervals to approach zero (i.e., define movement in continuous time), then Eq. 4.2 defines a *Weiner process* such that $\delta \sim \sigma W(\Delta_t)$. We can use this to define a stochastic partial differential equation that generalizes Eq. 4.1 in continuous time:

$$\partial S = \sigma \partial W \partial t \tag{4.3}$$

such that $\int_0^{\Delta_t} \sigma \partial W \partial t = \delta$ where this δ is the same as the one in Eq. 4.1, and Eq. 4.3 is sometimes called *Brownian motion*. The diffusion coefficient can then be related to the *mean free path* of a hypothetical particle, i.e., how far a particle (e.g., an individual animal) typically goes before randomly determining a new direction of travel.

In two or more dimensions, we then generalize this by defining location as vector $\mathbf{s}_t$, and this location evolves as:

$$\mathbf{s}(t + \Delta_t) = \mathbf{s}(t) + \delta$$
$$\delta \sim \text{MVN}(\mathbf{0}, \Delta_t \sigma^2 \mathbf{I}) \tag{4.4}$$

where $\Delta_t \sigma^2$ is the expected variance for each individual coordinate, and we could again write this using diffusion coefficient $D = \frac{\sigma^2}{2}$. We are defining σ^2 as the variance rate for each individual dimension, so the mean-squared displacement (MSD) is the sum of squares across n spatial coordinates, such that:

$$MSD = \mathbb{E}(|\mathbf{s}(t + \Delta_t) - \mathbf{s}(t)|) = 2nD\Delta_t \tag{4.5}$$

where n is the number of dimensions. This mean-squared displacement can then be compared with empirical data to infer mechanisms of animal dispersal [209, 167].

We next introduce taxis by defining a *habitat preference function* $h(s)$. This preference function is an imaginary property of each location, and can be defined based on local habitat, animal sensory capabilities, and other properties that are understood to affect animal movement [160]. It is analogous to potential functions that are widely used in physics, e.g., a gravitational potential that accelerates bodies with mass or an electric potential that accelerates bodies with electric charge [15]. For this reason, this preference function for describing taxis has been called a *potential function* [147, 182], although we prefer to use ecological vocabulary to describe the same process. Animals seek to move toward locations with higher preference, so they tend to move up a gradient of the habitat preference function. In the case of a single spatial dimension, we calculate this using the first derivative $h'(S) = \frac{d}{dS} h(S)$ of the preference function $h(S)$ at a given location S:

$$\partial S = h'(S)\partial t + \sigma W \partial t \tag{4.6}$$

However, we can write this more generally using a *gradient operator* $\nabla h(S)$, which describes taxis in any number of spatial coordinates:

$$\partial \mathbf{s} = \nabla h(\mathbf{s})\partial t + \sigma W \partial t \tag{4.7}$$

where $\nabla h(\mathbf{s})$ replaces the derivative in Eq. 4.6. For example, when defining movement in two dimensions such that location $\mathbf{s} = (x, y)$, we can write this gradient as $\nabla h(\mathbf{s}) =$

$(\frac{\partial}{\partial x}h(x,y), \frac{\partial}{\partial y}h(x,y))$. Gradient $\nabla h(\mathbf{s})$ then returns a vector pointing in the direction of greatest increase in preference $h(\mathbf{s})$, with vector length proportional to this gradient. Given this definition, taxis represents the tendency for animals to move up the gradient of this habitat-preference function.

In general, this SPDE can be solved exactly when preference h takes particular forms. As outlined in Section 1.1, however, we seek methods that we can apply regardless of the specific preference function being used. To do so, we introduce the *Euler method* to discretize the path that arises from this process, which results in a *directed random walk*. This involves replacing the differential equation (which must be solved via integration) with a difference equation (which can be solved via summation). In this case, taxis during the interval Δ_t is based only on habitat preference at the the begin of the interval:

$$\mathbf{s}(t + \Delta_t) - \mathbf{s}(t) = \nabla h(\mathbf{s}(t))\Delta_t + \delta \qquad (4.8)$$

Eq. 4.8 is conceptually identical to Eq. 4.7, but the continuous differentials have been replaced by discrete differences, the preference gradient $\nabla h(s(t))$ is now multiplied by Δ_t to account for the time resolution of the discretization, and δ is the time-integrated Weiner process described above. This discretization has larger errors when $h(\mathbf{s}_t)$ is not linear in the vicinity of $\mathbf{s}_t$ and $\mathbf{s}_{t+\Delta_t}$ (see Section B.6). In these cases it is helpful to decrease the time-interval Δ_t until this condition is approximately satisfied (or perhaps replacing the Euler method with a more accurate approximation like Runge-Kutta methods).

To illustrate, we simulate movement in two dimensions $\mathbf{s}_t = (x_t, y_t)$, where a *central-place forager* starts their foraging trip at location $s_0 = (0,0)$ and also ends the trip at approximately the same location. The distance from the start location is $\sqrt{x^2 + y^2}$, and we specify that preference is initially higher for locations that are distant from $\mathbf{s}_0$, such that the animal initially moves away from its starting location, but that preference later changes to be highest at $\mathbf{s}_0$ such that the animal later moves back toward that location (Fig. 4.1). In particular, we specify:

$$h(s,t) = \beta_t \sqrt{x^2 + y^2} \qquad (4.9)$$

where

$$\beta_t = 0.5 - \frac{t}{1000} \qquad (4.10)$$

CODE 4.1: R code for simulating central-place foraging trips and visualizing the vector-field representing taxis given a habitat preference function.

```
1  # Parameters
2  n_t = 1000
3  beta_t = seq(0.5, -0.5, length=n_t)
4  gamma_t = rep(0, n_t)
5  get_preference = function( beta, gamma, s ){
6    beta*sqrt(s[1]^2+s[2]^2) + gamma*(s[2])
7  }
8
9  # Simulation function
10 simulate_track = function( n_t, beta_t, gamma_t, get_preference ){
11   s_t = array( NA, dim=c(n_t,2), dimnames=list(NULL,c("x","y")) )
12   # Euler approximation
13   for( t in seq_len(n_t) ){
14     if(t==1) s_t[t,] = c(0,0)
15     if(t>=2){
16       gradient = grad( f=get_preference, x0=s_t[t-1,,drop=FALSE],
17                        beta=beta_t[t-1], gamma=gamma_t[t-1])
18       s_t[t,] = s_t[t-1,] + gradient + rmvnorm(n=1,sigma=diag(2))
19     }
```

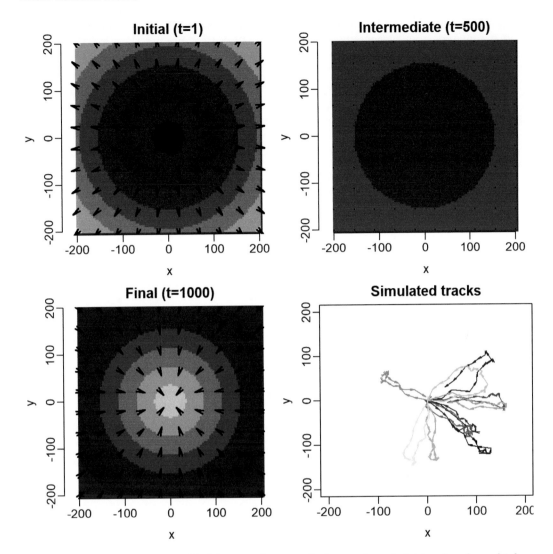

FIGURE 4.1: Illustration of habitat preferences (colors, where bright colors have higher preference) and resulting taxis (a vector-field generated as the gradient of the preference function, shown using arrows) in the initial, middle, and end time of the simulated foraging interval, as well as 10 simulated tracks for foraging trips of these central-place foragers.

```
20    }
21    return(s_t)
22 }
```

4.3   Track Reconstruction

In Section 4.2, we showed that we can simulate movement based on taxis (i.e., the gradient of a habitat preference function) and diffusion (otherwise unexplained changes in location).

However, applied ecologists often want to do this in reverse, i.e., infer movement processes and parameters from measurements of individual location, predict their location between measurements, and forecast future movement. We next show how to fit this as a bivariate state-space model, while treating location $\mathbf{s}$ as continuous valued and discretizing time t into intervals with duration Δ_t. We first specify a model that attributes all movement to "diffusion" (i.e., has no explanatory variables for passive drift or taxis). In this model, we specify that animal location $\mathbf{s}$ follows a multivariate random walk over time:

$$\mathbf{s}_{t+\Delta_t} \sim \mathrm{MVN}(\mathbf{s}_t, \Delta_t \mathbf{V}) \tag{4.11}$$

where $\mathbf{V} = \sigma_s^2 \mathbf{I}$ and $\mathbf{I}$ is an identity matrix, such that $\sigma_s^2 \Delta_t \mathbf{I}$ is the variance that accumulates over time Δ_t at rate σ_s^2. Specifying the same diffusion rate in both spatial coordinates is sometimes called *isotropic*, and we will later extend the model to specify different or correlated diffusion rates in different directions by replacing $\mathbf{I}$ with a different symmetric matrix.

We complete the model by specifying a distribution for measurement errors:

$$\mathbf{y}_t \sim \mathrm{MVN}(\mathbf{s}_t, \sigma_y^2 \mathbf{I}) \tag{4.12}$$

where we here assume that measurement error variance σ_y^2 is known (i.e., from the error of a Global Positioning System measurement calculated based on the number and position of GPS satellites during the measurement). This model is similar to the previous state-space Gompertz population model (Eq. 3.4), although it now involves a multivariate normal distribution for both process and measurement errors.

This model can be specified in TMB using a few new concepts and model components (Code 4.2). We estimate animal location x_iz at i times separated by intervals DeltaT_i , fitted to data y_iz . Similar to Code 3.2, we use function R_IsNA to allow the model to detect and skip data that are missing, while still predicting animal location during those times. Later, we will use this option to extract predictions of location between measurements (and associated standard errors), and it could also be used to forecast movement after the last available measurement. We use a function density::MVNORM that is included in a TMB library density to calculate a multivariate normal density from a specified covariance function. This *density library* will come in handy throughout the subsequent chapters as a compact way of specifying spatial and temporal correlations. We have also defined a function make_covariance which assembles the covariance matrix for diffusion. So far, we are defining $\mathbf{V} = \sigma_s^2 \Delta_t \mathbf{I}$ but we include this function to allow other, more complicated versions for the covariance matrix which we introduce in later sections. Finally, we also include covariates X_ij and covariate-response parameters beta_jz , which we also introduce later, and define a matrix Gsum_iz that calculates a cumulative sum of movement resulting from covariates.

CODE 4.2: TMB code for specifying a bivariate state-space model for animal movement from a Lagrangian perspective.

```
1  #include <TMB.hpp>
2  #include "make_covariance.hpp"
3  template<class Type>
4  Type objective_function<Type>::operator() ()
5  {
6    // Data
7    DATA_MATRIX( y_iz );        // NA for missing/dropped data
8    DATA_VECTOR( DeltaT_i ); // NA for first obs of each track
9    DATA_VECTOR( error2_i );
10   DATA_MATRIX( X_ij );
11   DATA_INTEGER( n_factors );
```

```
12   DATA_IMATRIX( RAM );
13
14   // Parameters
15   PARAMETER_VECTOR( sigma2_z );
16   PARAMETER_MATRIX( x_iz );
17   PARAMETER_MATRIX( beta_jz );
18
19   // Objective funcction
20   int n_i = y_iz.rows();
21   int n_z = y_iz.cols();
22   Type jnll = 0;
23   matrix<Type> I_zz(n_z,n_z);
24   I_zz.setIdentity();
25   matrix<Type> V_zz(n_z,n_z);
26   matrix<Type> S_zz(n_z,n_z);
27   matrix<Type> gamma_iz = X_ij * beta_jz;
28   matrix<Type> Vi_zz(n_z,n_z);
29
30   // Define process variance
31   V_zz = make_covariance( sigma2_z, RAM, sigma2_z, n_z, n_factors );
32
33   // Probability of random coefficients
34   for( int i=1; i<n_i; i++){
35     if( !R_IsNA(asDouble(DeltaT_i(i))) ){
36       Vi_zz = DeltaT_i(i) * V_zz;
37       jnll += density::MVNORM(Vi_zz)( x_iz.row(i) - (x_iz.row(i-1)+gamma_iz.
       row(i-1)) );
38     }
39   }
40
41   // Probability of data conditional on fixed and random effect values
42   for( int i=0; i<n_i; i++){
43     S_zz = I_zz * error2_i(i);
44     if( !R_IsNA(asDouble(y_iz(i,0))) ){
45       jnll += density::MVNORM(S_zz)( y_iz.row(i)-x_iz.row(i) );
46     }
47   }
48
49   // Cumulative sum of covariates
50   matrix<Type> Gsum_iz( n_i, n_z );
51   Gsum_iz.row(0) = gamma_iz.row(0) + x_iz.row(0);
52   for( int i=1; i<n_i; i++){
53     Gsum_iz.row(i) = Gsum_iz.row(i-1) + gamma_iz.row(i);
54   }
55
56   // Reporting
57   REPORT(gamma_iz);
58   ADREPORT(Gsum_iz);
59   REPORT( V_zz );
60   return jnll;
61 }
```

Before running the model, we check the sparsity of the inner Hessian matrix and confirm that every row of the Hessian is nonzero for at most three columns (Fig. 4.2). Intuitively, this makes sense, given that the location x_t for spatial coordinate x depends upon the values of (x_{t-1}, x_{t+1}, y_t), and is independent of all other values conditional upon these three.

We then fit this model in R only using data for 50 of the 1000 simulated time-steps (Code 4.3), which we accomplish by including NA values for 950 measurements. This model then allows us to predict location for all 1000 modeled times (Fig. 4.3). Given that we are fitting only 50 observations, the empirical Bayes prediction for movement follows 49 straight line-segments that connect each observation to the one before and after it (green lines). However, the model also estimates that uncertainty (i.e., the standard error on the

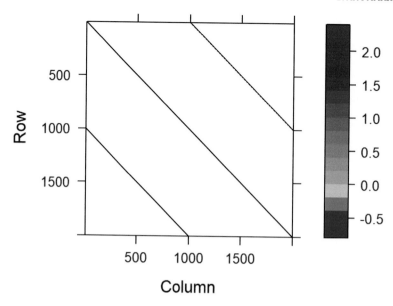

Dimensions: 2000 x 2000

FIGURE 4.2: Illustration of the sparsity of the inner Hessian matrix for the two-dimensional track-reconstruction using a first-order autoregressive process for diffusion.

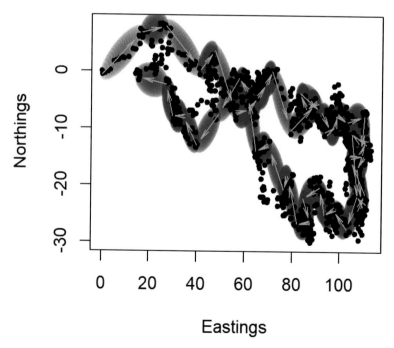

FIGURE 4.3: Illustration of a single simulation for the true location in each time (black dots), reconstructed tracks (green arrows), and uncertainty ellipse for each time (blue opaque circles).

random effect representing location) is lowest at the time of observations, and uncertainty increases and then decreases between any two observations. This is sometimes called a *Brownian bridge*, where a Brownian (random-walk) process is used to predict an increase and subsequent decrease in predictive variance between two observations.

CODE 4.3: R code fitting diffusive movement to location measured for a single individual.

```
1  # Define drift as a function of time
2  formula = ~ 0
3  X_ij = as.matrix( model.matrix(formula, data=data.frame("t_i"=t_i)) )
4
5  # Build inputs
6  Data = list( "y_iz"=y_iz,
7               "DeltaT_i"=DeltaT_i,
8               "error2_i"=rep(1,nrow(y_iz)),
9               "X_ij"=X_ij,
10              "n_factors"=0,
11              "RAM"=matrix(nrow=0,ncol=4) )
12 Parameters = list( "sigma2_z"=log(1),
13                    "x_iz"=Data$y_iz,
14                    "beta_jz"=matrix(0,nrow=ncol(Data$X_ij),ncol=2) )
15 Random = "x_iz"
16
17 # drop some data
18 which_include = seq(1, nrow(Data$y_iz), length=50 )
19 Data$y_iz[-which_include,] = NA
20
21 # Build and fit object
22 Obj = MakeADFun(data=Data, parameters=Parameters, random=Random )
23 Opt = nlminb( start=Obj$par, obj=Obj$fn, grad=Obj$gr )
24 Opt$SD = sdreport( Obj )
```

Next, we expand the model by approximating taxis (i.e., advection toward a preferred habitat) using a drift term (i.e., advection in a general direction without specifying a habitat preference function). To do so, we expand Eq. 4.11 to include a draft term in the mean of the multivariate normal distribution:

$$\mathbf{s}_{t+\Delta_t} \sim \text{MVN}(\mathbf{s}_t + \mathbf{z}_t\mathbf{B}, \Delta_t\mathbf{V}) \tag{4.13}$$

where $\mathbf{Z}$ is a matrix of covariates $\mathbf{z}_t$ applied to each time-interval t, and $\mathbf{B}$ is a matrix of coefficients that measure the response for each covariate (rows) for each locational coordinate (columns), such that $\mathbf{z}_t\mathbf{B}$ is the direction and magnitude of drift at time t. In the following, we specify a drift term as a cubic spline (using `splines::bs` to construct the spline) for the time since the foraging trip began (Code 4.4). This drift term then approximates the tendency for an animal to follow a general heading over a prolonged period. For a central-place forager, in particular, we assume that the drift term will tend to sum to zero over time, such that the animal has a tendency to "drift" back to it's initial location. To explore this, we have already defined `Gsum_iz` as a matrix that calculates the cumulative sum of advection that is predicted by the drift term $\mathbf{z}_t\mathbf{B}$, and we envision that `Gsum_iz` will tend to end at a location that is near the start location. We then compare this fit including diffusion and drift (Fig. 4.4), with the previous fit that only included diffusion (Fig. 4.3). This comparison shows that including the drift term in this single track results in a small reduction in the uncertainty ellipse (a smaller blue-shaded area), but otherwise has little impact on the overall estimate of the animal track.

CODE 4.4: R code to generate basis functions representing a cubic spline for the drift term and then rebuilding the TMB model.

```
1  # Define drift as a function of time
2  formula = ~ splines::bs(t_i,5)
```

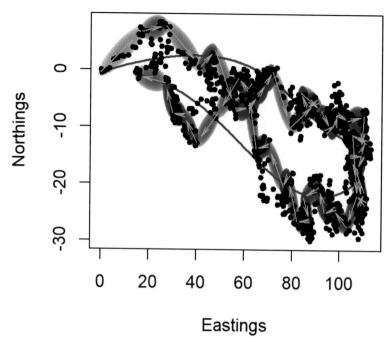

FIGURE 4.4: Same as 4.3, but also showing the cumulative sum of the drift term (red line), calculated by setting diffusion to zero in the fitted object. We include this to visualize the aggregate effect of the drift term on predicted location, where the start and end location of the red line are close to one another, as expected for a central-place forager.

```
3 Data$X_ij = as.matrix( model.matrix(formula, data=data.frame("t_i"=t_i)) )
4 Parameters$beta_jz = matrix(0,nrow=ncol(Data$X_ij),ncol=2)
5
6 # Refit model
7 Obj = MakeADFun(data=Data, parameters=Parameters, random=Random )
8 Opt = nlminb( start=Obj$par, obj=Obj$fn, grad=Obj$gr )
9 Opt$SD = sdreport( Obj )
```

Similarly, we can fit this same model to locational data for a northern fur seal (*Callorhinus ursinus*), specifically a lactating adult female, that was tagged at St. Paul Island in 2016 (Fig. 4.5). The data are obtained using a GPS tag and have been analyzed elsewhere [123]. We thin the 237 GPS location records to 25 measurements, and compare the predicted track with the withheld measurements. This analysis shows again that a basis spline for the time since the trip started can predict some but not all of the movement, and that the standard errors generally are similar to the withheld measurements.

4.4 Specifying Covariance Matrices

So far, we have explored a bivariate state-space model, which includes a state vector $\mathbf{s}_t = (x_t, y_t)$ which describes the location of a single animal using two spatial coordinates. However, many state-space models will involve more than two variables. For example, we might seek to describe:

- Abundance for multiple species in a community using a vector autoregressive model. For example, we might define $\log(\mathbf{n}_{t+1}) = \alpha + \mathbf{B}\log(\mathbf{n}_t)$ where $\mathbf{B}$ is a matrix of community

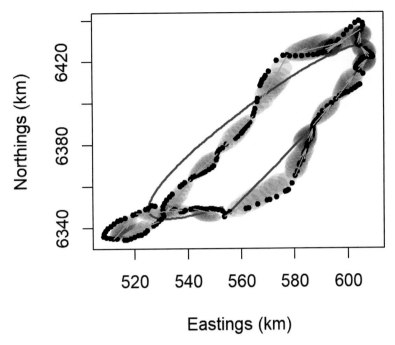

FIGURE 4.5: Same as 4.4, but fitted to data for a northern fur seal foraging trip from St. Paul Island in 2016 where x and y coordinates in the Universal Transverse Mercator (UTM) zone-2 projection.

interactions and $n_{c,t}$ is abundance for each species c. This results in a multivariate version of the Gompertz model for population dynamics [103, 228]. In this case, many properties of community dynamics can be calculated from the matrix **B** of species interactions;

- Size-at-age for multiple ages [212], used to identify environmental drivers that might affect individual growth rates;

- Juvenile survival rates for different habitats in a spatially structured meta-population [153], used to identify demographic stability for the entire species.

In these and other cases, we might want to estimate the correlation between different variables, and therefore need to replace the diagonal covariance $\mathbf{V} = \sigma_s^2 \mathbf{I}$ with a general covariance matrix. However, as the number of variables N increases, the number of correlation and variance parameters in an unstructured covariance matrix is $N(N+1)/2$, and this becomes prohibitive as $N >> 10$.

We first summarize four alternative parameterizations of the covariance matrix **V** before discussing them in detail subsequently:

1. *Diagonal and equal*: this is what we've been using so far, $\mathbf{D}_{equal} = \sigma^2 \mathbf{I}$, and it involves a single parameter, which is easiest to specify as $\sigma^2 = \exp(\theta)$;

2. *Diagonal and unequal*: alternatively, we might specify that each variable has a unique variance. For example, if modelling three variables this will result in:

$$\mathbf{D}_{unequal} = \begin{bmatrix} \sigma_1^2 & 0 & 0 \\ 0 & \sigma_2^2 & 0 \\ 0 & 0 & \sigma_3^3 \end{bmatrix} \tag{4.14}$$

This parameterization involves N parameters but still specifies that variables are independent;

3. *Factor model*: we might instead specify a factor model and use this to generate the covariance matrix. This factor model can range in complexity. For example, it could represent a single exogenous driver that explains variation in all variables. Alternatively, it can be used to represent an unstructured covariance matrix;

4. *Structural equation model*: finally, we might specify a structural equation model and use this to calculate the resulting covariance among factors. This approach allows correlations to be derived from parameters that are interpreted as regression slopes. However, it also requires specifying the structural relationship among variables a priori.

We next discuss computational details regarding Factor and Structural Equation Modelling approaches to covariance.

4.4.1 Factor Models

We might envision that the covariance among variables arises from a *factor model* with $N_{factors}$ axes of covariation.

$$\mathbf{V}_{factor} = \mathbf{\Lambda}\mathbf{\Lambda}^T \tag{4.15}$$

where $\mathbf{\Lambda}$ is a matrix with N rows and $N_{factors}$ columns. Eq. 4.15 is only identifiable given further restrictions on parameters [273, 274]. We choose to ensure identifiability by specifying that any element above the diagonal is fixed at zero. For example, when $N = 3$ and $N_{factors} = 2$ we get:

$$\mathbf{\Lambda} = \begin{bmatrix} \lambda_{1,1} & 0 \\ \lambda_{2,1} & \lambda_{2,2} \\ \lambda_{3,1} & \lambda_{3,2} \end{bmatrix} \tag{4.16}$$

This factor model can be interpreted in several ways:

1. *Trimmed Cholesky matrix*: as one interpretation, recall that covariance $\mathbf{V}$ is a symmetric matrix, and we can calculate a *Cholesky decomposition* of any symmetric matrix where $\mathbf{V} = \mathbf{L}\mathbf{L}^T$ and $\mathbf{L}$ is a lower-triangle matrix (i.e., has value 0 above the diagonal, see Section B.4.2 for details). In Eq. 4.16, we are therefore estimating N_{factor} columns of the Cholesky matrix, and we therefore call it a *trimmed Cholesky* matrix. When $N = N_{factors}$, the loadings matrix $\mathbf{\Lambda}$ is the full Cholesky matrix, $\mathbf{L} = \text{chol}(\mathbf{V})$, and $\mathbf{V}_{factor}$ is full rank. Estimating all parameters in this Cholesky matrix is then equivalent to estimating an *unstructured covariance*, and it involves estimating as many parameters as are identifiable. Alternatively, when $N_{factors} < N$, the resulting V_{factor} has rank $N_{factors} < N$ and it involves estimating fewer parameters;

2. *Loadings matrix*: we use the symbol $\mathbf{\Lambda}$ (the Greek letter for L) to indicate that it can be interpreted as a *loadings matrix* where, e.g., $\lambda_{2,1}$ represents how strongly variable 2 is loaded on (i.e., associated with) factor 1.

To illustrate these two interpretations of the factor-model covariance, consider that we have a vector $\mathbf{x}$ composed of standard normal random variables:

$$\mathbf{x} \sim \text{MVN}(\mathbf{0}, \mathbf{I}) \tag{4.17}$$

If we then take the product of this and the loadings matrix:

$$\mathbf{y} = \boldsymbol{\Lambda}\mathbf{x} \tag{4.18}$$

the resulting vector $\mathbf{y}$ can be written as:

$$\mathbf{y} \sim \mathrm{MVN}(\mathbf{0}, \mathbf{V}_{factor}) \tag{4.19}$$

where $\mathbf{V}_{factor} = \boldsymbol{\Lambda}\boldsymbol{\Lambda}^T$. We can therefore interpret the loadings matrix $\boldsymbol{\Lambda}$ either (1) as a projection matrix that transforms latent factors $\mathbf{x}$ into the response variable $\mathbf{y}$, or (2) as a tool to construct the covariance $\mathbf{V}_{factor}$. The former interpretation is useful, e.g., to decompose a multivariate process with N variables into a simpler process involving N_{factor} factors. For example, this is commonly done during species ordination, where a vector of species densities at different sites is reduced to two axes, and the loadings $\boldsymbol{\Lambda}^T$ and factors $\mathbf{x}$ are plotted on a two-dimensional scatterplot [148].

In practices, there are two caveats to using the factor-model covariance.

- *Ensuring full rank*: when using a rank-reduced factor model, $N_{factors} < N$, it may still be convenient to ensure that the covariance is full rank. This ensures that any possible vector $\mathbf{y}$ is within the support of the covariance (i.e., the null space is an empty set). To achieve this, we often use:

$$\mathbf{y} \sim \mathrm{MVN}(\mathbf{0}, \boldsymbol{\Lambda}\boldsymbol{\Lambda}^T + \mathbf{D}) \tag{4.20}$$

where $\mathbf{D}$ is diagonal and either equal or unequal;

- *Rotations*: similarly, we have specified that the loadings matrix $\boldsymbol{\Lambda}$ is lower-triangle; this constraint ensures that columns cannot be freely rotated, and hence ensures identifiability. However, factors cannot be interpreted independently of one-another. Ecologists therefore typically "rotate" factors prior to interpreting them:

$$\begin{aligned} \mathbf{x}^* &= \mathbf{H}\mathbf{x} \\ \boldsymbol{\Lambda}^* &= \boldsymbol{\Lambda}\mathbf{H}^{-1} \end{aligned} \tag{4.21}$$

where $\mathbf{H}$ is the *rotation matrix*. It is then easy to see that this rotation leaves the model unchanged, where $\mathbf{y} = \boldsymbol{\Lambda}\mathbf{x} = \boldsymbol{\Lambda}^*\mathbf{x}^*$ for any invertible matrix $\mathbf{H}$. Ecologists typically use a *varimax rotation*, which ensures that loadings $\boldsymbol{\Lambda}^*$ has a few large values and many small values (i.e., factors typically represent a small number of individual variables). However, we also use a *principal component analysis rotation* [237], where the first axis explains as much variance as possible, the 2nd explains as much as possible given this, etc. This rotation then ensures that the first axis is the "dominant mode of variability", and that displaying only the first two axes still captures as much of the total variance as possible.

4.4.2 Structural Equation Models

Alternatively, an analyst may have a priori knowledge about associations between variables. In the strongest case, the analyst might specify a *causal map*, e.g., specifying that changing variable X_1 causes a change in variable X_2, X_1 and X_2 both cause a change in X_3, and X_3 causes a change in X_4, which in turn affects X_1. These types of cycles are common in trophic relationships where species are often assembled in tightly interacting *trophic modules* with weak linkages between modules [95]. Additionally. this type of graphical model or causal map often arises from eliciting stakeholder input, where input can be used to generate a *conceptual model* of ecosystem function, which then be represented as a *structural equation model* [112]:

$$\mathbf{y} = P\,\mathbf{y} + \delta \tag{4.22}$$

where P is a matrix of path coefficients that represent the hypothesized linear relationships linking variables, and $\delta \sim \text{MVN}(\mathbf{0}, \boldsymbol{\Sigma})$ represents additional exogenous drivers. This then results in a covariance:

$$\text{Var}(\mathbf{y}) = \mathbf{B}\boldsymbol{\Sigma}\mathbf{B}^T \tag{4.23}$$

where $\mathbf{B} = (\mathbf{I} - P)^{-1}$ and $\boldsymbol{\Sigma} = \boldsymbol{\Lambda}\boldsymbol{\Lambda}^t$. Structural equation packages in R [59] typically combine this calculation with an inverse-Wishart distribution for the sample covariance from a set of measurements, but the procedure for converting mechanisms P and $\boldsymbol{\Gamma}$ to a covariance can then be used in any nonlinear model.

Specifying covariance using a structural-equation model has several benefits [236]:

1. *Parsimonious representation*: it allows detailed control over how many parameters N_{sem} are used to represent the relationship among variables. In particular, SEM can estimate any number of parameters from $N_{sem} = 1$ (i.e., $\boldsymbol{\Lambda} = \sigma\mathbf{I}$) to $N_{sem} = N(N+1)/2$ to represent the covariance, where N is the number of variables;

2. *Complex and cyclic ecological relationships*: the SEM can include cyclic dependencies (e.g., a trophic flow from resource to producer to consumer, but where the consumer increases the basal resource via excretion) as long as the total number of dependencies (N_{sem}) is less than the degrees of freedom in the covariance matrix, $N_{sem} \leq N(N+1)/2$;

3. *Conceptual modelling*: the structural equations can often be generated from existing conceptual or semi-quantitative models for a given ecological system;

4. *Path coefficients as regression slopes*: the estimated coefficients in P are interpreted similarly to regression slopes in a linear model, i.e., a change of Δ in y_1 causes a $\rho_{2,1}\Delta$ change in y_2.

However, the resulting estimates depend upon the structural equation model that is specified, so model-building requires additional information about the system that is explicitly represented by the SEM.

4.5 Comparing Factor and Structural Equation Models for Group Movement

We therefore have a wide range of methods available for specifying a covariance matrix among variables. These different options are implemented in the TMB custom function make_covariance (Code 4.5). We next demonstrate this using simulated data, representing 10 tracked individuals that pursue a similar foraging strategy that involves moving northward and then returning toward their starting location (Code 4.6). This shared foraging strategy results in correlated random walks that are correlated along the y-axis but not the x-axis across individuals (see Fig. 4.6).

CODE 4.5: Custom function make_covariance used to construct a covariance among species using a factor model or structural equation model.

```
1 template<class Type>
2 matrix<Type> make_covariance( vector<Type> s2_z,
```

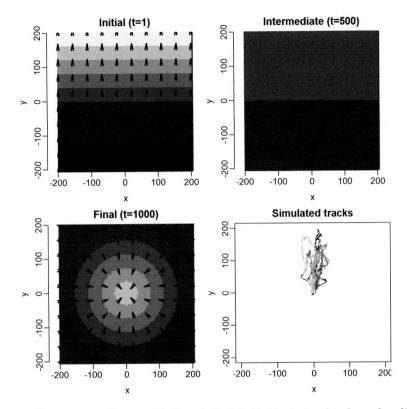

FIGURE 4.6: Same as 4.1, but modified such that individuals tend to leave heading north-ward before returning toward their original location.

```
3              matrix<int> RAM,
4              vector<Type> RAMstart_z,
5              int n_rows,
6              int n_cols ){

8   // Define temporary objects
9   matrix<Type> Cov_rr(n_rows, n_rows);
10  matrix<Type> I_rr( n_rows, n_rows );
11  I_rr.setIdentity();
12  if( RAM.rows()>0 ){
13    // Assemble Structural Equation Model covariance
14    matrix<Type> L_rc(n_rows, n_rows);
15    matrix<Type> Rho(n_rows, n_rows);
16    matrix<Type> Gamma(n_rows, n_rows);
17    Rho.setZero();
18    Gamma.setZero();
19    for(int zI=0; zI<RAM.rows(); zI++){
20      if( RAM(zI,3)>0 ){
21        if(RAM(zI,0)==1) Rho(RAM(zI,1)-1,RAM(zI,2)-1) = s2_z(RAM(zI,3)-1);
22        if(RAM(zI,0)==2) Gamma(RAM(zI,1)-1, RAM(zI,2)-1) = s2_z(RAM(zI,3)-1);
23      }else{
24        if(RAM(zI,0)==1) Rho( RAM(zI,1)-1, RAM(zI,2)-1 ) = RAMstart_z(zI);
25        if(RAM(zI,0)==2) Gamma( RAM(zI,1)-1, RAM(zI,2)-1 ) = RAMstart_z(zI);
26      }
27    }
28    L_rc = I_rr - Rho;
29    L_rc = atomic::matinv( L_rc );
30    L_rc = L_rc * Gamma;
```

```
31      Cov_rr = L_rc * L_rc.transpose();
32    }else{
33      // Assemble loadings matrix
34      matrix<Type> L_rc(n_rows, n_cols);
35      L_rc.setZero();
36      int Count = 0;
37      if( n_cols>0 ){
38        for(int r=0; r<n_rows; r++){
39        for(int c=0; c<n_cols; c++){
40          if(r>=c){
41            L_rc(r,c) = s2_z(Count);
42            Count++;
43          }else{
44            L_rc(r,c) = 0.0;
45          }
46        }}
47      }
48      // Diagonal and equal coariance matrix
49      Cov_rr = I_rr * exp( s2_z(Count) );
50      // Add factor-model covariance matrix
51      Cov_rr += L_rc * L_rc.transpose();
52    }
53    return Cov_rr;
54  }
```

CODE 4.6: R code fitting a multivariate state-space movement model to location measurements for multiple animals from a Lagrangian perspective.

```
1  # Parameters
2  gamma_t = beta_t = seq(0.5,-0.5,length=n_t)
3  beta_t = ifelse( beta_t>0, 0, beta_t )
4  gamma_t = ifelse( gamma_t<0, 0, gamma_t )
5
6  # Simulate three tracks
7  n_tracks = 4
8  set.seed(101)
9  x_iz = NULL
10 for( track in 1:n_tracks ){
11   s = simulate_track( n_t=n_t, beta_t=beta_t,
12                       gamma_t=gamma_t, get_preference=get_preference)
13   colnames(s) = paste0( c("x","y"), track )
14   x_iz = cbind(x_iz, s)
15 }
16 y_iz = x_iz + array(rnorm(prod(dim(x_iz))), dim=dim(x_iz))
17
18 # Build inputs
19 Data = list( "y_iz"=y_iz,
20              "DeltaT_i"=c(NA,rep(1,nrow(y_iz)-1)),
21              "error2_i"=rep(1,nrow(y_iz)),
22              "X_ij"=array(dim=c(nrow(y_iz),0)),
23              "n_factors"=0,
24              "RAM"=matrix(nrow=0,ncol=4) )
25 Params = list( "sigma2_z"=log(1),
26                "x_iz"=Data$y_iz,
27                "beta_jz"=array(0,dim=c(ncol(Data$X_ij),ncol(Data$y_iz))) )
28 Random = c("x_iz")
29
30 # drop some data
31 which_include = seq(1, nrow(Data$y_iz), length=20*n_tracks )
32 Data$y_iz[-which_include,] = NA
33
34 # Build and fit object
35 Obj = MakeADFun(data=Data, parameters=Params, random=Random )
36 Opt = nlminb( start=Obj$par, obj=Obj$fn, grad=Obj$gr )
```

We then fit this model again using the three alternative approaches to specifying the covariance in diffusion among animals (i.e., specifying $\mathbf{V}$ in Eq. 4.11):

1. The "diagonal-and-equal" covariance matrix $\mathbf{V} = \mathbf{D}_{equal}$ where $\mathbf{D}_{equal} = \sigma^2 \mathbf{I}$, involving one parameter representing covariance among individuals;

2. An alternative factor-model covariance, $\mathbf{V} = \mathbf{D}_{equal} + \mathbf{\Lambda}\mathbf{\Lambda}^T$ using $N_{factors} = 2$ under the assumption that there's potential correlations in x- and y-axis directions. This then results in 16 parameters representing covariance (15 in the factor model, and one more in the diagonal-and-equal matrix);

3. A structural equation-model covariance, under the assumption that animal #1 is leading the other animals and therefore movement for that animal then causes similar movement for the others. This specification results in 2 estimated parameters representing the impact of animal #1 on other animals, for each of two spatial coordinates.

We specify the structural equation model by writing a text file that is parsed by function specifyModel in R-package sem [59] (Code 4.7). This text file is used to specify which path coefficients should be estimated in P and these are indicated by using a single-headed arrow where, e.g., y1 -> y2 indicates that the model should estimate the magnitude of change in y2 expected from a change in y1 . This text file is also used to specify which parameters should be estimated in the Cholesky $\mathbf{\Lambda}$ of exogenous variance $\mathbf{\Sigma}$. These are indicated by using a double-headed arrow where, e.g., y1 <-> y1 indicates that the variance of y1 should be estimated, or y1 <-> y2 indicates that the covariance of y1 and y2 should be estimated. By specifying exog.variances=TRUE , specifyModel automatically adds a two-headed arrow for each variable, i.e., estimates a diagonal and unequal matrix for exogenous covariance E. Finally, the text file requires the user to specify a label for each parameter (i.e., in each row). When two or more parameters share a single label (e.g., all path coefficients are named y in this example), then those coefficients are all estimated to have the same value. Similarly, if a parameter is labeled NA , then they are fixed at a value *a priori* that is provided immediately after the parameter name. This text file is then parsed by a custom function build_ram that constructs a matrix RAM , which is used in turn by make_covariance to construct the covariance given model parameters (see Code 4.5).

CODE 4.7: R interface to build a Structural Equation Model for group movement.

```
# Specify SEM
text = "
    y1 -> y2, y
    y1 -> y3, y
    y1 -> y4, y
"
SEM_model = sem::specifyModel( text = text,
                               exog.variances = TRUE,
                               endog.variances = TRUE,
                               covs = colnames(y_iz) )
RAM = build_ram( SEM_model, colnames(y_iz) )
```

These three hypotheses therefore provide a wide range of model complexity, and illustrate different avenues to specify a covariance matrix. Similarly, they result in a range of different estimates of covariance among variables (Fig. 4.7), although the factor and structural-equation estimates of covariance both correctly identify a large covariance among individuals in movement along the y-axis. The Akaike Information Criterion identifies that the factor-model covariance is most parsimonious, despite having the most parameters (see bottom-right of each panel in Fig. 4.8). Despite these differences, however, all models estimate

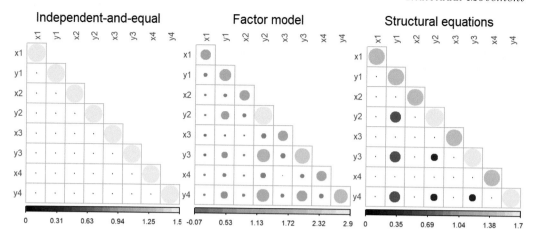

FIGURE 4.7: Covariance in movement **V** in Eq. 4.11 visualized using the function corrplot in the corrplot package [257], where the area of each circle is proportional to the covariance (and see the colorbar legend at the bottom of each panel), for the diagonal-and-unequal covariance (left panel), factor model (middle panel), or structural equation model (right panel), and the 8 variables are labeled x or y for movement in eastings or northings, respectively, and are also labeled 1 through 4 for the four animals.

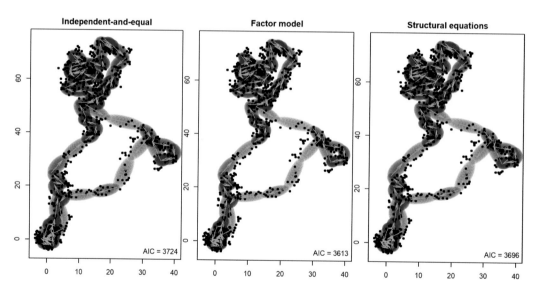

FIGURE 4.8: Depiction of reconstructed tracks for the 1st animal, following same plotting format as Fig. 4.3, using the diagonal-and-unequal covariance (left panel), factor model (middle panel), or structural equation model (right panel). We also list the Akaike Information Criterion (AIC) value for each model in the bottom-right of each panel, where the lowest value corresponds to the highest expected performance for new data (see Section 1.7). In this case, the lowest AIC occurs using the factor-model covariance. Differences in reconstructed track using these three covariance matrix structures were similarly small for the other three animals (results not shown here).

tracks that are almost indistinguishable (Fig. 4.8). This then suggests that the data are sufficiently informative that these small differences in model structure have little leverage over interpolated locations.

4.6 Chapter Summary

In summary, we have showed that:

1. Individual movement can be described explicitly (by tracking individual position over time) or implicitly (by deriving properties arising from aggregating movement across time or individuals). It is possible to calculate ecological patterns arising from explicit consideration of organism movement, but harder to go in the reverse direction and derive organism movement processes from observed ecological patterns;

2. Individual movement can be decomposed into three components representing taxis (movement toward preferred habitat), drift (movement in a specified direction), and diffusion (otherwise unexplained movement). Taxis and drift are sometimes combined using the term advection. These processes can be described in continuous space and time using a stochastic differential equation, or discretized in continuous space and discrete time as a difference equation;

3. Individual tracks arise from a movement in two dimensions, and can be simulated using taxis, drift, and diffusion. Similarly, these tracks can be reconstructed using intermittent location measurements using a multivariate state-space model. In these cases, taxis can be approximated using a drift term if the covariates driving habitat preferences are unknown or unavailable;

4. Tracks for multiple individuals can be reconstructed jointly using a multivariate state-space model with a larger number of position variables. Large multivariate models involve a covariance matrix. If correlations are ignored, this can be approximated as "diagonal-and-equal" or "diagonal-and-unequal" matrices.

5. Covariance among multiple time-series (representing animal locations or otherwise) can be estimated using a factor-model covariance. This factor model involves estimating a few columns of the Cholesky decomposition of covariance. This can be interpreted as a loadings matrix, which measures the association of each variable with each latent factor. The factor model requires additional restrictions to ensure that results are identifiable, and results can be interpreted after applying a varimax or PCA rotation.

6. Alternatively, covariance among multiple time-series can be estimated using a structural-equation covariance matrix. This approach yields path coefficients that are interpreted like standard regression slopes. The structural-equation covariance allows detailed control over the number of estimated parameters, and can represent ecological assumptions about the causal mechanisms that operate among variables.

4.7 Exercise

In Section 4.2 and Code 4.1, we simulate tracks for a central-place forager such that preference initially increases with distance from the starting point but then reverses, such that the individual tends to move away and then back to the centroid of its range. First, please modify this simulation to match an alternative ecological scenario of your choice, e.g., simulating daily movement where the animal has a normally distributed preference in summer and winter, but where the centroid differs between these seasons, such that they move from one to other range seasonally. Then, calculate the expected habitat utilization given this preference function, e.g., by calculating 1000s of animal tracks, discretizing the spatial domain, and then calculating the proportion of time spent in each grid cell. Consider changing the relative magnitude of diffusion vs. taxis (where the latter is controlled by the height of the preference function), and seeing how this affects the resulting habitat utilization.

5

Spatial Models

5.1 Exogenous and Endogenous Drivers of Spatial Patterns

Ecologists often seek to identify the processes that underlie patterns that they observe in nature. For example, community ecologists have classified vegetation communities based on local temperature and precipitation, and subsequently identified tradeoffs in plant structure and metabolic constraints that underlie these [189]. In this case, an ecologist might construct a regression predicting vegetation densities as a function of local precipitation and temperature, and might seek to develop a statistical model that explains a large portion of observed variance. As another example, meta-population theory arose from describing how butterfly movement between habitat patches can "rescue" local extirpation arising from small population sizes in each individual patch [80]. In this latter case, the variable of interest (abundance in each habitat patch) is expected to vary stochastically (due to low sample sizes for individual birth and death) such that environmental covariates are not sufficient in isolation to explain demographic variation.

These two examples illustrate that spatial patterns arise both from:

- *Exogenous drivers*, i.e., where physical habitat constraints (e.g., temperature and precipitation) drive population densities via their effect on demographics (birth, death, growth, movement, etc.);

- *Endogenous drivers*, i.e., where age-structured movement, density-dependent habitat selection, selective social information usage [139], and other properties of a given population or community might cause a particular form of spatial aggregation to arise.

This distinction becomes important because analysts might hope to eventually identify and measure the exogenous drivers that are associated with spatial variation in population density. However, even in this perfect world where all exogenous drivers have been identified and measured, there will always be endogenous drivers of spatial distribution that cause local densities to not reflect long-term averages.

When endogenous drivers have a large role, or when important exogenous drivers have not been measured, regressing population density on available covariates is likely to result in large residuals (i.e., instances where observations are much larger or smaller than model predictions), where these residuals are then correlated for nearby samples. Neglecting these spatially correlated residuals will cause predicted densities to be a poor description of underlying patterns, and therefore cause any downstream ecological interpretation to be suspect.

In practice, ecologists generally recognize the importance of accounting for residual patterns that are correlated across space, time, or among variables (e.g., between ages, species, etc.), both when fitting models [48] and when evaluating their performance [190]. Ecologists have therefore developed a wide range of techniques to account for residuals spatial variation [47]. In the following chapter, we contrast widely used methods to estimate spatially

DOI: 10.1201/9781003410294-5

correlated residual patterns, while emphasizing those that can combined with other types of ecological dynamics in later chapters.

5.2 Basis Expansion and Splines

In Section 1.4, we introduced Generalized Linear Models (GLMs), which we described as a model having a linear predictor μ_i, link function g, and probability distribution f defining the likelihood of data given parameters:

$$Y_i \sim f(\mu_i) \tag{5.1}$$
$$g(\mu_i) = \mathbf{x}_i \beta \tag{5.2}$$

We first focus on approaches that address residual spatial variation by increasing the flexibility of available covariates[1].

To define model "flexibility" we first introduce the concept of *degrees of freedom*, which measures the the number of dimensions that are necessary to describe the predictions from a model. For example, consider a model where covariate matrix $\mathbf{X}$ includes a column of $1s$ (representing an intercept) and then a single covariate. In this case, there's (at most) two degrees of freedom for the predicted density surface $\mu = \mathbf{X}\beta$, represented by the vector of coefficients β (Fig. 5.1). Put another way, the potential densities μ could be described as the sum of two *basis functions*. To increase model flexibility, we could therefore expand the number of basis functions that are needed to describe the predicted response μ.

To increase the flexibility of this GLM, we therefore apply *basis expansion* to create an expanded set of covariates $\mathbf{X}^*$, and an associated increase in the degrees of freedom β. Basis expansion involves specifying a function $b : X \mapsto \mathbf{X}^*$ that transforms the original covariates $\mathbf{X}$ to an expanded set of covariates $\mathbf{X}^*$ where the rank of $\mathbf{X}^*$ is greater than that of $\mathbf{X}$. In practice, basis expansion often takes as input one or two covariate vectors, and as output gives an expanded set of covariate vectors. The simplest example is quadratic expansion, where a covariate vector (e.g., elevation) is replaced with two vectors, representing elevation and elevation-squared (Fig. 5.2). This then allows the potential response function $\mathbf{X}^*\beta$ to generate a dome-shaped response, potentially identifying the value of a given covariate where densities are expected to be highest. Quadratic basis expansion is one example of polynomial basis expansion, where a covariate X is combined with its polynomials, e.g., its square X^2, cube X^3, etc.

However, higher-order polynomials tend to have extreme slopes and perform poorly when extrapolating outside the observed values for a given covariate. We therefore next introduce *splines*. We will typically use the splines2 package [253], which can construct a conventional B-spline as well as a wide range of alternatives having useful and specialized characteristics. Constructing a spline involves at least two choices:

- *Order*: splines are designed to vary smoothly, such that the linear predictor $\mu = \mathbf{X}\beta$ is also a smooth function. Technically, this involves specifying the spline order, where all derivatives of each basis up to this order are continuous. By default, software involving splines often uses a 3rd-order (cubic) spline by default, meaning that the 1st and 2nd derivatives vary smoothly, such that the human eye typically can't identify any discontinuity in the resulting response function;

---

[1]See https://github.com/james-thorson/Spatio-temporal-models-for-ecologists/Chap_5 for code associated with this chapter.

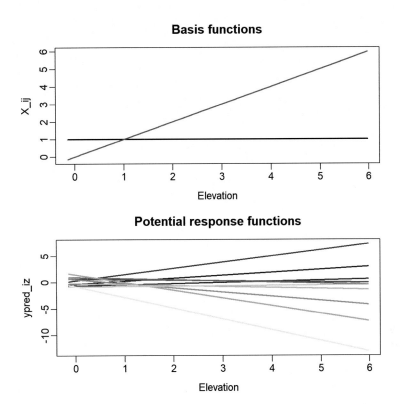

FIGURE 5.1: Illustration of the basis functions resulting from a single intercept and co-variate (top panel) representing elevation in km in China, and ten random simulations of the density function that could arise from using this basis (bottom panel).

- *Knots*: many types of spline also result in a change in some derivative at specific values of a given covariate. These changes occur at specified locations (often called *knots*) where a discontinuity occurs in the specified derivative.

The total degrees of freedom for many basic types of splines is the Order plus the number of knots, such that a 3rd-order spline with 7 knots would then have 10 degrees of freedom. Said another way, using a 3rd-order basis-spline with seven knots for basis expansion would transform a variable that provides one degree of freedom when fitted in a GLM into 10 separate variables that collectively provide 10 degrees of freedom when fitted in a GLM (Fig. 5.3). If we define the basis expansion as $\mathbf{X}^* = b(\mathbf{X})$, we then calculate the linear predictor for a generalized linear model as $g(\mu_i) = \mathbf{x}_i^* \beta$.

To illustrate, we envision that an ecologist might seek to describe the distribution or density of an alpine flower in Mainland China. This analyst might obtain the elevation throughout mainland China but also want additional flexibility to estimate whether highest densities occur at intermediate elevations. They might therefore apply spline basis expansion to calculate an expanded set of covariates. Illustrating this using an intercept and a 3rd-order basis spline with four degrees of freedom (i.e., one internal knot), we see that the first basis function corresponds to lowland areas in eastern and northern China, while the third basis function corresponds largely to the Tibetan plateau (Fig. 5.4). These five covariates could then be used to provide more flexibility when describing the plant distribution in this hypothetical example.

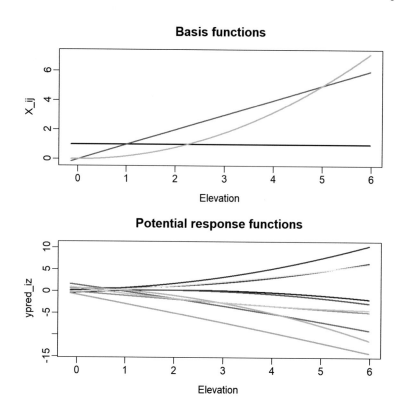

FIGURE 5.2: Illustration of the basis functions resulting from a single intercept and covariate (top panel) that is transformed using a quadratic polynomial basis expansion, and ten random simulations of the density function that could arise from using this basis (bottom panel).

5.3 Tensor Splines and Generalized Additive Models

Importantly, splines can be easily defined over multiple dimensions. For example, vegetation communities depend on the interaction of temperature and precipitation [189], such that an analyst might want a spline for the interaction of these two variables. This then requires a two-dimensional (2D) spline. A 2D *tensor spline* can easily be created as:

$$
\mathbf{z}_i^* \beta = \sum_{j=1}^{n_x} \sum_{k=1}^{n_y} \beta_{j,k} \mathrm{x}_{i,j}^* \mathrm{y}_{i,k}^*
$$
$$
= (\mathrm{x}_i^* \otimes \mathrm{y}_i^*) \beta
$$
(5.3)

where $\otimes$ represents a Kronecker product. If matrix $\mathbf{M}_1$ has dimension $m \times n$ and matrix $\mathbf{M}_2$ has dimension $p \times q$, then $\mathbf{M}_1 \otimes \mathbf{M}_2$ has dimension $mp \times nq$. Restating Eq. 5.3, a tensor-spline is created as the outer product of each individual dimension, where $b(X, Y) = b(X) \otimes b(Y)$. If basis $b(x)$ has n_x degrees of freedom, and basis $b(y)$ has n_y then tensor spline basis $b(X, Y)$ has $n_x n_y$ degrees of freedom.

In particular, ecologists often find that fitting a GLM to data that are associated with spatial information (which we call *spatial data* in the following) results in spatially correlated

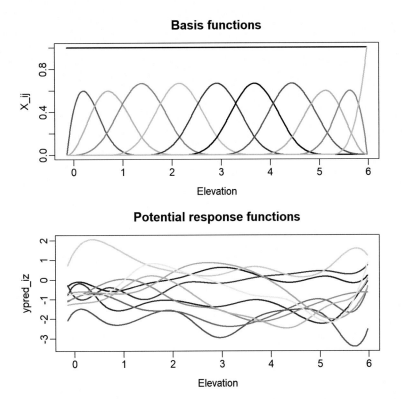

FIGURE 5.3: Illustation of the basis functions resulting from a single intercept and covariate (top panel) that is transformed using a 3rd-order basis-spline with seven knots (i.e., 10 degrees of freedom) for basis expansion, and 10 random simulations of the density function that could arise from using this basis (bottom panel).

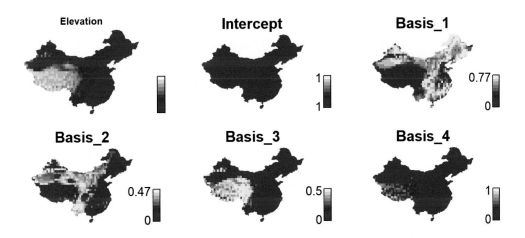

FIGURE 5.4: Illustation of the elevation in Mainland China (top-left panel, in meters above sea level), the intercept, and four basis functions resulting from a basis-spline with four degrees of freedom.

residuals. In these cases, it is feasible to start by including a spline basis-expansion of two-dimensional location. This is useful for several reasons, e.g.:

- *Using available information*: given the low cost of global position systems (GPS), most ecologists already record the location for each sample. It is therefore easy and cheap to augment many models with additional information representing geographic location;

- *Improved uncertainty estimates*: GLMs that have spatially correlated residuals violate the assumption that the log-likelihood can be calculated as the sum of log-likelihoods for each datum (see Section 1.5). This violation generally causes standard errors to be too small, and unimportant variables will be identified as statistically significant more often than they should (sometimes called Type-1 error) [48];

- *Conditional inference*: GLMs that include a basis-expansion of geographic location can then improve predictions. For example, when predicting density at a location near a positive residual, it might be useful to predict a similarly positive residual at that nearby location. This involves conditioning the prediction at a given location upon the statistical residuals of nearby locations.

For example, expanding our previous discussion of predicting distribution for an alpine plant in mainland China, an analyst might construct a tensor-spline using geographical coordinates. Using a basis spline with four degrees of freedom for each dimension then results in 16 spline basis functions that each represent a different section of a given spatial domain (Fig. 5.5). Fitting this model then requires estimating 16 additional parameters. However, using this tensor spline as a covariate allows a GLM to identify whether the hypothetical plant is associated with broad geographic areas beyond what is explained using elevation.

Next, imagine that we use this 2D basis spline but also choose to shrink the coefficients β towards zero by specifying them as a random effect:

$$g(\mu_i) = \mathbf{X}\alpha + \mathbf{Z}\beta$$
$$\beta_k \sim \text{Normal}(0, \sigma^2) \tag{5.4}$$

where $\mathbf{Z} = b(\mathbf{x}, \mathbf{y})$ is the 2D tensor-spline basis expansion of spatial coordinates and $\mathbf{X}$ is additional covariates with estimated slopes α. To illustrate the consequences of this specification, we note that specifying:

$$\mathbf{Y} = \mathbf{L}\mathbf{X}$$
$$\mathbf{X} \sim \text{MVN}(\mathbf{0}, \mathbf{\Sigma}) \tag{5.5}$$

is equivalent to specifying:

$$\mathbf{Y} \sim \text{MVN}(\mathbf{0}, \mathbf{L}\mathbf{\Sigma}\mathbf{L}^T) \tag{5.6}$$

so returning to Eq. 5.4, we can re-write our GLM as:

$$g(\mu) \sim \text{MVN}(\mathbf{X}\alpha, \sigma^2 \mathbf{Z}\mathbf{I}\mathbf{Z}^T) \tag{5.7}$$

where the distribution on spline coefficients β implies a covariance $\sigma^2 \mathbf{Z}\mathbf{I}\mathbf{Z}^T$ for all locations, which depends upon the exact basis functions $\mathbf{Z}$ being used.

We can therefore specify a generalized linear mixed model (GLMM) using 2D tensor-spline basis functions as covariates, and treating their response coefficients as random effects such that they are shrunk towards zero. This then allows the smoothness to be estimated as σ^2. A sufficiently large value for σ^2 results in estimates of β that approach those that would arise from estimating basis functions freely in a GLM (analogous to what we saw

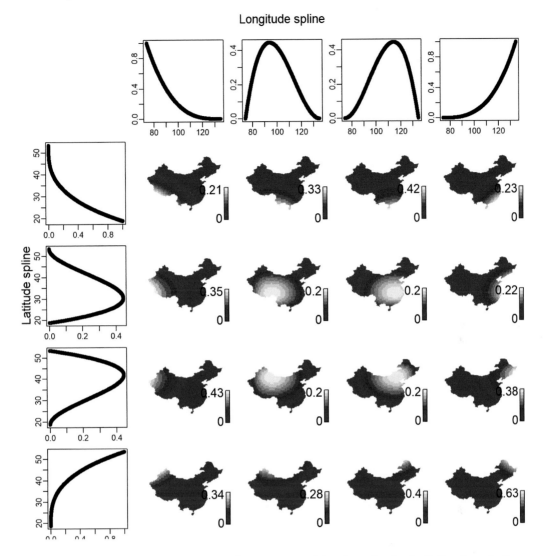

FIGURE 5.5: Illustation of the basis functions resulting from Latitude (left column), Longitude (top row), or the two-dimensional tensor-spline basis expansion of Latitude and Longitude arising from the outer product of these two individual dimensions.

in Fig. 2.5). Extending our discussion from Section 2.3, estimating spline coefficients as random effects is conceptually similar to specifying a GAM, where the log-determinant of the Hessian replaces the penalty term that is using in GAM to restrict the wiggliness of the resulting response function.

Given this link between GLMMs with spline basis functions and GAMs, ecologists often use widely available software for GAMs to implement spatial models. By default, package `mgcv::gam` includes 3rd order splines with many degrees of freedom (e.g., >1000 per dimension), and then avoids overparameterization by shrinking coefficients towards zero. This approach is sometimes called a *smoothing spline* (or specifically a *thin-plate regression spline*), and the actual placement of knots has little influence on results as the GAM penalty increases [268].

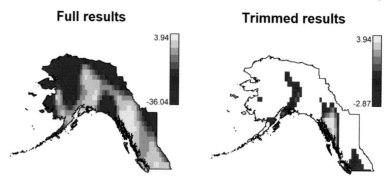

FIGURE 5.6: Estimated summertime log-density of bald eagles in Alaska, British Columbia, and the Yukon Territory, estimated using a GAM fitted to Breeding Bird survey counts (see colorbar for scale of log-densities in units numbers per survey station), showing either the full results (ranging widely from $e^4 = 55$ to $e^{-40} \leq 10^{-15}$ birds per station) as well as results that trim densities less than 0.1% of the maximum estimated density (to better show contrast in results in areas with non-zero densities).

To demonstrate, we download data from the Breeding Bird Survey [202] using the 2020 release [172][2]. This survey started in 1966 and has expanded to include thousands of fixed survey sites across the US and Canada, where these sites are visited annually during peak seasons of May-June. Each site includes 50 stops located 0.5 miles apart, for a total length of 24.5 miles per survey site. At each stop within a site, observers spend 3 minutes counting all birds within 0.25 miles of that stop . We specifically download data for bald eagles *Haliaeetus leucocephalus* in 2018, and restrict data to Alaska, British Columbia and the Yukon (where densities are generally high). We also combine all stops within a site to calculate the total density at that site.

We fit these data by specifying a log-linked GAM and using a Poisson distribution. This GAM is easily fitted using R package mgcv [267]:

CODE 5.1: R code to fit a spatial model using R-package mgcv .

```
1 library(mgcv)
2 Gam = gam( SpeciesTotal ~ 1 + s(X,Y,bs="gp",m=1), data=samples, family=
    poisson )
```

where the formula 1 + s(X,Y,bs="gp",m=1) specifies a GAM with an intercept and a tensor-product smoothing spline on Latitude and Longitude. We specifically use a first-order penalty and calculate the GAM penalty using a Gaussian process. We specify a Gaussian process for the penalty on the 1st derivative to better match later results, and invite readers to explore the sensitivity to different smoother options available in package mgcv . This then results in smoothed density estimates, where bald eagles have the highest densities in coastal areas of Alaska and British Columbia (Fig. 5.6). This GAM can be easily extended to include spatio-temporal variation (i.e., spatial variation that changes over time) by including additional terms in the function s(.) used to construct the smoothing spline.

---

[2]Data are publicly available, and we downloaded data from https://www.sciencebase.gov/catalog/item/5ea04e9a82cefae35a129d65 on June 14, 2022.

5.4   Separable Correlation and Sparsity

As an alternative to specifying 2D tensor-product basis functions as covariates to estimate residual spatial variation, we next extend autoregressive models to two dimensions. Specifically, we seek to define the probability distribution for a random effect $\omega_{x,y}$, where coordinates (x, y) are defined on an evenly spaced grid.

As we saw in Section 3.5, we can fit an autoregressive model by specifying either a series of conditional distributions, or by constructing the precision of their joint distribution. One way to generalize autoregression to two dimensions on an evenly spaced grid involves combining these:

$$\omega_y \sim \begin{cases} \text{MVN}\left(\mathbf{0}, \frac{\sigma^2}{1-\rho^2}\mathbf{R}\right) & \text{if } y = 1 \\ \text{MVN}\left(\rho\omega_{y-1}, \sigma^2\mathbf{R}\right) & \text{if } y > 1 \end{cases} \tag{5.8}$$

where this defines the joint distribution for each row of ω_{y+1} conditional upon its value in the row before ω_y. We can write a custom function conditional_distribution that evaluates this probability density in TMB (Code 5.2), where we first construct the precision matrix Q_yy for each row of the matrix epsilon_xy of random, and loop through rows applying a custom function dmvnorm to evaluate the conditional probability.

CODE 5.2: TMB code defining a custom function to assemble a precision matrix for a one-dimensional autoregressive process and looping that across a second dimension.

```
 1  template<class Type>
 2  matrix<Type> conditional_distribution( array<Type> epsilon_xy,
 3                                         Type rho,
 4                                         Type sigma2,
 5                                         Type &jnll_pointer ){
 6
 7    int n_x = epsilon_xy.rows();
 8    int n_y = epsilon_xy.cols();
 9
10    // Make precision matrix for Q_yy
11    matrix<Type> Q_yy(n_y, n_y);
12    Q_yy.setZero();
13    for(int y=0; y<n_y; y++) Q_yy(y,y) = (1+pow(rho,2));
14    for(int y=1; y<n_y; y++){
15      Q_yy(y-1,y) = -rho;
16      Q_yy(y,y-1) = -rho;
17    }
18
19    // Calculate probability
20    vector<Type> Tmp_y(n_y);
21    matrix<Type> Q0_yy(n_y,n_y);
22    Q0_yy = Q_yy * ( 1-pow(rho,2) ) / sigma2;
23    matrix<Type> Q1_yy(n_y,n_y);
24    Q1_yy = Q_yy / sigma2;
25    for(int x=0; x<n_x; x++){
26      for(int y=0; y<n_y; y++){
27        if(x==0) Tmp_y(y) = epsilon_xy(0,y);
28        if(x>=1) Tmp_y(y) = epsilon_xy(x,y) - rho*epsilon_xy(x-1,y);
29      }
30      if(x==0) jnll_pointer -= dmvnorm( Tmp_y, Q0_yy, true );
31      if(x>=1) jnll_pointer -= dmvnorm( Tmp_y, Q1_yy, true );
32    }
33    return(Q1_yy);
34  }
```

Alternatively, we can write a custom function joint_distribution (Code 5.3) that evaluates the same probability density by constructing the Kronecker product of the 2-dimensional autoregressive process:

$$\text{vec}(\omega) \sim \text{MVN}(\mathbf{0}, \sigma^2 \mathbf{R}_1 \otimes \mathbf{R}_2) \tag{5.9}$$

This function constructs the two precision matrices Q_xx and Q_yy and then uses the internal function kronecker to construct the precision Q_zz . Any correlation or precision matrix that can be constructed as the Kronecker product of two smaller matrices is called a *separable* process, and constructing the precision in this way is the key to constructing computationally efficient distributions for high-dimensional processes. Again extending what we saw in Section 3.5, we can equivalently write this as:

$$\text{vec}(\omega) \sim \text{MVN}(\mathbf{0}, \sigma^{-2} \mathbf{Q}_1^{-1} \otimes \mathbf{Q}_2^{-1}) \tag{5.10}$$

where $\mathbf{Q}$ is the precision matrix for the autoregressive process in one dimension. As we saw before, each individual precision matrix can be formed analytically (Eq. 3.15) and is sparse (i.e., nonzero only for adjacent cells).

CODE 5.3: TMB code defining a custom function to assemble a precision matrix for a two-dimensional autoregressive process.

```
1   template<class Type>
2   matrix<Type> joint_distribution( array<Type> epsilon_xy,
3                                     Type rho,
4                                     Type sigma2,
5                                     Type &jnll_pointer ){
6
7     int n_x = epsilon_xy.rows();
8     int n_y = epsilon_xy.cols();
9
10    // Make precision matrix for Q_xx
11    matrix<Type> Q_xx(n_x,n_x);
12    Q_xx.setZero();
13    for(int x=0; x<n_x; x++) Q_xx(x,x) = (1+pow(rho,2));
14    for(int x=1; x<n_x; x++){
15      Q_xx(x-1,x) = -rho;
16      Q_xx(x,x-1) = -rho;
17    }
18    // Make precision matrix for Q_yy
19    matrix<Type> Q_yy(n_y,n_y);
20    Q_yy.setZero();
21    for(int y=0; y<n_y; y++) Q_yy(y,y) = (1+pow(rho,2));
22    for(int y=1; y<n_y; y++){
23      Q_yy(y-1,y) = -rho;
24      Q_yy(y,y-1) = -rho;
25    }
26
27    // Calculate probability
28    int n_z = n_x * n_y;
29    matrix<Type> Q_zz(n_z, n_z);
30    Q_zz = kronecker( Q_yy, Q_xx );
31    Q_zz = Q_zz / sigma2;
32    vector<Type> epsilon_z(n_z);
33    int Count = 0;
34    for( int y=0; y<n_y; y++){
35    for( int x=0; x<n_x; x++){
36      epsilon_z(Count) = epsilon_xy(x,y);
37      Count++;
38    }}
39    jnll_pointer -= dmvnorm( epsilon_z, Q_zz, true );
```

```
40    return Q_zz;
41 }
```

Combining two precision matrices in a separable model then has several useful properties:

- *Sparsity*: if we define the proportion of nonzero elements in matrix $\mathbf{M}_1$ as its sparsity p_1, and the proportion in correlation $\mathbf{M}_2$ as p_2, then the sparsity of $\mathbf{M}_1 \otimes \mathbf{M}_2$ is $p_1 p_2$. Said another way, a separable precision matrix is more sparse than each dimension individually. For example, in the case of this 2D equally spaced grid, the sparsity pattern of the inner Hessian (Fig. 5.7) matches the sparsity of the precision matrices $\mathbf{Q}_1^{-1} \otimes \mathbf{Q}_2^{-1}$;

- *Modularity*: if we learn how to construct the precision or covariance matrix arising from a few elementary processes (e.g., an equally spaced grid in Eq. 3.15, or a structural equation model in Eq. 4.23), we can then mix-and-match which matrix is the best for describing covariance over space vs. over time vs. among variables. We can then construct the joint covariance from the Kroenecker product of these individual matrices.

Alternatively, TMB includes a namespace density that includes functions that can automate many of these steps when constructing a separable spatial model (see Table 5.1 for a list). We next show how to fit the two-dimenstional spatial model using these functions. Specifically:

- AR1(rho) constructs the precision matrix for an evenly spaced autoregressive process with conditional standard deviation of 1;

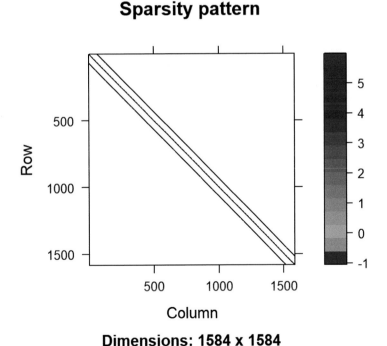

Sparsity pattern

Dimensions: 1584 x 1584

FIGURE 5.7: Sparsity pattern arising from a 2D equally spaced grid with 21 rows and 52 columns (1092 elements total), showing the banded pattern that is typical for a separable correlation.

TABLE 5.1: A brief summary of some functions available in namespace density , which can be used to simplify code when implementing a separable spatial or spatio-temporal model using TMB.

Function in density namespace	What it does
GMRF_t⟨Type⟩gmrf	Constructs an object named gmrf of class GMRF_t , where users can later call jnll += gmrf(user_vector) to evaluate the negative log-likelihood of a user-supplied vector user_vector given a distribution defined by gmrf
gmrf = GMRF(precision)	Given a specified precision matrix precision , define a GMRF_t object with covariance defined as the inverse of precision
gmrf = AR1(rho)	Defines a GMRF_t object such that it evaluates the probability distribution arising from a regularly spaced, first-order autoregressive process with autocorrelation rho and unit variance
gmrf = SCALE(gmrf1, sd)	Given a specified GMRF_t object, rescales it to have standard deviation sd
gmrf = SEPARABLE(gmrf1, gmrf2)	Given two GMRF_t objects, gmrf1 and gmrf2 , construct a new GMRF_t object with covariance that arises as the Kronecker product of the covariance for gmrf1 and gmrf2

- SEPARABLE(AR1(rho),AR1(rho)) constructs the separable precision arising from two AR1 processes;

- SCALE(SEPARABLE(AR1(rho),AR1(rho)), scale) takes a separable precision with a standard deviation of 1 and rescales it to have a standard deviation of scale ;

- SCALE(SEPARABLE(AR1(rho), AR1(rho)), scale)(epsilon_xy) takes this rescaled and separable precision, and calculates the negative log-likelihood of epsilon_xy arising from that distribution.

In summary, a few statements in TMB then implements the entire calculation for the separable covariance function (Code 5.4), while internally tracking the inverse covariance ("precision") matrix for efficient computation.

CODE 5.4: TMB code comparing three alternative ways to implement a two-dimensional autoregressive process for a spatial generalized linear mixed model.

```
1    //// Probability of random effects
```

TABLE 5.2: Comparing the estimated intercept, natural-log of spatial variance, the logit-transformed spatial autocorrelation, and the runtime in seconds for three alternative implementations of a log-linked Poisson-GLM fitted using TMB. These three specify the joint distribution for each row of the spatial process conditional upon the previous row, construct the joint precision as the Kronecker product of one-dimensional autoregressive terms, or using the built-in density package.

	Conditional	Simultaneous	Density_package
beta0	–2.187	–2.209	–2.393
ln_sigma2	0.299	0.279	0.505
logit_rho	0.973	1.002	0.718
Time_seconds	2.103	325.967	2.192

```
2   // Conditional in dimension-Y, joint in dimension-X
3   if( Options_vec(0)==1 ){
4     matrix<Type> Q_yy = conditional_distribution( epsilon_xy, rho, sigma2,
      jnll );
5     REPORT( Q_yy );
6   }
7
8   // Kroenekcer product of precision in both dimensions
9   if( Options_vec(0)==2 ){
10    matrix<Type> Q_zz = joint_distribution( epsilon_xy, rho, sigma2, jnll );
11    REPORT( Q_zz );
12  }
13
14  // Calculate using built-in TMB functions
15  using namespace density;
16  if( Options_vec(0)==3 ){
17    // Include "pow(1-pow(rho,2),0.5)" twice for 2D unit variance
18    Type scale = pow(sigma2,0.5) / pow(1-pow(rho,2),0.5) / pow(1-pow(rho,2)
      ,0.5);
19    jnll += SCALE( SEPARABLE(AR1(rho), AR1(rho)), scale )( epsilon_xy );
20  }
```

We again fit these models to data for bald eagles in Alaska, British Columbia, and Yukon in 2019. Comparing these three implementations of a 2D autoregressive for these data shows that they give similar parameter estimates (Table 5.2), where differences arise from different treatments of boundary effects. However, the implementation that constructs the joint precision matrix takes much longer to run, because it does not efficiently use information about the sparcity of the inner Hessian matrix when calculating the log-determinant as used in the Laplace approximation (Eq. 2.26). Similarly, these three models all estimate density maps that are generally similar (Fig. 5.8). We note that the GAM results included density estimates of less than one bird per quadrillion stations, which intuitively seems like an extreme value to predict. By contrast, the various two-dimensional autoregressive models predict low densities of approximately one bird per 100 stations, which seems like a more reasonable estimate for the lower range of densities.

We can further visualize the properties of the covariance matrix resulting from a 2D autoregressive process $\Sigma_{full} = \sigma^{-2} \mathbf{Q}_1^{-1} \otimes \mathbf{Q}_2^{-1}$ via it's *eigendecomposition*. Specifically, any covariance can be decomposed as:

$$\Sigma = \mathbf{V}\Lambda\mathbf{V}^T \tag{5.11}$$

where the eigendecomposition is defined as eigen $: \Sigma \mapsto (\mathbf{V}, \lambda)$, such that eigenvectors $\mathbf{V}$ are orthogonal to one-another and have length of one, and eigenvalues λ measure the variance

FIGURE 5.8: Estimated summertime log-density of bald eagles in Alaska, British Columbia, and the Yukon Territory, estimated using a GLMM fitted to Breeding Bird survey counts using three alternative versions of a 2D autoregressive process that estimates density across an entire grid but only showing those cells that overlap with an intended spatial domain.

associated with each eigenvector (see Section B.4.1 for more details). As a consequence of Eq. 5.4, we can rewrite 5.9 as:

$$\text{vec}(\omega) = \mathbf{V}\mathbf{\Lambda}^{0.5}\text{vec}(\omega^*)$$
$$\text{vec}(\omega^*) \sim \text{MVN}(\mathbf{0}, \mathbf{I})$$

where $\mathbf{\Sigma}_{full} = \mathbf{V}\mathbf{\Lambda}^{0.5}(\mathbf{V}\mathbf{\Lambda}^{0.5})^T$. In a sense, then, $\mathbf{V}\mathbf{\Lambda}^{0.5}$ are the basis functions associated with covariance $\mathbf{\Sigma}_{full}$. If we map the first basis functions (Fig. 5.9), we see that the largest basis functions capture correlations over broad spatial scales, e.g., where v.2 represents variation from west to east and therefore distinguishes British Columbia and the Yukon Territory from central and western Alaska. Although these basis functions differ from the tensor-spline basis functions constructed previously (Fig. 5.5), the dominant eigenvectors serve a similar role in representing broad-scale spatial patterns. Furthermore, estimating this model as a GLMM shrinks coefficients towards zero based on the marginal likelihood, and this shrinkage is conceptually similar to the penalty term that occurs in a GAM. We therefore see that the GLMM using a 2D autoregressive correlation is similar to a GAM that is specified using the same basis functions.

5.5   Models for Irregular Shaped Spatial Domains

Despite the speed and convenience of specifying a 2D autoregressive process using separable functions via package density , this approach wastes computational resources estimating random effects for grid cells that are outside of the spatial domain (in this case over water), which we then ignore when interpreting the output. This issue is analogous to defining state-space models over a single temporal dimension, but defined for uneven time intervals, as we discussed previously in Section 3.6. It is therefore helpful to define a precision matrix for unevenly spaced locations in two-dimensional space. There are many different approaches to do so [250], but we here focus on two methods: the conditional autoregressive model, and the stochastic partial differential equation (SPDE) method.

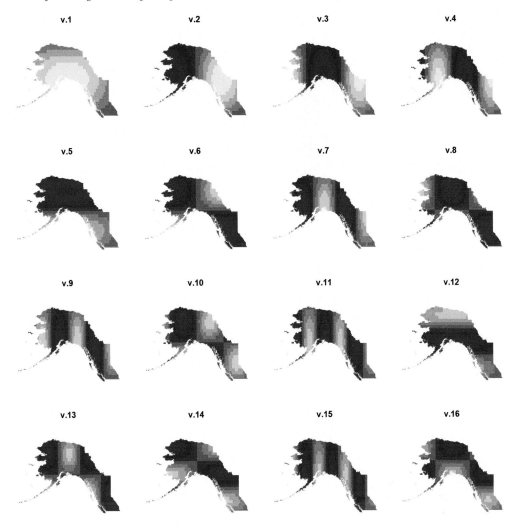

FIGURE 5.9: First 16 basis functions for the bald eagle species distribution model, showing $\mathbf{V}\mathbf{\Lambda}^{0.5}$ from the eigendecomposition of the covariance matrix resulting from a separable 2D autoregressive process.

5.5.1   Conditional Autoregressive Model

Spatial modelling has typically emphasized *conditional autoregressive* and *simultaneous autoregressive* approaches that can directly construct the precision matrix for a spatially correlated variable [36]. We previously introduced the simultaneous autoregressive process in Section 3.4, and see Appendix B.5 for a derivation of a SAR for a one-dimensional autoregressive process. We here introduce the alternative *conditional autoregressive process* or CAR model.

To define the CAR model, we divide a spatial domain into a set of n_g polygons (a.k.a. cells) that are non-overlapping and contain the entire domain. We then introduce the *adjacency matrix* $\mathbf{A}$ with dimension n_g by n_g. The adjacency matrix $\mathbf{A}$ has a value $a_{i,j} = 1$ if cell i is adjacent to cell j and $a_{i,j} = 0$ otherwise. We here will divide a spatial area into square grid cells, and define adjacency as grid cells that share an edge (not just a vertex).

CAR method

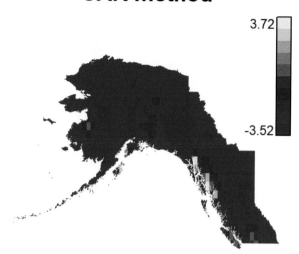

FIGURE 5.10: Estimated bald eagle densities, using a GLMM fitted to Breeding Bird survey counts using a conditional autoregressive model that only requires fitting to those grid cells that are within the intended spatial domain.

This implies that a square grid cell is adjacent to at most four other cells; this is sometimes called *rook adjacency*, similar to how a rook is allowed to move along rows or columns in the game of chess. The adjacency matrix is typically sparse and depends entirely on the spatial configuration and shape of modeled cells. Alternative definitions of adjacency are possible which also result in a sparse adjacency matrix. For example, *queen adjacency* defines two polygons as adjacent if they share either an edge or a vertex, such that a square grid cell could be adjacent to at most 8 neighboring cells. The following methods can be applied to these different definitions of adjacency, although we will use rook-adjacency for simplicity of presentation.

Given evenly sized grid cells and adjacency matrix $\mathbf{A}$, a CAR model constructs the inverse covariance $\mathbf{Q}$ as:

$$\mathbf{Q} = \frac{1}{\sigma^2}(\mathbf{I} - \rho\mathbf{A}) \qquad (5.12)$$

where this construction is nonzero only on the diagonal or for adjacent cells [251]. However, this simple construction comes with the drawback that ρ has a less intuitive interpretation than other methods. In particular, this equation only results in a valid precision matrix if $\frac{1}{\lambda_{min}} < \rho < \frac{1}{\lambda_{max}}$, where λ_{min} and λ_{max} are the minimum and maximum eigenvalues of adjacency matrix $\mathbf{A}$. We specifically use R-package igraph [38] to calculate these minimum and maximum eigenvalues efficiently given that $\mathbf{A}$ is sparse.

Mapping estimated log-density using the CAR model (Fig 5.10), confirms that it estimates high densities in southeast Alaska in agreement with the previous autoregressive methods (Fig. 5.8). Similarly, the estimated basis functions for the CAR (Fig 5.11) show patterns that are similar to those of the autoregressive methods (Fig. 5.9), i.e., the first several basis functions represent broad spatial patterns followed by finer-scale variation for subsequent basis functions.

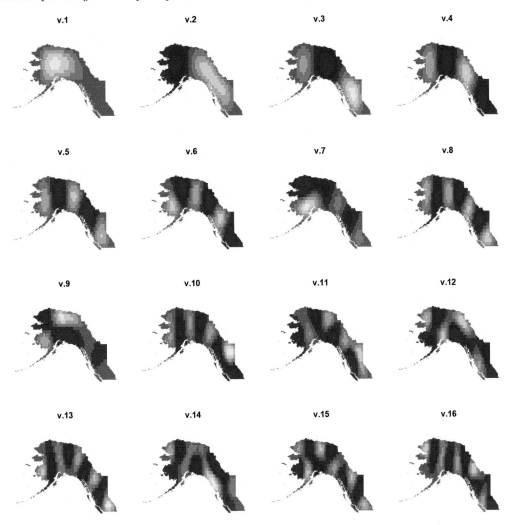

FIGURE 5.11: First 16 basis functions for the bald eagle species distribution model, showing $\mathbf{V}\mathbf{\Lambda}^{0.5}$ from the eigendecomposition of the covariance matrix resulting from a CAR model.

5.5.2 SPDE Method

We next introduce an alternative stochastic partial different equation or *SPDE approach* to modelling a spatial variable measured at any set of locations in two dimensions [6, 136]. This approach has several noteworthy characteristics:

- *Customized resolution*: the SPDE approach can be adapted to include a denser arrangement of random effects in some areas than others. For example, in cluster sampling designs, it might be convenient to represent spatial variation at a high resolution near clustered sampling, while still using lower resolution elsewhere;

- *Nonstationarity*: it is relatively straightforward to include some types of nonstionarity, i.e., where the correlation rate varies as a function of covariates;

Sample locations   **Mesh composed of triangles**

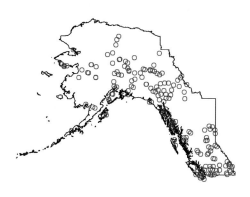
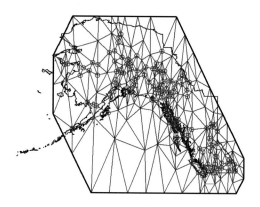

FIGURE 5.12: The location of samples (blue circles in left panel) and the resulting finite element analysis mesh composed of triangles and with samples overlayed (right panel), showing the higher resolution (smaller triangles) arising in areas with many samples (blue points).

- *Analytic and sparse precision matrix*: similar to the CAR model, the SPDE method allows the precision matrix to be constructed directly for a 2D field, and the SPDE method also results in a precision matrix that is sparse. This sparsity remains even when adding additional properties, e.g., an oscillatory covariance function [136];

- *Bounded or spherical domain*: we can easily adopt an irregular spatial domain. For example, we can modify the model to avoid correlations that proceed across a boundary (e.g., over land when modelling correlations for a marine fish [5]) or include correlations arising on a sphere (e.g., when modelling correlations for an atmospheric measurement over the entire globe [136]);

- *Geometric anisotropy*: finally, the SPDE method can be easily adapted to represent *geometric anisotropy*, defined as a correlation function that differs based on cardinal direction [136, 235]. This property is useful in ecological applications, e.g., when spatially correlated residuals (representing unmodeled exogenous or endogenous drivers) differ more greatly in one direction than another.

These characteristics could also be added to the CAR method shown previously, but the SPDE method provides a convenient user interface to do so.

To introduce the SPDE approach, first recall that an autoregressive process for two locations s_1 and s_2 separated by distance $d = |s_1 - s_2|$ follows an exponential correlation function, $C_{exponential}(d) = e^{-\theta d}$, and that the precision matrix is zero for nonadjacent locations (see Eq. 3.15). To generalize this for two dimensions, however, we have to define which locations are "adjacent". To do so, the SPDE method defines a *finite element analysis* (FEA) mesh. This involves defining a set of triangles that cover the entire spatial domain, where every location is within exactly one triangle. For example, the mesh for the bald eagles example has a higher density of triangles in southern British Columbia where samples are also dense (Fig. 5.12).

This FEA mesh has several characteristics:

- *Vertices*: each triangle of the FEA mesh has three vertices, and the value of the spatial variable ω_s at each of these vertices is treated as a random effect;

- *Edges*: similarly, each triangle of the mesh has three edges. Two vertices are called *adjacent* if and only if they are connected by a single edge.

We also introduce the Matérn covariance function [77]:

$$V_{matern}(d) = \frac{1}{\tau^2 2^{\nu-1} \Gamma(\nu)} (\kappa d)^\nu K_\nu(\kappa d) \tag{5.13}$$

where κ is an estimated parameter representing the decorrelation rate (with units $distance^{-1}$), such that a high value of κ corresponds to a process where nearby locations have a low correlation, and τ is the estimated parameter that controls the variance of the Matérn covariance function. Finally, ν is a parameter that is fixed a priori and controls how the smoothness of the Matérn covariance function, $\Gamma(\nu)$ is the gamma function and K_ν is a Bessel function. This Matérn covariance function is useful in part because it reduces to the exponential covariance function when smoothness $\nu = 0.5$, and approaches a Gaussian covariance function as smoothness $\nu \to \infty$.

Usefully, if we specify a smoothness that is intermediate between the exponential and Gaussian correlation functions (i.e., $\nu = 1$ for a two-dimensional model), we can then approximate the precision matrix as:

$$\mathbf{Q}_{spde} = \tau^2 (\kappa^4 \mathbf{M}_0 + 2\kappa^2 \mathbf{M}_1 + \mathbf{M}_2) \tag{5.14}$$

where $\mathbf{M}_0$ is a diagonal matrix, $\mathbf{M}_1$ is nonzero only vertices that are connected by an edge (i.e., the same sparsity as the adjacency matrix), and $\mathbf{M}_2$ is nonzero only for vertices that are connected by two edges (i.e., the sparsity of the adjacency matrix squared). In this example, $\mathbf{M}_2$ still only has <9000 nonzero elements, such that 95% of vertices are conditionally independent (Fig. 5.13). Using the Matérn correlation function (Eq. 5.13) and $\nu = 1$, we also get simple expressions for the distance at approximately 10% correlation (termed the *geostatistical range*, $\sqrt{8}/\kappa$), and the variance as distance increases asymptotically (called the *geostatistical sill* or pointwise variance, $\frac{1}{4\pi\tau^2\kappa^2}$). These properties require us to fix ν a priori, while only estimating τ and κ. In general, we suspect that having a computationally efficient approach that is extensible (e.g., allows adding geometric anisotropy, covariate-based nonstationarity, or a customized domain) is more important than estimating ν to bridge performance between the exponential and Gaussian correlation functions [207].

This SPDE method then allows the value of a spatial variable ω_s at vertices $s \in 1, 2, ..., S$ to be computed as a multivariate normal distribution, $\omega \sim \text{MVN}(\mathbf{0}, \mathbf{Q}_{spde}^{-1})$. We then can interpolate the value ω^* at any other locations as:

$$\omega^* = \mathbf{A}\omega \tag{5.15}$$

where $\mathbf{A}$ is a sparse matrix that represents bilinear interpolation. Specifically, for any location g we identify the triangle of the finite-element mesh that contains it, identify the value ω_s at its three vertices, and calculate ω_g^* as a weighted average of these three values based on the distance between g and those three vertices. The bilinear interpolation matrix $\mathbf{A}$ represents this process by being nonzero for only three values in each row, with values representing weights used in that weighted average.

CODE 5.5: R code to construct objects required for the SPDE method using the fmesher package.

```
1  # create mesh
2  xy_i = st_coordinates(out)
```

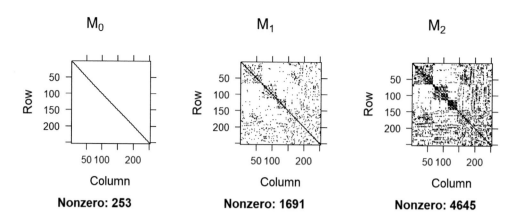

FIGURE 5.13: Depiction of the three sparse matrices used to assemble the precision matrix using the SPDE method, showing the number of non-zero elements in each.

```
3  mesh = fm_mesh_2d( xy_i, refine=TRUE, cutoff=0.5)
4
5  # Create matrices in fmesher / INLA
6  spde <- fm_fem(mesh, order=2)
7
8  # create projection matrix from vertices to samples
9  A_is = fm_evaluator( mesh, loc=xy_i )$proj$A
10
11 # Create extrapolation grid
12 cellsize = 1
13 grid = st_make_grid( sf_states, cellsize=cellsize )
14
15 # create projection matrix from vertices to grid
16 A_gs = fm_evaluator( mesh, loc=st_coordinates(st_centroid(grid)) )$proj$A
```

Usefully, these matrices M_0, M_1, M_2, and A can be computed automatically using the R-package fmesher [135] (Code 5.5). This package provides simplified access to functions originally provided by R-package INLA [134], although we will use package fmesher to simplify software dependencies. We will typically export these matrices to TMB and evaluate the probability of random effects using the density::GMRF function (Code 5.6), and therefore do not otherwise need to install the R-package INLA . In this case, we have defined a vector of random effects omega_s , and we also show how to project from the vector at SPDE vertices to the vector omega_i at all sample locations using projection matrix A_is (constructed using function fmesher::fm_evaluator in Code 5.5 and then passed to TMB). This workflow allows us to combine the SPDE approach with other customized model components. However, we note that some models explored throughout the textbook can instead be fitted using the Integrated Nested Laplace Approximation [200], after which R-package INLA is named. The details of this nested Laplace approximation are beyond our intended scope, but detailed comparison suggests that estimates using INLA or the default Laplace approximation in TMB are quite similar, while TMB allows more customized control over model configuration [168].

CODE 5.6: TMB code used to evaluate the probability of a random effect using the SPDE method and also project from the vector of random effects at SPDE vertices to the vector of sample locations.

SPDE method

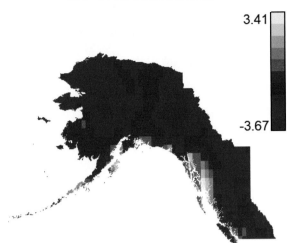

3.41

-3.67

FIGURE 5.14: Estimated summertime density of bald eagles using a GLMM and the SPDE method (see Fig. 5.8 for details).

```
1   // Probability of random effects
2   Eigen::SparseMatrix<Type> Q = (exp(4*ln_kappa)*M0 + Type(2.0)*exp(2*
    ln_kappa)*M1 + M2) * exp(2*ln_tau);
3   jnll += GMRF(Q)( omega_s );
4
5   // Project using bilinear interpolation
6   vector<Type> omega_i( A_is.rows() );
7   omega_i = A_is * omega_s;
```

We then fit the bald eagle dataset using this SPDE method, again interpolating densities to an evenly spaced grid to display results using a second projection matrix $\tilde{\mathbf{A}}$ (labeled A_gs in Code 5.5). As expected, this shows the same spatial patterns in estimated log-densities as using the 2D autoregressive grid (Fig. 5.14) or the CAR model (Fig. 5.10). Similarly, we can calculate the covariance at the set of locations that we use for plotting. The covariance at those locations is then $\Sigma = \tilde{\mathbf{A}} \mathbf{Q}_{spde}^{-1} \tilde{\mathbf{A}}^t$. If we then calculate the basis functions implied by that covariance matrix, we again see that the dominant basis functions represent broad-scale spatial patterns (Fig. 5.15), similar to the previous CAR and 2D autoregressive methods. We therefore conclude that the SPDE method provides a convenient generalization of the 2D autoregressive process for unequally spaced locations.

5.6   Chapter Summary

In summary, we have showed that:

1. Ecologists need a variety of approaches to identify spatially correlated residuals that remain when predicting population densities based on measured covariates, where these residuals arise from a variety of exogenous and endogenous mechanisms;

2. Spatially correlated residuals can be addressed by using a basis-expansion of covariates, where basis expansion takes one or more covariates (e.g., elevation,

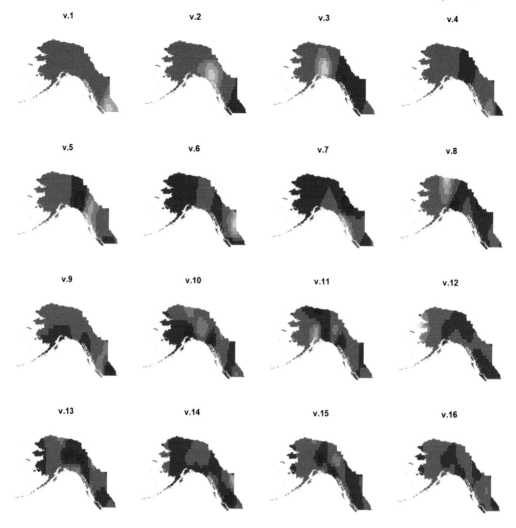

FIGURE 5.15: First 16 basis functions for the bald eagle species distribution model and the SPDE method (see Fig. 5.9 for details).

temperature, location, etc.) and creates additional vectors that can more flexibly describe a nonlinear relationship between covariates and population response. Two-dimensional basis expansion (i.e., for two spatial coordinates) then requires a tensor product of the basis-expansion for each individual covariate;

3. Generalized Additive Models (GAMs) use splines for basis expansion, and penalize the coefficients for each spline towards zero. This then shrinks the resulting function towards some smooth shape. Similarly, generalized linear mixed models (GLMMs) using splines as covariates will use the log-determinant of the inner Hessian matrix as a penalty term, and with similar behavior to the penalty used by GAMs;

4. We can define a two-dimensional (2D) distribution for a spatial variable by using a separable correlation function, i.e., the outer product of the correlation in each dimension individually. If each dimension uses a first-order autocorrelation with even spacing, then the resulting 2D grid has a precision matrix that is highly sparse;

5. Alternatively, we can define a 2D distribution for an irregular grid using a conditional autoregressive process, or for any set of locations that approximates a Matérn correlation function by defining a finite element mesh and using the SPDE method. All matrices involved in the SPDE approach can be produced easily using the fmesher package in R.

5.7   Exercise

Download elevation for samples and gridded locations in the bald eagle example (e.g., using package `elevatr` used in Section 5.3). Then, please adapt the code presented for either the CAR model or SPDE method, and add a quadratic or spline basis expansion of elevation to the linear predictor for that model. After fitting, visualize the resulting response curve (e.g., adapting code from Section 1.7 to use the `marginaleffects` package). Are the coefficients for this estimated effect of elevation statistically significant?

6

Spatial Sampling Designs and Analysis

6.1 Ecological Inference and Field Sampling

Like all scientists, ecologists seek to understand the world through an iterative process of reduction and subsequent composition. This typically involves decomposing a complicated phenomenon into smaller processes that can be studied experimentally and in isolation (*reduction*), and then some effort to identify how much of the original phenomenon can be explained via those mechanisms that have been identified via previous experiments (*composition*). However, ecologists are faced with many unique challenges in this scientific process including:

1. *Multicausality*: ecologists seek to predict a wide range of dynamics, including individual behavior, population/community dynamics, and ecosystem function (to name a few). However, most ecological variables are impacted by many other variables simultaneously, and those in turn are impacted by many others. Analysts can represent these causal mechanisms using a *causal map* [173], and the causal maps associated with real-world ecosystems are very complex relative to a controlled experiment in classical mechanics (e.g., a pendulum operating under drag and gravity) or chemistry (e.g., reactions occurring under well-mixed concentrations and known physical principles);

2. *Tapering effects*: most ecological problems are defined at a level of organization that involves hundreds or thousands of mechanisms that have been observed in isolated settings. For example, the tendency for communities to persist through time is called *community regulation*, and modern coexistence theory seeks to enumerate and categorize the many ecological processes that promote community persistence into a smaller set of mathematical expressions [54]. However, essentially all potential mechanisms are likely to have some non-zero (if vanishingly small) role in dynamics for a given system. If we plotted the importance (e.g., variance explained) by different possible measurements or mechanisms in rank order from most to least important, this plot would likely have a longer tail than other types of science (this phenomenon is called *tapering effects* [19]). This contrasts, for example, with classical mechanics (studied in physics) where a small number of mechanisms can describe essentially all properties of solid objects and their movement within space;

3. *Lack of experimental control*: finally, ecologists often address questions at a scale that does not permit experimentation, either because it would be unethical (i.e., experiments on humans as elements of an ecosystem without their consent), impractical (i.e., multi-generational experiments on long-lived populations), technically impossible (i.e., experimental control of individual behavioral mechanisms

DOI: 10.1201/9781003410294-6

in complex organisms), or context-dependent (i.e., require repeating experiments for every described species and habitat). In these cases, ecologists must instead use insights derived from experiments in analogous systems and identify dominant mechanisms based on incomplete information from lower levels of organization.

These challenges all complicate the traditional scientific process of reduction and composition when applied to ecological systems.

For these three reasons (among others), applied ecologists have adopted a scientific process that prioritizes repeated, direct measurements of important variables in real-world ecosystems (*ecological monitoring*) that can directly inform the management of real-world systems. These ecological monitoring programs (e.g., the International Long-Term Ecological Research Network [249]) then provide data sets that are described using a shared set of dynamical models, and ecological theory seeks in part to define this set of shared dynamical models. However, the design and analysis of ecological monitoring programs comes with additional challenges. In the following, we specifically address the following topics related to the design and analysis of ecological monitoring programs:

1. *Spatial integration*: how to estimate a spatially integrated total (i.e., total abundance) by fitting a spatial model to local measurements;

2. *Non-ignorable sampling designs*: how to analyze data that arise non-experimentally, and where the probability of data being available may be correlated with the variable of interest;

3. *Muli-stage sampling*: how to analyze data that involves multiple levels of subsampling.

These problems all involve some background in sampling theory, and we note that other textbooks provide an alternative perspective on these and other topics from a design-based perspective [30].

6.2 Spatial Integration and Weighting

Ecologists often seek to estimate the average or total value for a given variable when aggregating across space. Many governments mandate a change in regulation based on the aggregated total for a system variable. For example, environmental agencies might want to maintain a population sizes for the endangered North Atlantic right whale (*Eubalaena glacialis*) above some threshold, or ensure that wetland habitats maintain a total area above some target. As we will see, calculating this total leads to several different algorithms for spatial integration, and each is suitable for analyzing different ecological variables, attributes, or processes[1].

To illustrate these principles, we introduce a data set of samples of air pollution compiled by AirNow [244]. AirNow includes daily measurements of particles suspended in the atmosphere that are smaller than a specified size (PM2.5 and PM10) as well as ozone, using standardized measurement systems that are approved by the United States Environmental Protection Agency. We specifically downloaded 8-hour average *ozone concentrations* in southeastern and mid-Atlantic states (Florida through New York)[2] and note that the EPA

---

[1]See https://github.com/james-thorson/Spatio-temporal-models-for-ecologists/Chap_6 for code associated with this chapter.

[2]obtained from the EPA Outdoor Air Quality Data center on July 19, 2022.

in 2015 defined a standard for human ozone exposure of 0.07 parts per million (ppm). We then combine these air pollution measurements with global estimates of human population density at 1-degree resolution available in 2000, 2005, 2010, 2015, and 2020, using the UN WPP-Adjusted Population Density, v4.11 [24][3].

We first fit a spatial linear mixed model to ozone measurements occurring on July 1, 2019, using the SPDE method to define the distribution of a spatial variable ω and using population density as a covariate:

$$\log(\mu) = \beta_0 + \mathbf{A}\omega + \beta_1\mathbf{d}$$
$$\omega \sim \text{MVN}(\mathbf{0}, \mathbf{Q}^{-1}) \tag{6.1}$$
$$y_i \sim \text{Gamma}(\sigma^{-2}, \mu_i\sigma^2)$$

where $\mathbf{Q}$ is the precision matrix defined using the SPDE method, ω is the vector of spatial random effects at each SPDE vertex, $\mathbf{A}$ is the matrix representing bilinear interpolation from SPDE vertices to the locations of data, $\mathbf{d}$ is the vector of population density at each sample, and Gamma is a gamma disribution with shape σ^{-2} and scale $\mu_i\sigma^2$ where σ is the estimated coefficient of variation.

6.2.1 Estimating Uncertainty in Derived Quantities

This model can then be used to predict ozone concentrations at any other location $s \in \mathcal{D}$ within the spatial domain $\mathcal{D}$ where covariate $d(s)$ is also known. These predictions therefore define a function $\mu(s)$ over that same spatial domain and we start by mapping the value $\mu(s)$ at evenly spaced locations (Fig. 6.1). In general, we will also want to calculate, visualize, and communicate our uncertainty when predicting $\hat{\mu}(s)$. Recall from Eq. 2.28) that we can construct the *joint precision matrix* from the inner and outer Hessian matrices, as well as the gradient of predicted random effects given fixed effects (the outer Jacobian matrix). This joint precision $\mathbf{Q}_{joint}$ is often sparse, due to the sparsity of the inner Hessian matrix when using a sparse precision matrix (i.e., a CAR, SAR, or SPDE method to define spatial autocorrelation). We could further calculate the joint covariance, $\mathbf{V}_{joint} = \mathbf{Q}_{joint}^{-1}$, although taking the matrix inverse will then typically result in a dense covariance matrix, so computing the joint covariance $\mathbf{V}_{joint}$ is computationally infeasible for large models.

We then approximate the uncertainty in fixed and random effects in two ways:

1. *Sample-based*: we can estimate uncertainty by taking samples of both fixed and random effects while assuming that they are normally distributed with mean equal to their empirical Bayes prediction (for random effects) and maximum likelihood estimates (for fixed effects) and variance of their inverse precision $\mathbf{Q}_{joint}$. Given that $\mathbf{Q}_{joint}$ is often sparse, we specifically use a custom R-function rmvnorm_prec to generate multivariate-normal samples from this precision matrix without having to compute or store the matrix inverse in memory (Code 6.1). This code specifically uses the Matrix package [8] to decompose a sparse precision matrix $\mathbf{Q} = \mathbf{P}^t\mathbf{L}\mathbf{L}^t\mathbf{P}$, where $\mathbf{L}$ is a sparse lower-triangle matrix and $\mathbf{P}$ is a permutation matrix. This generalizes the Cholesky decomposition (see Section B.4.2) for efficient use with sparse matrices by including a permutation matrix $\mathbf{P}$. This decomposition then allows us to use Matrix to efficiently calculate $\mathbf{z}^* = \mathbf{P}^t(\mathbf{L}^t)^{-1}\mathbf{z}$ where $\mathbf{z}$ are independent samples from a standard normal distribution, and $\mathbf{z}^* \sim \text{MVN}(\mathbf{0}, \mathbf{Q}_{joint}^{-1})$ [199];

---

[3]accessed from the NASA Socioeconomic Data and Applications Center on July 22, 2022.

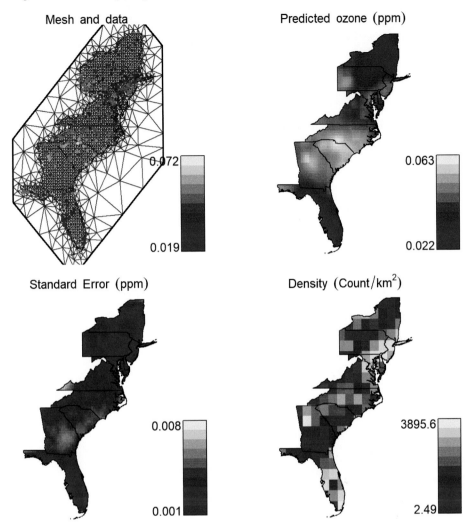

FIGURE 6.1: Measured ozone concentrations (top-left panel) along with the SPDE mesh arising from a uniform grid overlaid on the spatial domain, the predicted ozone densities (top-right panel, in ppm), predictive standard errors (bottom-left, in ppm), and human population density in 2020 extracted from a global data product with 1-degree resolution (bottom-right panel and plotted on log-scale).

For each sample from the joint precision matrix, we then plug these into our calculation of μ (using Eq. 6.1). We specifically do this using custom function `sample_var` (Code 6.2) by passing the vector new fixed and random effects to the compiled TMB object `obj$report`, so that we can use the same code as was used during parameter estimation. We can use this function `sample_var` generically for any fitted TMB model (e.g., in Code 6.3). We then interpret the standard deviation across samples as a sample-based approximation to the predictive standard error;

2. *Generalized delta method*: alternatively, we can can define a function ϕ for calculating a derived quantity, in this case $\mu_i = \phi(\theta, \epsilon)$, where θ and ϵ are fixed and

random effects, respectively. We then calculate the matrix of gradients of this function ϕ with respect to fixed and random effects, $\mathbf{G} = \nabla\phi$, and then calculate the covariance of derived quantities as $\mathbf{G}\mathbf{Q}_{joint}^{-1}\mathbf{G}^t$.

The generalized delta method requires calculating gradients for each derived quantity, and thus becomes computationally expensive when calculating the standard error for a large number of quantities (e.g., predicting ozone concentrations at a high spatial resolution). By contrast, the sample-based method approximates the uncertainty using a specified number of samples, and therefore introduces some sampling error in the calculated standard errors. When to use each of these two approximations depends somewhat on context, including the number of random effects and the need for a precise calculation.

CODE 6.1: R function `rmvnorm_prec` demonstrating how to efficiently sample from a multivariate normal distribution when supplying a sparse precision matrix.

```r
 1  rmvnorm_prec <-
 2  function( mu, # estimated fixed and random effects
 3            prec, # estimated joint precision
 4            n.sims) {
 5
 6    require(Matrix)
 7    # Simulate values
 8    z0 = matrix(rnorm(length(mu) * n.sims), ncol=n.sims)
 9    # Q = t(P) * L * t(L) * P
10    L = Cholesky(prec, super=TRUE)
11    # Calcualte t(P) * solve(t(L)) * z0 in two steps
12    z = solve(L, z0, system = "Lt") # z = Lt^-1 * z
13    z = solve(L, z, system = "Pt") # z = Pt     * z
14    return(mu + as.matrix(z))
15  }
```

CODE 6.2: R function `sample_var` demonstrating how to efficiently sample from the joint precision matrix of fixed and random effects estimated by TMB using a mixed-effects model.

```r
 1  # Function to generate samples
 2  sample_var <-
 3  function( obj, # TMB object
 4            var_name, # name of REPORTed variable
 5            mu, # Estimated mean
 6            prec, # Estimated precision
 7            n_samples = 500, # Number of samples
 8            fun1 = function(x,...) x, # optional samples transform
 9            fun2 = sd, # calculate fun2 across samples
10            ... ){
11
12    require(abind)
13    # Take samples using precision matrix
14    u_zr = rmvnorm_prec( mu=mu, prec=prec, n.sims=n_samples )
15    # Calculate REPORTed variable for each sample
16    for( rI in 1:n_samples ){
17      Var = obj$report( par=u_zr[,rI] )[[var_name]]
18      # Transform REPORTed variable using fun1
19      Var = fun1( Var, ... )
20      # Bind samples together into array
21      if(is.vector(Var)) Var = as.array(Var)
22      if(rI==1) Var_zr = Var
23      if(rI>=2){
24        Var_zr = abind( Var_zr, Var, along=length(dim(Var))+1 )
25      }
26    }
27    # summarize across samples using fun2
28    out = apply(Var_zr, MARGIN=1:(length(dim(Var_zr))-1), FUN=fun2)
```

```
29    # Return value
30    return( out )
31 }
```

CODE 6.3: R code demonstrating interface to efficiently sample from joint precision of fixed and random effects.

```
1 # Optimize
2 Opt = nlminb( start=Obj$par, obj=Obj$fn, gr=Obj$gr )
3 Opt$SD = sdreport( Obj, bias.correct=TRUE, getJointPrecision=TRUE, bias.
      correct.control=list(sd=TRUE) )
4
5 # Sample standard errors
6 SE_g = sample_var( obj = Obj,
7                    var_name = "yhat_g",
8                    mu = Obj$env$last.par.best,
9                    prec = Opt$SD$jointPrecision )
```

Using the sample-based method to approximate and map the standard errors for ozone concentrations, this model predicts concentrations approaching 0.06 ppm (but still below the 0.07 ppm EPA standard) near Atlanta and along the boundary of North and South Carolina, with the lowest concentrations in central Florida and northern New York (Fig. 6.1). We also see comparatively low standard errors throughout the spatial domain.

6.2.2 Expansion Weights

Ecological inference often involves integrating an estimated function across some spatial domain $s \in \mathcal{D}$.

$$\hat{y} = \int_{\mathcal{D}} \hat{\mu}(s)w(s)\mathrm{d}s \qquad (6.2)$$

where $w(s)$ is a expansion term when calculating this integral. In practice, this integral is typically approximated using *Monte Carlo integration* [252] (Table 2.1) by defining a set of n_g *integration points*, $g \in \{1, 2, ..., n_g\}$ where $s_g \in \mathcal{D}$. This then replaces the integral in Eq. 6.2 with a summation:

$$\hat{y} = \sum_{g=1}^{n_g} \hat{\mu}(s_g)w_g a_g \qquad (6.3)$$

where a_g represents the area associated with each integration point s_g. These integration points can be chosen in many ways, perhaps with higher densities in areas of particular interest, but in the following, we will generally define a set of evenly spaced grid cells, so that the same locations can be used for plotting and spatial integration.

We highlight four different weighting methods that define the weight w_g and area a_g associated with each integration point, noting that other methods will arise in practice:

- *Sample weighted total*: most simply, an analyst might calculate density function $\hat{\mu}$ at the set of locations where samples were taken. This specifies the same number of integration points and samples $n_g = n_i$ and $s_g = s_i$, and presumably also an equal area assigned for each integration point $w_g = a_g = 1$. Sample weighting is useful, e.g., when sampling locations were chosen as part of a panel design, they were chosen at locations that are somehow representative of the sampled variable, or otherwise have specific ecological or stakeholder interest;

- *Area-weighted average*: instead, an analyst might calculate the total over some irregularly shaped domain $\mathcal{D}$. In this case, quadrature points near a boundary might represent a smaller area than quadrature points that are entirely enclosed by the domain, such that a_g varies among integration points while $w_g = 1$ for all of them;

- *Covariate-weighted average*: in other cases, an analyst might weight the estimated variable μ based on a quantity that is assumed to be known without error. For example, to calculate total ozone exposure, we might calculate the weighted average of predicted ozone concentrations $\hat{\mu}(s_g)$, weighted by human population densities $d(s_g)$ such that $w_g = \frac{d(s_g)}{\sum_{g=1}^{n_g} d(s_g)}$;

- *Multivariate-weighted average*: finally, it might be helpful to use covariate-weighting but using other modeled variables in place of a fixed covariate for weighting w_g. For example, an analyst might seek to measure the average body size of individuals in a population. To do so, they could jointly model a measurement of body size $B(s)$ as well as population density $N(s)$ at different locations $s \in \mathcal{D}$. The analyst could then calculate the average body size for each integration point $\hat{\mu}_g = B(s_g)$, weighted by the proportion of numerical abundance $w_g = \frac{a_g N(s_s)}{\sum_{g=1}^{n_g} a_g N(s_g)}$ associated with that location [74].

In the following, we contrast the first three weighting methods when applied to the ozone concentration example. However, we first introduce a potential source of bias called *retransformation bias*, and also introduce a practical solution called *epsilon bias-correction*.

6.2.3   Epsilon Bias-correction

First recall that hierarchical models treat random effects as random variables that follow some specified distribution, and that the marginal likelihood is calculated by integrating across the value of random variables (Section 2.1.2). For illustration, we introduce a single random variable ϵ but seek to estimate a variable $Z = \phi(\epsilon)$ that is calculated via some transformation ϕ that might be nonlinear. To illustrate, let's imagine that ϵ follows a normal distribution and is then exponentiated, and we seek an unbiased pointwise estimator $\hat{Z}$, defined as the expected value for the distribution Z:

$$\epsilon \sim \text{Normal}(\mu, \sigma^2)$$
$$Z = e^\epsilon \tag{6.4}$$
$$\hat{Z} = \mathbb{E}(Z)$$

In this simple case, we know that Z follows a lognormal distribution, and we can calculate the mean of this lognormal distribution analytically $\hat{Z} = e^{\mu + 0.5\sigma^2}$, where an unbiased estimator will have this same value. However, the simplest "plug-in" estimator involves estimating the mode of the random effect, $\hat{\epsilon} = \mu$, and then plugging that value into the transformation, such that $\hat{Z} = e^\mu$. This plug-in estimator obviously has a negative bias relative to the known expected value $\mathbb{E}(Z)$. This bias in the plug-in estimator is called *retransformation bias*.

   The plug-in estimator may also be biased due to *posterior skewness bias*. To see this, imagine that we replace the normal distribution $\epsilon \sim \text{Normal}(\mu, \sigma^2)$ with a shape-scale parameterization of the Gamma distribution $\epsilon \sim \text{Gamma}(k, \theta)$ and assuming that $k \geq 1$ for illustration purposes. Let us further imagine that we want an unbiased estimator for $\hat{Z} = \mathbb{E}(\epsilon)$, i.e., using an identity link instead of the previous log-link function. In this case, the plug-in estimator involves calculating the mode of the random effect, $\hat{\epsilon} = (k-1)\theta$ and then calculates $\hat{Z} = (k-1)\theta$. However, the mean of the gamma distribution is actually $\mathbb{E}(\epsilon) = k\theta$, where the mean is greater than the mode due to the positive skewness of the Gamma distribution when $k \geq 1$. We can see that the skewness of the random effect ϵ

contributes to a bias when using the plug-in estimator to calculate the expected value for some transformation $\mathbb{E}(\phi(\epsilon))$.

From these thought-experiments, we can deduce that the magnitude of bias for the plug-in estimator increases whenever:

1. random effect ϵ has a large standard deviation. In a mixed effects model, this will occur when the empirical Bayes prediction for a random effect used in the calculation of a given derived quantity has a large standard error; or

2. available data result in an estimate of random effects that has non-zero skewness; and

3. the function used to compute the derived quantity is highly nonlinear with respect to random effects. The bias therefore arises in generalized linear mixed models for all link functions except the identity link.

Many ecological applications for spatial and spatio-temporal models have a response variable that is constrained to be positive (such that analyses often specify a nonlinear link function like the log-link), and also have limited data (hence having large imprecision for estimated random effects). Furthermore, many population-dynamics models resulted in non-zero skewness for random effects. We therefore conclude that the plug-in estimator will be biased in many practical applications.

To correct for plug-in estimator bias when using the Laplace approximation, we therefore introduce the *epsilon bias-correction estimator*. Skipping the derivation [227, 243], recall that the Laplace approximation involves defining the joint log-likelihood of fixed θ and random effects ϵ given data y, $f(\theta, \epsilon; y)$, such that the marginal likelihood $\mathcal{L}(\theta; y) = \int e^{f(\theta, \epsilon; y)} \mathrm{d}\epsilon$. We then augment the joint log-likelihood by including an additional term, and calculate the Laplace approximation of this augmented likelihood $\mathcal{L}^*(\theta, \delta; y)$:

$$f^*(\theta, \epsilon, \delta; y) = f(\theta, \epsilon; y) - \delta\phi(\theta, \epsilon) \tag{6.5}$$

$$\mathcal{L}^*(\theta, \delta; y) = \int e^{f^*(\theta, \epsilon, \delta; y)} \mathrm{d}\epsilon \tag{6.6}$$

$$\mathbb{E}(\phi(\theta, \epsilon)) = \hat{\phi}(\theta, \epsilon) = \frac{\partial}{\partial \delta} \log(\mathcal{L}^*(\theta, \delta; y))|_{\delta=0} \tag{6.7}$$

where we generalize the previous notation by noting that derived quantity $\phi(\theta, \epsilon)$ might be a nonlinear function of both fixed and random effects. We then evaluate the gradient of the log-marginal augmented likelihood with respect to δ evaluated at $\delta = 0$. This gradient is then a high-quality approximation to the unbiased estimator for the derived quantity $\phi(\theta, \epsilon)$.

When applying the epsilon estimator to a single random effect (e.g., Eq. 6.5), it can be decomposed into the following three terms:

$$\hat{\phi}(\epsilon) = \underbrace{\phi(\hat{\epsilon})}_{\text{Plug-in}} - \underbrace{\frac{1}{2}\frac{\phi''(\hat{\epsilon})}{f''(\hat{\epsilon})}}_{\text{Nonlinearity in } \phi} - \underbrace{\frac{1}{2}\phi'(\hat{\epsilon})\frac{f'''(\hat{\epsilon})}{f''(\hat{\epsilon})^2}}_{\text{Skewness in } f} \tag{6.8}$$

where $\phi''(\cdot)$ is the second derivative of transformation $\phi(\cdot)$ (which is zero when ϕ is linear, i.e., using an identity link function), $f''(\hat{\epsilon})$ is the second derivative of the joint likelihood (i.e., the Hessian, such that $\frac{1}{f''(\hat{\epsilon})}$ is the variance of f), and $f'''(\hat{\epsilon})$ is the third derivative of the joint likelihood and therefore measures skewness. The second term (labeled *Nonlinearity in g*) accounts for the issues raised in the lognormal example, where the variance in random effects ($\frac{1}{f''(\hat{\epsilon})}$) and the nonlinearity in the transformation ($\phi''(\hat{\epsilon})$) combine to cause bias.

The third term (labeled *Skewness in f*) is calculated from the third-derivative of the joint log-likelihood and therefore corrects for skewness.

Alternatively, it is feasible to correct for bias resulting from nonlinear transformations and skewness by taking samples from the estimated distribution of random effects, transforming these samples, and taking the mean of these transformed values. For example, this sample-based method occurs naturally when using *Markov Chain Monte Carlo* to sample from a posterior distribution, and Bayesian hierarchical models therefore provide a natural way of computing the posterior mean for any derived quantity. However, it is difficult in general to obtain samples from the joint log-likelihood f for use in calculating $\phi(\theta, \epsilon)$ for large and nonlinear hierarchical models; this is precisely why we instead introduced the Laplace approximation in the previous chapters. Alternatively, we saw in Sec 6.2.1 that we can take samples from a multivariate normal approximation to the joint log-likelihood (Eq. 2.28). This multivariate normal approximation corrects for the second term of Eq. 6.8. However, the multivariate-normal distribution has zero skewness, so this multivariate-normal approximation does not account for the skewness of random effects that is included as the third term of Eq. 6.8. We therefore expect samples from the joint precision (e.g., using Eq. 2.28) to perform poorly when the random effects have substantial skewness.

We demonstrate these concepts for our example where ϵ is normally distributed and $Z = e^\epsilon$, and add code to calculate the epsilon-correction manually or sample directly from the known distribution in a TMB function (Code 6.4). In this case, we manually add the additional term delta(0) * Z to the joint negative log-likelihood (see the first line of Eq. 6.5). We also compare this manual implementation with the automated version that can be accessed by outputting any variable using the ADREPORT function in TMB, and then calling the function sdreport(..., bias.correct=TRUE) in R (Code 6.5). This code allows us to explore either normal or gamma distributions, using either an identity, square-root, or exponential transformation, and compares the epsilon estimator against the mean of samples drawn from the known distribution.

CODE 6.4: TMB code demonstrating how the epsilon bias-correction estimator is implemented.

```
1  #include <TMB.hpp>
2  template<class Type>
3  Type objective_function<Type>::operator() ()
4  {
5    // Data
6    DATA_IVECTOR( options_z );
7    // options_z(0) = 0:  eps ~ Normal( mean=mu, sd=sigma )
8    // options_z(0) = 1:  eps ~ Gamma( shape=mu, scale=sigma )
9    // options_z(1) = 0:  Y = eps
10   // options_z(1) = 1:  Y = sqrt(eps)
11   // options_z(1) = 2:  Y = exp(eps)
12   DATA_SCALAR( mu );
13   DATA_SCALAR( sigma );
14
15   // Parameters
16   PARAMETER( epsilon );
17   PARAMETER_VECTOR( delta );  // calculate gradient of derived quantity Z
18
19   // Define distribution for random variable epsilon
20   Type jnll=0, Z=0;
21   if(options_z(0)==0) jnll = -1 * dnorm( epsilon, mu, sigma, true );
22   if(options_z(0)==1) jnll = -1 * dgamma( epsilon, mu, sigma, true );
23
24   // Define derived quantity, Y = phi(X)
25   if(options_z(1)==0) Z = epsilon;
```

```
26   if(options_z(1)==1) Z = sqrt(epsilon);
27   if(options_z(1)==2) Z = exp(epsilon);
28
29   // Add delta*Z for manual epsilon correction
30   if( delta.size() > 0 ){
31     jnll += delta(0) * Z;
32   }
33
34   // Sample-based estimator for true distribution
35   SIMULATE{
36     Type Zmean_sampled = 0;
37     for( int i=0; i<10000; i++ ){
38       if(options_z(0)==0) epsilon = rnorm( mu, sigma );
39       if(options_z(0)==1) epsilon = rgamma( mu, sigma );
40       if(options_z(1)==0) Zmean_sampled += epsilon / 10000;
41       if(options_z(1)==1) Zmean_sampled += sqrt(epsilon) / 10000;
42       if(options_z(1)==2) Zmean_sampled += exp(epsilon) / 10000;
43     }
44     REPORT( Zmean_sampled );
45   }
46
47   REPORT( Z );
48   ADREPORT( Z ); // Built-in TMB SE and epsilon correction
49   return jnll;
50 }
```

CODE 6.5: R code demonstrating the equivalence of the manual and automated implementation of the epsilon bias-correction estimator.

```
1  # Define true distribution and transformation
2  Dist = 0   # 0=Normal; 1=Gamma
3  Trans = 2  # 0=Identity; 1=sqrt; 2=exp
4  if( Dist==0 ){
5    mu = 1
6    sigma = 0.5
7  }else{
8    mu = 2
9    sigma = 1
10 }
11
12 # Build object
13 Data = list( "options_z"=c(Dist, Trans), "mu"=mu, "sigma"=sigma )
14 Params = list( "epsilon"=1.2, "delta"=vector() )
15 Obj = TMB::MakeADFun( data=Data, parameters=Params, random="epsilon" )
16 Obj$fn(Obj$par)   # optimize random effects
17
18 # Compute SEs and apply epsilon estimator
19 SD = TMB::sdreport( Obj, bias.correct=TRUE )
20
21 # Extract various estimators
22 BiasCorr = summary(SD,"report")['Z','Est. (bias.correct)']
23 Plugin = summary(SD,"report")['Z','Estimate']
24
25 # Manually calculate epsilon estimator
26 parhat = Obj$env$parList()
27 parhat[["delta"]] = 0
28 Obj = TMB::MakeADFun( data=Data, parameters=parhat, random="epsilon" )
29 (BiasCorr_manual = Obj$gr(Obj$par)[1])
30 (BiasCorr_sampled = Obj$simulate()$Zmean_sampled)
```

Finally, we compare the plug-in and epsilon bias-corrected estimators with the known expectation for this simple lognormal case (Fig. 6.2), and this shows that the epsilon estimator is very close to (but not quite) identical to the known value. This epsilon bias-correction

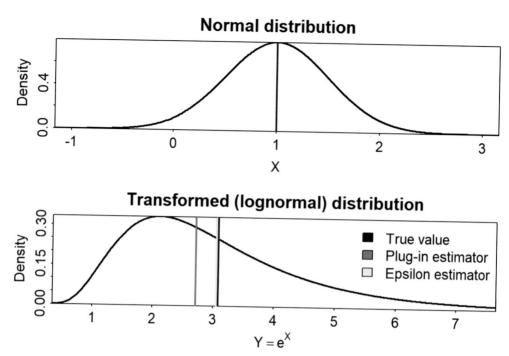

FIGURE 6.2: The density for a normally distributed random variable (top panel), and the lognormal density that arises after it is exponentiated (bottom panel), showing the analytical calculation of it's expected value (which is the true value for an unbiased estimator), the plug-in estimator (which is negatively biased), and the epsilon bias-corrected estimator (which has very minimal bias).

estimator therefore serves as a generic approach to deal with retransformation and skewness bias without requiring samples from the joint log-likelihood, either using MCMC at high computational cost or a multivariate-normal approximation that ignores the skewness of random effects. Despite its convenience, however, we note that calculating the sum across a set of random effects and including them in the calculation of a joint likelihood often causes these random effects to no longer be conditionally independent (see the graphical representation of sparsity from Section 2.4). Therefore, the epsilon estimator will sometimes break the sparsity of the inner Hessian matrix, and thereby require long computation when running sdreport(..., bias.correct=TRUE) in R. We therefore use it for only those variables where it is most important that a high accuracy bias-corrected estimator be used. In other cases where a lower accuracy approximation is sufficient (e.g., for high-resolution maps of results), we then tend to revert to the multivariate-normal approximation knowing that it will ignore the skewness of random effects.

We next return to a more complicated example of retransformation bias and epsilon bias-correction. In section 6.2, we defined a log-linked linear predictor involving a spatial variable and a single covariate (Eq. 6.1) and then calculated the total $\hat{y}$ by summing across the exponentiated value of the linear predictor (Eq. 6.3). The plug-in estimator y^* involves taking empirical Bayes predictions of random effects $\hat{\omega}$, and plugging them directly into the formula for spatial expansion (Eq. 6.3):

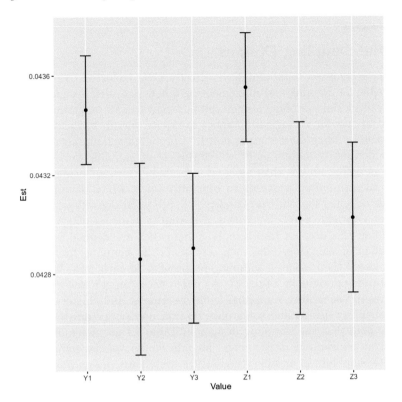

FIGURE 6.3: The sample weighteed (Y1), area-weighted (Y2), or population-density-weighted (Y3) average 8-hour ozone concentration on July 1, 2019 in parts per million using the plug-in estimator, or showing the same weighting methods but using the epsilon bias-correction estimator (Z1, Z2, and Z3 respectively).

$$y^* = \sum_{g=1}^{n_g} w_g a_g e^{\hat{\beta}_0 + \mathbf{A}\hat{\omega} + \hat{\beta}_1 \mathbf{d}} \tag{6.9}$$

where $\hat{\omega}$ are predicted as the mode of the joint likelihood given the estimated value of fixed effects. However, ω are estimated as random effects, and we would like to calculate $\hat{y}$ as the expected value across these random variables:

$$\hat{y} = \mathbb{E}\left(\sum_{g=1}^{n_g} w_g a_g e^{\hat{\beta}_0 + \mathbf{A}\omega + \hat{\beta}_1 \mathbf{d}} \right) \tag{6.10}$$

We therefore calculate the sample-weighted, area-weighted, and population-density-weighted average ozone, either using the plug-in or epsilon bias-correction estimators. In this example (Fig. 6.3), the sample-weighted average is slightly higher than either area or population-density weighted averages. As expected the epsilon bias-correction results in a slight increase for each estimator, where this positive correction results because the derivative of the transformation (in this case, exponentiation) is positive.

6.3   Spatial Sampling Designs

So far, we have analyzed data that have already been collected. However, ecologists are often responsible for defining and implementing a new sampling design, so we next turn to a brief review of spatial sampling designs and how they can affect spatio-temporal inference.

Sampling designs for ecological inference are quite varied, but can be classified as following a *probability sampling* design or not. In a probability sampling design, a target population is divided into sampling units, where this set of sampling units is called a *sampling frame*. Each sampling unit is then assigned a probability π_u of being included in a given sample, and then sampling proceeds to randomly select which sampling units to sample based on their assigned sampling probability [30]. Spatial designs typically define the target population as a variable occurring within a spatial domain $s \in \mathcal{D}$. A spatial design then involves dividing this domain into sampling units that are defined geographically, where this set or union of sampling units then represents the sampling frame. A spatial probability design would then assign a sampling probability to each unit. We further categorize ecological sampling designs in Table 6.1. As noted there, simple and stratified random designs are types of probability sampling, while systematic and cluster sampling sometimes involve a probability design and sometimes do not, and opportunistic sampling generally does not. We note that the sampling designs in Table 6.1 are not mutually exclusive. For example, a stratified random design might include simple random sampling in one stratum and systematic sampling in another. Similarly, a systematic design might involve sampling fish densities, and these densities might then be subsampled using a simple or stratified random design based on attributes of the captured individuals to measure characteristics like sex, age, and body size. Multi-stage designs such as this then includes elements of multiple designs.

We will illustrate these alternative sampling designs by simulating new data from the fitted ozone-concentration model analyzed previously, and then refitting the estimation model to simulation replicates. We specifically apply three spatial sampling designs, each involving $n_i = 50$ measurements of ozone:

1. *Simple random sampling*, where each modeled location has equal chance of being sampled;

2. *Systematic sampling* based on the numbered sequence of grid cells, such that samples are distributed evenly across the spatial domain;

3. *Opportunistic sampling*, where inclusion probability $\pi_g = a_g d_g$ is higher in grid cells with higher human population sizes (calculated as the product of population density and grid-cell area) in this example.

We then simulate data for 100 simulation replicates. For each simulation replicate, we simulate a new value of the spatial variable ω from its distribution when using maximum-likelihood estimates of fixed effects obtained when fitting to real-world data. We then simulate new data at selected sampling units according to their probability in each sampling design. An example of sampling locations (Fig 6.4) arising under each design confirms that systematic sampling results in samples that are distributed broadly across space, but also results in diagonal bands of closely packed samples where the sample spacing is similar to the number of cells in the dimensions of a given state, while opportunistic sampling results in dense sampling near New York City, Atlanta, Orlando, and other major urban areas.

For each replicate and design, we record the true area-weighted ozone concentration. We then refit the estimation model while applying epsilon bias-correction to estimate area-weighted ozone concentrations. We calculate error as the difference between

TABLE 6.1: A brief summary of common ecological sampling designs

Sampling design	Description
Simple random	A type of probability sampling where all sampling units S_i in the sampling frame have an equal inclusion probability. In this case, a design-based estimator for the population mean simply involves calculating the mean of those samples, and similarly, the squared standard error is the sample variance divided by the number of samples
Stratified random	An extension of simple random sampling, where the spatial domain is divided into multiple non-overlapping sub-populations, defines a sampling frame for each sub-population (called *sampling strata*), and then applies simple random sampling within each sampling stratum. A design-based index is then calculated for each stratum individually, where the total is the sum of design-based indices for each stratum, and the squared standard error is the sum of squared SEs for each stratum
Systematic	Defining an order for sampling units and then sampling at some pre-defined frequency along that order. If sampling units are ordered spatially, then systematic sampling ensures that samples are evenly spread across space and thereby can improve sampling precision in some cases. However, a different set of design-based estimators are appropriate to calculate the population mean and standard error
Fixed station (a.k.a. *panel*)	Identifying a set of sampling units that are repeatedly sampled over time. In this case, inference from samples to the population presumably requires that the fixed stations are representative, or that they are rotated periodically following a probability design (sometimes called a *rotating panel design*). In other cases, ecologists might select sites such that they can be intensively studied over time, and processes at each site are assumed to be representative of ecological processes in general, e.g., how inference is made from Long-Term Ecological Research network sites to global ecological processes
Two-stage (or multi-level)	When sampling the density of a population that is highly clustered, ecologists often conduct some probability or systematic sample for primary sampling units across the entire sampling frame, and then conduct secondary sampling in secondary sampling units within those primary units. For example, secondary sampling might be higher in the vicinity of individuals detected in the first stage. Design-based estimators typically involve calculating the mean and variance in each primary sampling unit based on secondary sampling, and applying an estimator across primary sampling units
Opportunistic	Analyzing data that arise from some process that the analyst does not control and hence cannot randomize (or perhaps even fully document). For example, continuous-plankton records [31] have been collected opportunistically on ocean-going vessels for many decades, and the Christmas Bird Count has involved citizen-scientists sampling birds [12]. In general, there is no design-based estimator for these data that do not follow a design, so inference has historically involved proposing a model and using model-based inference [22]

simple_random   **systematic**   **opportunistic**

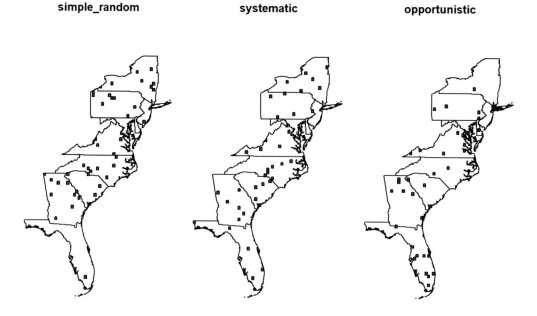

FIGURE 6.4: The location of samples for a single replicate of the simulation design, arising from simple-random, stratified-random, or opportunistic sampling, where the latter has inclusion probability that is proportional to human population density.

estimated and true area-weighted ozone concentration, and plot the range of errors for each sampling design and also when comparing plug-in and epsilon bias-corrected estimates.

This experiment illustrates several points (Fig. 6.5). First, the bias-corrected estimator is slightly higher than the plugin estimator for all designs, as expected given that we apply a log-linked linear model. Second, analyzing data from the simple-random sampling design and applying the epsilon estimator results in estimates that are (approximately) unbiased. Meanwhile, the stratified design results in estimates that have a slight negative bias, while the opportunistic sampling design has a substantial positive bias. The systematic design has smaller variation among replicates (i.e., more precision) than simple-random sampling, presumably due to the more even spread of sampling across space, and this increased precision appears to come at the cost of a slight negative bias. Finally, the opportunistic design has a large positive bias, e.g., it generally estimates a higher area-weighted ozone concentration than actually holds in a given simulation replicate. We discuss this bias in detail next.

6.4   Preferential Sampling

As seen in the simulation experiment (Fig. 6.5), the opportunistic sampling-design has a substantial positive bias. This phenomenon is called *preferential sampling*, and it arises whenever the spatial variable $\omega(s)$ at each location s is correlated with inclusion probability $\pi(s)$ [34, 43]. To see this, consider a case where inclusion probability and the target function

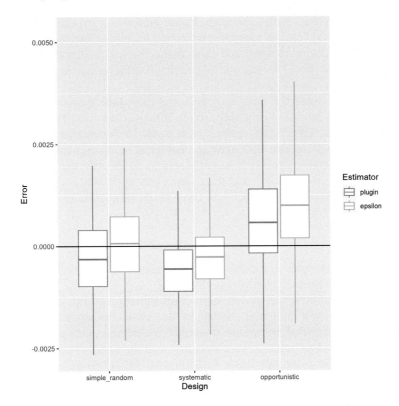

FIGURE 6.5: Boxplots showing errors when estimating area-weighted average ozone concentrations using a log-linked spatial model fitted to data arising from simple-random, stratified-random, or opportunistic sampling and using either a plug-in estimator or epsilon bias-correction, where a well-performing model would have average error approaching zero (horizontal grey line). Note that the opportunistic sampling design has inclusion probability that is proportional to human population abundance (*preferential sample*), which is expected to result in bias.

$\mu(s)$ both vary based on the set of covariates $\mathbf{Z}$:

$$\begin{aligned}
\text{logit}(\pi_i) &= \mathbf{Z}_i \gamma \\
\log(\mu_i) &= \mathbf{X}_i \beta + \mathbf{Z}_i \lambda \\
y_i &\sim g(\mu_i)
\end{aligned} \tag{6.11}$$

This can then result in $\text{logit}(\pi_i)$ being correlated with $\log(\mu_i)$. In this case, the sample average $\sum_{i=1}^{n_i} y_i$ will have expectation equal to the weighted average of μ_i weighted by π_i, rather than the area-weighted average. Similarly, we might then fit a spatial model without covariates:

$$\begin{aligned}
\log(\mu) &= \beta_0 + \mathbf{A}\omega \\
\omega &\sim \text{Normal}(\mathbf{0}, \mathbf{Q}_{spde}^{-1}) \\
y_i &\sim g(\mu_i)
\end{aligned} \tag{6.12}$$

where $\mathbf{A}$ is the projection matrix representing bilinear interpolation and $\mathbf{Q}_{spde}$ is the precision matrix calculated using the SPDE method (see Section 5.5.2). In this case, the estimated intercept β_0 will reflect the sample average, and the prediction of the target variable $\log(\mu)$

will be shrunk towards that sample average. In our simulation experiment involving ozone concentrations, the opportunistic sampling scenario had higher inclusion probability in locations with high population density, and these areas also had higher ozone concentrations, hence fitting the spatial model for the opportunistic sampling scenario resulted in a positive bias.

The easiest way to avoid this bias is by ensuring that data arise from some probability sampling design where inclusion probabilities are independent from the target variable, or to intercalibrate opportunistic data against such data. However, these solutions are often not feasible at the spatial scales that arise in ecological analyses without larger resources than are available. Therefore, research has also sought to identify model-based approaches to mitigate bias from preferential sampling. One such approach is simply to extend the spatial model to include in the spatial model those variables that drive variation in inclusion probability:

$$\log(\mu) = \beta_0 + \mathbf{A}\omega + \mathbf{Z}(s)\lambda \tag{6.13}$$

Including this term then controls for the effect of covariate $\mathbf{Z}$, and results in spatial variable ω being uncorrelated with inclusion probability $\pi(s)$.

To demonstrate, we refit the ozone simulation experiment but including population density as a covariate in the spatial model. In this case, the simulation and estimation models are perfectly matched, so we are unsurprised that the epsilon estimator results in an (essentially) unbiased estimate of area-weighted average ozone concentrations (Fig. 6.6). By contrast, the plug-in estimator has a small negative bias, presumably due to a failure to correct for retransformation bias. Comparing these results with the other scenarios (Fig. 6.5) shows that the model-based correction for opportunistic sampling results in somewhat larger interquartile range (i.e., boxplot width) and whiskers than the original model fitted to data arising from simple-random sampling. This increased imprecision presumably arises due to the spatially clustered sampling arising from the opportunistic sampling scenario, as well as any uncertainty in the estimated preferential-sampling parameter λ.

6.5   Multi-stage Sampling

Finally, multi-stage sampling designs arise frequently in ecological systems, so we discuss them in greater detail here. As an illustrative example, consider monitoring surveys in fisheries science, where ecologists typically use an established sampling protocol and gear consistently over large areas and several decades (i.e., a bottom trawl towed at a known speed for a fixed time) to obtain samples of fish abundance using a standardized unit of sampling effort. Analysts then typically assume that this protocol catches a proportion of local abundance, where this proportion is called *detectability* or *catchability*. If this detectability or catchability is then assumed to be constant over time and space, then spatial and temporal variation in expected catch is proportional to changes in local densities.

However, ecologists then often want to estimate population-level attributes of those fishes, and proceed by measuring those attributes for sampled individuals. For those individuals that are detected during a sampling design, an ecologist might conduct a subsample and for those subsampled individuals measure their age, body size, sex, maturation state, or their physiological condition (energy content, metabolic rates, stomach contents, etc.). Fisheries monitoring therefore involves subsampling the total catch from each sampling event (i.e., dividing into randomized buckets on deck) and enumerating the species, size,

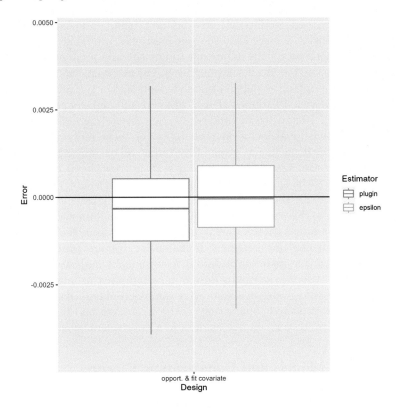

FIGURE 6.6: Boxplots showing errors when estimating area-weighted average ozone concentrations using a log-linked spatial model fitted to data arising from opportunistic sampling where inclusion probability is proportional to human population abundance, but now fitted with population abundance as a covariate in the spatial model (and again showing either a plug-in estimator or epsilon bias-correction and a grey line at zero errors). The decreased bias relative to the opportunistic design in Fig. 6.5 illustrates the benefits of the model-based approach to preferential sampling.

age, or attributes of individuals in that subsample.

6.5.1   Estimators for Multi-stage Sampling

In this illustrative example, we call the original bottom-trawl set a *primary sampling unit*, where the sampling locations in a monitoring survey are typically randomized using a simple or stratified-random design with known inclusion probability. The subsampled individuals are then *secondary sampling units*. In our example, these secondary units are obtained using a simple-random sampling design while using the outcome of the primary unit as the sampling frame; the inclusion probability π_j for each individual in the secondary sampling is then equal to the fraction of the primary unit that is subsampled. This multi-level design might continue onward to a tertiary sampling design, wherein individual fish have their stomachs removed, and only a simple random subsample of stomach contents is then enumerated to species.

The key insight in multi-stage sampling designs is that a secondary (or tertiary etc.) sampling unit has an inclusion probability that arises from the inclusion probabilities of each higher level in a given design. For example, a primary sampling unit might have inclusion probability π_i, and conditional on that unit being sampled a given fish might have inclusion probability π_j, such that the unconditional probability of that fish being sampled is $\pi_i \pi_j$. For this reason, it is important to identify and account for mechanisms for preferential sampling occurring at all sampling level. For example, subsampling fractions in a secondary sampling design often depend upon the outcome of the primary sampling unit; a bottom trawl sample with 10 tonnes of fish will typically subsample a smaller fraction than a sample with only 10 kg (due to time-constraints for sorting each sample). This then results in a lower unconditional inclusion probability for those individuals in high-density areas, and this variation in inclusion probability might be systematically correlated with species composition or size. Ignoring this variation in unconditional inclusion probability then gives rise to preferential sampling bias, and as a result, it could yield biased inference about average age or size for the population as a whole (e.g., [223]).

One simple solution to this scenario is to apply a separate and unbiased estimator to each level of the design. This might involve taking a measurement from the secondary sampling unit, estimating the expected measurement if the entire primary sampling unit had been subsampled, and treating that estimate as data in a subsequent analysis of primary sampling units. In our multi-level fisheries sampling design, for example, this might involve recording the count-at-age for a given species in a secondary sample, dividing this by the subsampling fraction to estimate the expected count-at-age if the entire tow had been subsampled, and then treating this expanded count-at-age as data in a subsequent model [225].

6.5.2   Occupancy Models and Demographic Closure

However, these concepts are more complicated when ecologists create a design where a sampling unit is repeatedly sampled over a short time-scale, such that the variable in each sampling unit is unlikely to change between samples. When sampling abundance this assumption is often called *demographic closure*, because the sampling unit is closed to demographic changes like births, deaths, and migration during secondary sampling. It is possible to interpret these multiple replicated measurements as secondary sampling units, average them to reduce sampling variance for the primary sampling unit, and proceed with spatial analysis as normal. However, replicated samples given demographic closure can often yield additional useful information about detection probabilities.

For example, consider the case that a given site has latent abundance N, and we obtain J measurements of animal abundance $\{C_1, C_2, ..., C_J\}$. Let's make several further assumptions: (1) that the sampling unit is demographically closed during these replicated counts; (2) that each individual is randomly distributed and has probability q of being counted in each sample, where this probability is constant for all individuals and samples; and (3) each count does not mistakenly enumerate the same individual multiple times such that each count C_j follows a Binomial distribution with the same latent abundance and detectability. These assumptions result in an *N-mixture model* [198], and we can then define the probability for these counts as:

$$\Pr(C_1, C_2, ..., C_J | N, q) = \prod_{j=1}^{J} \mathrm{dBinomial}(C_j | N, q) \qquad (6.14)$$

where $\mathrm{dBinomial}(C | N, p) = \frac{N!}{(N-C)!C!} p^C (1-p)^{N-C}$ is probability mass function for the Binomial distribution. The marginal likelihood of detection probability is then approximated by marginalizing across latent abundance N from a specified lower to upper limit (e.g., lower limit $L = \max(C)$ to upper limit $U = 200$):

$$\mathcal{L}(q; C_1, C_2, ..., C_J) = \sum_{N=L}^{U} \Pr(C_1, C_2, ..., C_J | N, q) \qquad (6.15)$$

where the simulation below illustrates that this yields a biased but potentially useful estimator for detection probability q.

CODE 6.6: R code simulating an N-mixture estimate of detection probability.

```
1  # settings
2  N = 20
3  q = 0.8
4  n_count = 10
5
6  # simulate counts
7  C = rbinom(n=n_count, size=N, prob=q)
8
9  # define negative log-marginal likelihood
10 jnll = function(q, C, Nmax=200){
11   prob_n = NULL
12   for( N in max(C):Nmax ){
13     prob_n = c( prob_n, prod(dbinom(x=C, size=N, prob=q, log=FALSE)) )
14   }
15   return( -1*log(sum(prob_n)) )
16 }
17
18 # optimize marginal likelihood
19 optimize( f=jnll, interval=c(0.001,0.999), C=C )$minimum
```

Under the specified assumptions, the variance among samples can be used to define the variance of measurement errors. To provide intuition, note that:

- The sample variance of replicated counts $\text{Var}(C_1, C_2, ..., C_J)$ provides a measurement of the theoretical variance $\text{Var}(C)$ arising from a Binomial distribution $C \sim \text{Binomial}(N, q)$;

- As sampling captures all individuals in the demographically closed sampling unit ($q \to 1$), each sample becomes a census and the sample variance among replicated measurements will vanish $\text{Var}(C_1, C_2, ..., C_J) \to 0$;

- As sampling captures a miniscule proportion of local individuals (i.e., $q \to 0$ and $N \to \infty$), then each sample will follow a Poisson distribution with intensity $\mathbb{E}(C) = \lambda = qN$, such that the variance among replicates also approaches this value $\text{Var}(C_1, C_2, ..., C_J) \to \text{mean}(C_1, C_2, ..., C_J)$;

- Intermediate values for detection probability will also result in intermediate values for among-sample variance (i.e., if $0 < p < 1$ then $0 < \text{Var}(C_1, C_2, ..., C_J) < \text{mean}(C_1, C_2, ..., C_J)$, where the function relating among-sample variance $\text{Var}(C_1, C_2, ..., C_J)$ to detection probability is monotonic and decreasing.

The variance among replicated samples $\text{Var}(C_1, C_2, ..., C_J)$ therefore provides information that is useful to estimate detection probabilities q. More formally, we can compute the expected variance and expected mean:

$$\mathbb{E}(C) = qN$$
$$\text{Var}(C) = q(1 - q)N \tag{6.16}$$

such that their ratio $\frac{\text{Var}(C)}{\mathbb{E}(C)} = 1 - q$. Therefore the ratio of the sample variance and sample mean is an estimator for the probability of not detecting each individual. We note that a similar model can be derived when measuring whether each sampling unit is occupied or unoccupied, and this is conventionally called an *occupancy model* [141]. Both occupancy and N-mixture models can then be modeled with abundance or occupancy varying between primary sampling units, although this still requires that each primary sampling unit is demographically closed during replicated sampling.

This N-mixture model [198] can then be used in place of other estimators for secondary sampling units, and a spatial model used to extrapolate density based on primary sampling units [67, 231]. However, we recommend extreme caution when using the among-sample variance of secondary sampling units to infer detection probabilities, because the estimate of detection probability can be biased due to even mild violations of the assumptions listed previously [195].

6.6   Chapter Summary

In summary, we have showed that:

1. Inference about regional patterns involves spatial integration. This can be implemented by identifying integration points, predicting variables at those locations, and then aggregating across integration points. This spatial integration can be weighted in many different ways, including sample, area, covariate, and multivariate-weighting methods, and these typically correspond to different types of ecological inference. For example, population abundance typically requires area-weighting, calculating per-capita exposures requires covariate weighting, and measuring average demographic rates requires density-weighting (a type of multivariate weighting);

2. Ecologists often estimate quantities that result from both fixed and random effects (e.g., total abundance where densities arise from spatially correlated random effects). In these cases, a simple "plug-in" estimator will be subject to *retransformation bias*, where this bias increases with the nonlinearity of the transformation as well as the uncertainty and skewness of random effects. These biases can be largely mitigated using the *epsilon estimator*, which involves calculating the gradient of an expanded form of the marginal likelihood with respect to the derived quantity;

3. Analysts should estimate, visualize, and communicate uncertainty in density predictions. Uncertainty in derived quantities can be calculated from the joint precision matrix of fixed and random effects, either using the generalized delta method or a sample-based approximation. The sample-based method introduces additional sampling error, but does not substantially increase computational time even for a large number of uncertainty calculations. Therefore, choosing between these two methods depends upon analytical goals and context;

4. Ecologists often specify a sampling design that results in new data. Designs can be categorized as following probability sampling or not, and further categorized as simple or stratified-random sampling, systematic or fixed-station sampling, two-stage (cluster) sampling, and opportunistic sampling. These designs represent a trade-off between sampling efficiency and potential bias where, e.g., systematic sampling can spread data more evenly and reduce standard errors when applying a spatial model. Each of these can be specified to include replicated sampling during a short time interval that is closed to demographic changes, and in this case the sampling variance is sometimes informative about detection probabilities;

5. If sampling rates ("inclusion probabilities") are correlated with the variable being modeled (termed "preferential sampling"), then a standard spatial model will

result in biased inference about regional averages. This bias can sometimes be mitigated by including additional covariates or random effects that explain the variation in sampling intensity within the spatial model for a target variable.

6.7 Exercises

1. In Section 6.5.2, we defined an N-mixture model by specifying a Bernoulli distribution for replicated samples of a primary sampling unit that is demographically closed. We then provided Code 6.6, which showed how to simulate samples from this model for a single primary sampling unit, and then marginalize across local abundance to calculate the likelihood function for estimating detectability. Please expand this code to simulate abundance and replicated samples for 20 primary sampling units (sites), where site each has local abundance drawn from a Poisson distribution $N_i \sim \text{Poisson}(\lambda = 20)$ and has three replicated samples. Then, expand the estimation model (either in R or TMB) to include the same Poisson distribution for local abundance, now estimating two parameters, detectability q and average density λ. Please use a simulation experiment to explore the performance for estimating detectability and average density when varying four inputs: the true values for these two parameters, as well as the number of sites and the number of replicated samples per site. For each of these four experimental axes, please identify whether increasing its value increases or decreases the precision for estimates of detectability and average density.

2. In Section 6.4, we showed that we can mitigate bias arising from opportunistic sampling when the inclusion probability $\pi(s)$ is correlated with the random variation in the target variable $\omega(s)$. Please expand the code used for the simulation experiment in that section to also apply the preferential sampling estimator (Eq. 6.11) to simple random and systematic scenario sampling scenarios. What is the distribution for estimates of the parameter λ linking inclusion probability to target variable density? How does estimating this extra parameter affect the precision and bias for the total area-weighted ozone concentration?

7

Covariates Affecting Densities and Detectability

7.1 Reasons to Include Covariates

In previous chapters, we illustrated how an individual-based process where each individual has marks representing demographic variables (i.e., a marked point process from a Lagrangian viewpoint) can be approximated as gridded densities (i.e., an Eulerian viewpoint) and fitted as a generalized linear model (GLM). We have also shown how residual spatial variation can be modeled by estimating the response to spatial basis functions while integrating across the value of random effects in a generalized linear mixed model (GLMM). This then allowed us to develop basic types of species distribution models (SDMs), where species encounters, counts, or biomass densities are treated as response variables in a GLMM. However, we haven't previously emphasized the role of covariates in a GLMM and SDM.

Covariates might be included in an SDM for a variety of different reasons including:

- *Mitigating bias*: the SDM could be fitted to data that differ substantially in measurement methods or sampling protocols. For example, an ecologist might fit samples of population density using historical (visual) and contemporary sampling methods (environmental DNA), where the mean of each sample is affected both by population density and the characteristics of each sampling method. In this case, including information about the sampling methods can be used to filter out a portion of variance that is explained by sampling method, and thereby eliminate what would otherwise cause a bias in resulting estimates of population density;

- *Interpolation*: similarly, the SDM can be fitted and then used to predict species densities at new locations. If the location of samples is representative of the area being sampled (e.g., they arise from a probability sampling design), then the estimated model is likely representative of other locations in the same spatial domain. We call this predictive task *interpolation*;

- *Attribution*: the SDM can be fitted with a variety of covariates representing environmental or experimental conditions. If these covariates are truly exogenous (as known strongly in the case of randomized treatments, or sometimes believed in the case of observed covariates), then an estimated response curve can be treated as measuring a causal mechanism. This then allows researchers to conduct experiments *in silico* to see how a response would have been different under different environmental or experimental conditions, and thereby attribute real-world outcomes to observed conditions;

- *Counterfactual evaluation*: finally, the SDM could be used to predict species densities under novel conditions. As we will see, this differs from interpolation because it may involve predicting the response to ecological conditions that have never been observed,

DOI: 10.1201/9781003410294-7

either because these conditions reflect some new ecological context, human pressure, or experimental treatment.

Both attribution and counterfactual evaluation are tasks that require predicting a response under covariate values that were not directly observed. Both could perhaps be called *extrapolation*, and it is often observed that extrapolation is more difficult than interpolation [190]. However, we distinguish two different types of extrapolation, where it is easier to extrapolate to conditions that are fundamentally similar to those previously observed (attribution) than extrapolating to conditions that are in some way different from previous observations (counterfactual predictions). SDM performance when extrapolating to new ecological conditions (which we call counterfactual prediction) is sometimes called *transferability*, and we will see that ecological knowledge is typically required to hope for transferability.

In this chapter, we will explore how covariates can aid in all of these tasks. To do so, we will first discuss causal inference and structural equation models in detail. We will use this discussion to illustrate when conventional regression models will result in biased estimates of the effect of covariates (i.e., misleading attribution) and poor performance under new conditions (i.e., poor counterfactual prediction), and also show when these will be unbiased. We will then emphasize a distinction between detectability and density covariates, and show how detectability covariates can be used to intercalibrate data from multiple sources.

7.2 Predicting the Effect of System Changes

To begin, imagine a study focused on how the abundance of soil invertebrates N in a standardized sample is affected by soil temperature T and tree cover C. For illustration purposes, we will imagine that soil invertebrate densities increase with increasing soil temperature (due to increased foraging and growth rates), and that soil temperature decreases in areas with high tree cover (due to decreased solar heating and increased evapotraspiration)[1]. For illustration, we specify parameter values and assume that soil invertebrate densities follow a log-linked Poisson process:

$$C \sim \text{Normal}(0, 1)$$
$$T \sim \text{Normal}(-C, 0.1^2) \tag{7.1}$$
$$N \sim \text{Poisson}(e^T)$$

such that a change in temperature results in a proportional change in log-density. We visualize these effects using a *causal map* that represents the causal mechanisms operating in our hypothesized ecosystem (Fig. 7.1 left panel). This causal map includes:

- One box for each variable, without differentiating between those that might be considered response or predictor variables;

- One arrow for each direct effect of one variable on another.

We specifically construct a causal map with linear effects for each mechanism, and show the *path coefficient* that represents the strength of each linear effect. The total effect of one variable A on another variable B can then be computed by taking the sum across paths flowing from A to B, where the effect from each path is computed as the product of path

---

[1]See https://github.com/james-thorson/Spatio-temporal-models-for-ecologists/Chap_7 for code associated with this chapter.

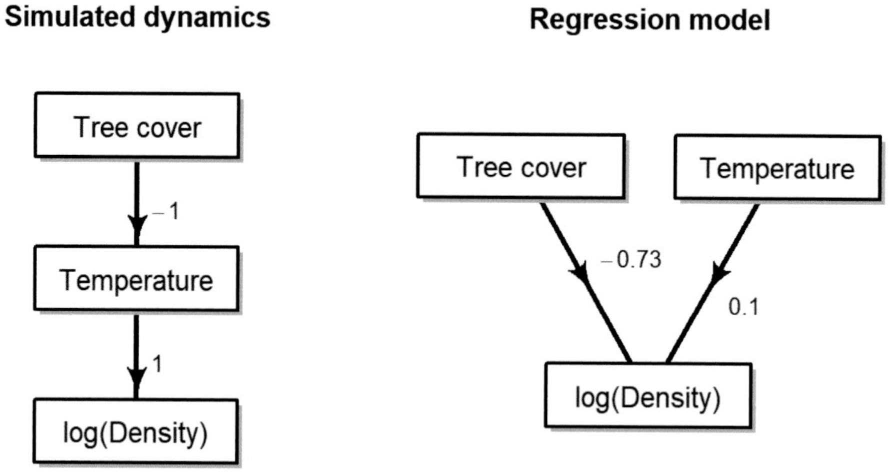

Simulated dynamics   **Regression model**

FIGURE 7.1: The true causal map (left panel) representing mechanisms in a hypothesized soil community with three system variables (boxes) where an arrow indicates a linear causal effect and the number next to each arrow shows the magnitude of response. For example, the arrow from Temperature to log(Density) indicates that a 0.1-degree increase in Temperature would cause a 10% increase in Density. We also show the estimated effect (right panel) when erroneously analyzing data from this ecosystem as a standard generalized linear model.

coefficients for each arrow along that path [271]. In this simplified model, for example, we have only a single path linking tree cover C to log-densities $\log(D)$, i.e., $C \to T \to \log(D)$, which implies that C has a linear effect on $\log(D)$ with a net magnitude of $1 \times -1 = -1$.

To analyze data resulting from this hypothetical ecosystem, many ecologists have been trained to use a regression model that treats an ecological variable of interest (e.g., soil invertebrate densities) as a response and available ecosystem characteristics (e.g., tree cover, temperature, etc) as covariates. A sophisticated analyst might then acknowledge that invertebrate counts are non-negative counts and specify a log-linked Poisson distribution, or might additionally use random effects to address pseudoreplication resulting from a nested sampling design. We therefore fit 100 measurements from this causal map with a generalized linear model of the form glm(N $\sim$ C + T, family="poisson") .

Fitting this GLM results in poor estimates for both covariates (Fig. 7.1 right panel), where e.g. the estimated effect of Temperature is highly imprecise. The measured response N could be correctly modeled using a GLM of the form glm(N $\sim$ T, family="poisson") that is nested within the fitted GLM. Therefore, we are not suprised to see that the estimated Temperature effect is approximately unbiased when we replicate the experiment 100 times (Fig. 7.2 left panels).

Despite being unbiased, estimates of the Temperature effect are highly variable among replicates (e.g., the Temperature effect ranges from -1 to 3.5), and standard errors are extremely large in any single experiment. Standard errors are large because the two modeled covariates (Tree Cover and Temperature) are highly (-0.99) correlated, and this is called *collinearity* when both are included in the same regression model. It might surprise some readers, however, to see that collinearity does not degrade extrapolation-performance when predicting soil invertebrate densities for 100 new locations with temperature and tree cover drawn from the same underlying process (Fig. 7.2, top right panel). However, we also

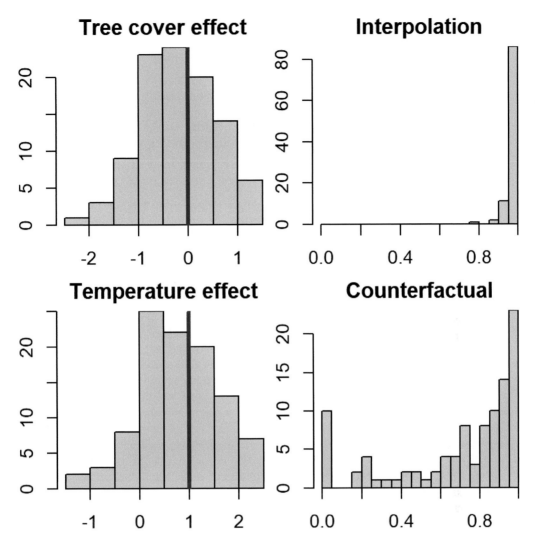

FIGURE 7.2: Estimates of the effect of tree cover (top left) and temperature (bottom left) using a GLM glm(N ∼ C + T, family="poisson") , along with the true values (blue line) from the data-generating process (Eq. 7.1), and the performance (proportion variance in true log-density explained by predicted log-density) when predicting new data arising from the same data-generating process (top-right panel) or predicting densities under a counterfactual where the temperature has increased by 1 degree Celcius without any change in tree cover (bottom-right panel).

simulate new data from the same model, but where temperatures have increased by 1 degree Celcius relative to their average in the fitted data without any change in tree cover (e.g., due to climate change that was not represented in the historical data available to estimate parameters). In this counterfactual scenario, the fitted GLM then has substantially degraded performance for predicting soil invertebrate densities under novel conditions (Fig. 7.2, bottom right panel). One interpretation for this degraded performance is that the structure of collinearity has changed between the fitted and predicted samples [47].

From this simple example we can draw several lessons that are also true of causal inference in general:

1. Inference about the effect of covariates is hugely benefited by (or depends upon) a correct understanding of any causal mechanisms that link those covariates;

2. Fitting to multiple covariates that are correlated (*collinearity*) will typically result in imprecise estimates of slopes in a regression model, but predictions can still be accurate when using collinear variables as long as the relationship among covariates during predictions are similar to those occurring in the fitted data;

3. Predicting the consequences of counterfactual changes in covariates is highly sensitive to the specified relationship among covariates (the *causal map*).

Given these conclusions, we next introduce *do-calculus* as an general and comprehensive vocabulary for thinking about the effect of covariates in ecological models, and use this to discuss structural equation models (SEM) as a potential solution to these issues.

7.2.1   Do-calculus and Causal Inference

All scientists are trained to understand that *correlation does not imply causation*, i.e., that identifying two correlated variables does not imply that a change in one will cause a change in the other. Similarly, scientists are typically trained to understand that randomly assigning sampling units to alternative experimental treatments and measuring the outcome (a *randomized controlled experiment* and its variants) is the gold-standard for inferring mechanisms (which we will call *causal inference*). However, there is less general understanding and agreement about when observational data can be used to infer causality, or how to identify those variables that must be measured when a system cannot be manipulated experimentally due to logistical or ethical constraints.

To address these and other questions about causal inference, we briefly review the literature regarding *do-calculus* [173, 174]. In essence, this literature defines a *do-operator*, where this do-operator then defines the value or distribution of variables B as if other variables A have been exogenously fixed at some hypothetical value or distribution. In particular, we can write the distribution for B as $B|\mathrm{do}(A = a_{new})$ when covariates are fixed at value a_{new} or $B|\mathrm{do}(A = A_{new})$ when covariates are fixed at distribution A_{new}. We can fit a probabilistic model, and then use these estimated distributions for three types of analysis [173]:

1. *Policy evaluations*: We can calculate the distribution for a given response $B|\mathrm{do}(A = A_{policy})$ under alternative assumptions about an exogenous policy choice A_{policy}. This allows us to identify an efficient policy by optimizing the outcome with respect to that policy;

2. *Attribution*: We can calculate the probability of an observed event under scenarios that differ from those that really occurred, and use this to calculate how real-world conditions either increased or decreased this probability relative to some baseline probability. This then allows us to identify what past conditions were responsible for an increased or decreased probability of a past event. This type of probabilistic attribution has seen increased use in policy discussions of climate change and its effect on extreme weather events [161];

3. *Mediation analysis*: Returning to our soil community example, let us imagine that $C \to T \to \log(D)$ and also $C \to \log(D)$. In this expanded example, we see that tree cover C has a direct effect on log-density $\log(D)$ and also has an indirect effect that is mediated by Temperature T. We might then want to know what fraction of the effect of C on $\log(D)$ is due to mediator variable T. This *mediation*

analysis then provides insight into the proximal mechanisms that underlie large-scale correlations.

In our previous example, we defined a joint distribution for three variables $\Pr(C, T, \log(D))$ where sampled abundance was a noisy measure of log-density $\log(D)$. In the causal map defined for the simulation experiment, we then decompose this joint distribution:

$$\Pr(C, T, \log(D)) = \Pr(\log(D)|T) \times \Pr(T|C) \times \Pr(C) \tag{7.2}$$

i.e., where log-density $\log(D)$ is independent of tree cover C conditional upon the value of temperature T (and see Section 2.4 for more discussion of conditional independence and graphical models). However, this probability notation is ambiguous about how to interpret an experiment where temperature T is changed to a new value. This change in temperature could perhaps represent:

- An *endogenous* change, wherein tree cover C is changed exogenously (i.e., due to changing land management or fire regimes), and this causes an endogenous change in temperature T and then a resulting change in soil densities $\log(D)$;

- An *exogenous* change, wherein some un-modeled variable (i.e., an experimental treatment, or global climate change) causes a change in temperature T even though tree cover C is unchanged, via a causal pathway that is not explicitly represented in the causal map.

Importantly, the endogenous change in T is then represented using do-calculus as:

$$\Pr(C, T, \log(D)) = \Pr(\log(D)|T) \times \Pr(T|\mathrm{do}(C = C_{new})) \tag{7.3}$$

where the distribution $\Pr(T|\mathrm{do}(C = c_{new}))$ results from an exogenous change in C to value c_{new}. Alternatively, the exogenous change in Y is represented using do-calculus as:

$$\Pr(C, T, \log(D)) = \Pr(\log(D)|\mathrm{do}(T = T_{new})) \times \Pr(C) \tag{7.4}$$

where T_{new} represents some new distribution that arises from mechanisms that are not represented in the probability distribution (i.e., an experiment, or some other exogenous change), and hence does not affect the distribution for C.

As we have seen in Fig 7.2, fitting a regression model to covariates that have a complicated mechanistic relationship can result in biased or imprecise predictions when covariates are then changed exogenously. Unfortunately, this scenario is precisely what ecologists are often asked to do, i.e., to predict changes resulting from exogenous changes in global climate, human behavior, and introduced species. We therefore present one approach to addressing this issue, i.e., using a structural equation model that explicitly represents mechanisms among covariates.

7.2.2 Structural Equation Models

We next demonstrate these concepts in detail by introducing structural equation models (SEM). We demonstrate how SEM can be used to model the correlation among covariates as well as their effect on a response variable, and also how this can be used during extrapolation to conduct both attribution and counter-factual prediction. Recall that in Section 4.4.2, we defined a structural equation model as providing a distribution for a vector of variables $\mathbf{y}$, e.g., $\mathbf{y} = (C, T, \log(D))$. Unlike a standard regression, none of these variables are defined as predictor or response variables; instead, we define endogenous and mechanistic relationships among variables using a matrix of path coefficients P, and also define exogenous variation δ which follows covariance $\boldsymbol{\Sigma} = \boldsymbol{\Lambda}\boldsymbol{\Lambda}^t$.

$$\mathbf{y} = \mathrm{P}\,\mathbf{y} + \delta$$
$$\delta \sim \mathrm{MVN}(\mathbf{0}, \boldsymbol{\Sigma}) \tag{7.5}$$

This then defines the covariance among variables $\mathbf{y}$:

$$\mathrm{Var}(\mathbf{y}) = \mathbf{V} = \mathbf{B}\boldsymbol{\Sigma}\mathbf{B}^T \tag{7.6}$$

where $\mathbf{B} = (\mathbf{I} - \mathrm{P})^{-1}$. We now can see that this SEM equation and the resulting covariance (Eq. 7.5 and 7.6) has similar structure to the simultaneous autoregressive (SAR) process that can be used to calculate the covariance resulting from a spatial or temporal model (e.g., Eq. B.8 and B.12 in Section B.5).

Alternatively, we could fit a model that explains the covariance among variables $\mathbf{v}$ but instead estimates the covariance directly $\mathrm{Var}(\mathbf{y}) = \mathbf{V}$ as an unstructured matrix, i.e., by estimating parameters in the lower-triangular Cholesky matrix L where $\mathbf{V} = \mathbf{L}\mathbf{L}^t$ (see Section 4.4). After estimating parameters for this unstructured covariance matrix, we could then make predictions about values that are missing at random, and summarize relationships between variables by calculating the eigendecomposition of $\mathbf{V}$ and visualizing the slope of the first couple eigenvectors (called *major axis regression*) [238, 255]. However, SEM offers several advantages relative to estimating this unstructured covariance among parameters. We introduced some of these advantages previously in Section 4.4.2, and here emphasize in particular the justifications labeled *Parsimonious representation* and *Path coefficients as regression slopes*. However, we now have sufficient context to add a fifth justification, i.e., that SEM provides a natural approach to implementing calculations involving the do-operator (Sec. 7.2.1). Specifically, if we want to implement a policy evaluation, attribution exercise, or mediation analysis that involves changing a variable A to a new value or distribution A_{new}, we specify that distribution $\mathrm{do}(A = A_{new})$ and then trace impacts along arrows pointing from that variable to other modeled variables [173, 271]. We therefore add this to the list from Sec 4.4.2:

5 *Clarify how to do counter-factual analyses*: fitting an SEM provides a natural way to conceptualize and compute the expected impact of changing one variable on downstream variables.

In particular, we might seek to predict the consequences of an exogenous change in the vector of state variables from $\mathbf{y}_0$ to $\mathbf{y}_0 + \Delta$. Exogenous change Δ might be an indicator vector (i.e., a vector of zeros except one nonzero element corresponding to the single variable that is changed) or it might represent some simultaneous change in all variables (i.e., all elements of Δ are nonzero). This change Δ then causes a first-order change of $\mathrm{P}\,\Delta$ in other variables (see the first line in Eq. 7.5), then a second-order change of $\mathrm{P}^2\,\Delta$, and so on. The total effect of change Δ is then calculated by summing across these first, second, and higher-order effects. Assuming that this power-series is stationary (i.e., the effect of an exogenous change does not grow exponentially as it moves through the estimated system of equations), we can calculate the total effect as:

$$\sum_{n=0}^{\infty} \mathrm{P}^n\,\Delta = (\mathbf{I} - \mathrm{P})^{-1}\Delta \tag{7.7}$$

To understand this, let us first assume that the system of equations is not *recursive*, i.e., there are no loops like $A \to B \to C \to A$ (such that P can be re-ordered as a lower-triangle matrix). In this case, Eq. 7.7 reduces to calculating the total effect of A on C by summing across paths flowing from A to C, where each path is the product of path coefficients for arrows along that path. In the notation of do-calculus for non-recursive systems, we can therefore use Eq. 7.7 to compute $\Pr(C|\text{do}(A = A + \Delta_A, B = B + \Delta_B))$ where Δ_A and Δ_B are the exogenous changes to variables A and B. However, Eq. 7.7 can also be computed for recursive systems, where computing the total effect is otherwise less straightforward.

CODE 7.1: R code to specify parameters in a SEM using the arrow notation that is then parsed by sem::specifyModel and subsequently passed to TMB.

```
# Define model and convert to RAM
text = "
  C -> T, b1
  T -> logD , b2
  logD <-> logD , NA , 0.01
"
SEM_model = sem::specifyModel( text=text , exog.variances=TRUE ,
                               endog.variances=TRUE , covs=c("C","T","logD") )
RAM = build_ram( SEM_model , c("C","T","logD") )
```

To estimate parameters for the causal map in Fig. 7.1, we first define the model using the textual interface from R-package sem [59], which is itself based upon common notation for defining path-analysis models [270]. We used this notation previously to illustrate parsimonious specification of the covariance in movement among animals (Section 4.5) but review the notation again here. This text interface requires the user to specify:

1. A one-headed arrow, e.g., A -> B to specify that a change in variable A from $A = A_0$ to $A = A_0 + \Delta$ is expected to cause B to change from $B = B_0$ to $B = B_0 + \rho\Delta$, where ρ is an estimated parameter in path matrix P;

2. A two-headed arrow to specify variance (e.g., A <-> A) or covariance (e.g., A <-> B) parameters that are then estimated in matrix $\mathbf{\Lambda}$;

3. A name for the parameter associated with each one-headed or two-headed arrow, e.g., b1 or b2 , and any two parameters that have the same name are assumed to also have the same value;

4. Which parameters are fixed at some value a priori and not subsequently estimated, where the user can name any parameter NA , in which case they are assigned the value that immediately follows that name.

We then use the custom function build_ram to convert this input format to a matrix that lists all structural parameters in the SEM, where this matrix can then be interpreted in TMB to build covariance $\mathbf{V}$ from estimated parameters. We specify endog.variances = TRUE as a shortcut to also estimate a variance parameter in $\mathbf{\Lambda}$ for every variable that was not explicitly assigned one. For example, to specify the SEM on the left-hand-side of Fig 7.1, we specify (Code 7.1) a path coefficient from C to T named b1 and a second path coefficient from T to $\log(D)$ named b2 . In this model, we will specify a log-linked Poisson distribution for measured response N, and this Poisson distribution represents all exogenous variation in the response. We therefore "turn off" exogenous variation for $\log(D)$ estimated in the covariance term $\mathbf{V}$ by fixing the estimate of additional variation at 0.01 a priori (i.e., using the line logD <-> logD, NA, 0.01). In this case, this text file results in the following RAM matrix, which prescribes the parameters to be estimated:

```
  heads to from parameter start
1     1  2    1         1 <NA>
2     1  3    2         2 <NA>
3     2  3    3         0 0.01
4     2  1    1         3 <NA>
5     2  2    2         4 <NA>
```

where column heads indicates whether a parameter is in Γ or P, to and from indicate the row and column numbers of those matrices defined by a given row, parameter lists the parameter number, and start provides a user-specified starting value. The latter is particularly important for any parameters where parameter=0 in the RAM matrix, indicating that this parameter is turned off.

In this case, we are estimating:

$$P = \begin{bmatrix} 0 & 0 & 0 \\ \theta_1 & 0 & 0 \\ 0 & \theta_2 & 0 \end{bmatrix} \tag{7.8}$$

and:

$$\Lambda = \begin{bmatrix} \theta_3 & 0 & 0 \\ 0 & \theta_4 & 0 \\ 0 & 0 & 0.01 \end{bmatrix} \tag{7.9}$$

where this specification involves estimating a vector of four parameters $\theta = \{\theta_1, \theta_2, \theta_3, \theta_4\}$, representing the two arrows of the causal map as well as the exogenous variance in Tree Cover and Temperature. We assign an arbitrarily low value (e.g., 0.01) to the standard deviation of exogenous variation in variable $\log(D)$, to match the generalized linear model which does not include any variance beyond the effect of the two covariates.

To fit this SEM using a log-linked Poisson distribution for soil invertebrate counts, we:

1. specify a multivariate normal distribution for three variables; and

2. specify how those three variables are linked to measurements, e.g., where C and T are assumed to be measured without error (and hence are mapped to their observed values), while N is specified to follow the a Poisson distribution with log-intensity $\log(D)$.

Fitting this model to the same data set as Fig. 7.1, we see that a generalized SEM results in an accurate estimate of the link between tree cover and temperature, as well as between temperature and densities (Fig. 7.3). Replicating this experiment one-hundred times, we also see that the model has highly precise estimates of each effect, and also has similar performance for both extrapolation and counter-factual prediction (Fig. 7.4). This similar

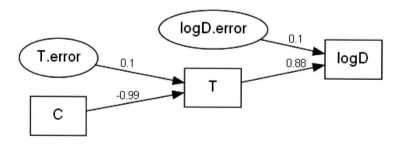

FIGURE 7.3: Path diagram (a.k.a., causal map) estimated using a structural equation model fitted to the same data as Fig. 7.1, using graphics generated automatically using R-package sem , where boxes are variables, ellipses are exogenous errors that follow a standard normal distribution, arrows connecting variables have a number listing the estimated path coefficient, and arrows connecting exogenous errors to variables have a number listing the estimated standard deviation for that exogenous error.

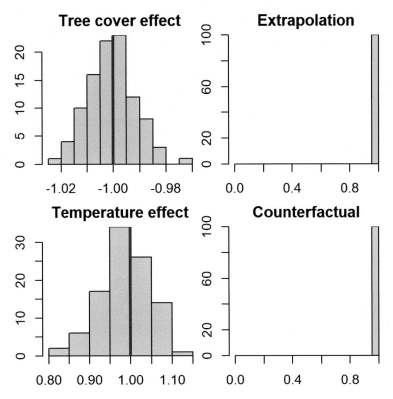

FIGURE 7.4: Estimates of the covariate effects and predictive performance; see Fig. 7.2 caption for details and note the smaller x-axis range for tree cover and temperature-effect estimates.

performance shows that the SEM has addressed the changing structure of collinearity during counter-factual prediction by correctly specifying the relationship among variables [47].

We therefore conclude that properly specifying the mechanistic associations among covariates permits accurate predictions both when covariates follow the same process as sampling data, and when exogenous drivers cause covariates to differ systematically from their relationship in available sampling data. Usefully, this SEM reduces to a familiar GLM when the user defines one variable as a response and the others as predictors, i.e., by specifying that all mechanisms involve an effect of each predictor on the response. This confirms that a GLM will perform well for counterfactual prediction if there is no mechanistic association among covariates.

However, we note that SEM does not resolve all problems regarding causal inference. We instead use SEM as a simple example of the larger universe of models that can be constructed and interpreted using do-calculus. In particular, SEM has the following limitations:

- *Linearity*: SEM assumes that causal mechanisms are well approximated by a linear response. However, ecological systems often have thresholds and saturating relationships, and these are presumably not well described by SEM. These circumstances then require developing a nonlinear approach, e.g., using a path diagram that involves generalized additive models;

- *Dependence on specified causal map*: more generally, both SEM and do-calculus generally will only yield unbiased estimates if the causal map represents real-world mechanisms. If additional mechanisms occur in nature but are not modeled, then the estimates of direct and indirect effects will likely be both biased and inconsistent.

In summary, we recommend more development of causal methods in statistical ecology. We also acknowledge that fundamental challenges remain when analyzing observational data to infer mechanistic relationships among variables. In particular, the dependence on the specified causal map implies that ecologists must continue to use experimental field and laboratory methods to inform the structure among covariates that is assumed when analyzing observational data [104, 241].

7.3   Benefits of Density Covariates

We next demonstrate the potential benefit of including covariates when estimating total abundance from spatially representative sampling designs (a.k.a., interpolation). As we saw in Section 7.2, this predictive task is less dependent upon correctly specifying the association among covariates than counter-factual prediction. We therefore proceed with the conventional regression-model assumption that covariates are exogenous and independent. We also explore the case of a well-known model, e.g., the log-Gaussian Cox process from Section 2.3.

   We specifically use the mainland territory of Finland as the spatial domain, and overlay a square grid with dimensions 0.5 latitude by 0.5 longitude, which results in 271 grid-cells. We use this domain to define a Conditional Autogressive (CAR) model assuming rook-adjacency (see Section 5.5.1), and simulate two latent variables $\mathbf{x}_1$ and $\mathbf{x}_2$ that each follow a multivariate normal distribution:

$$\begin{aligned}
\mathbf{x}_1 &\sim \mathrm{MVN}(\mathbf{0}, \sigma_1^2 \mathbf{Q}^{-1}) \\
\mathbf{x}_2 &\sim \mathrm{MVN}(\mathbf{0}, \sigma_2^2 \mathbf{Q}^{-1}) \\
\log(\lambda_s) &= \beta_0 + x_{i,1} + x_{i,2} \\
c_s &\sim \mathrm{Poisson}(\lambda_s)
\end{aligned} \tag{7.10}$$

where $\mathbf{Q} = \mathbf{I} - \rho\mathbf{A}$ and $\mathbf{A}$ is the rook adjacency matrix and we simulate ρ at 75% of the maximum admissible value (i.e., 0.75 times the maximum eigenvalue for $\mathbf{A}$).

   We then estimated parameters by fitting a similar model structure as was used to simulate data:

$$\begin{aligned}
\omega &\sim \mathrm{MVN}(\mathbf{0}, \sigma_\omega^2 \mathbf{Q}^{-1}) \\
\log(\lambda_s) &= \beta_0 + \omega_i + \mathbf{x}_i \gamma \\
c_s &\sim \mathrm{Poisson}(\lambda_s)
\end{aligned} \tag{7.11}$$

where ω is treated as a random effect, and we can supply some combination of $\mathbf{x}_1$ and $\mathbf{x}_2$ as covariate matrix $\mathbf{X}$ while estimating the appropriate number of response coefficients γ.

   To illustrate simple properties of this model, we first test performance under the scenario that one sample is available from each grid-cell, that we have no covariates available (i.e., $\mathbf{X}$ is a matrix with zero columns), and explore performance for estimating the area-weighted total density, $\sum_s^{n_s} \lambda_s$. We have two hypotheses about model performance:

- *Unbiasedness*: given that the same model is used for simulation data and estimating parameters, we hypothesize that the epsilon bias-correction estimator will result in an unbiased estimate of area-weighted total abundance;

- *Poisson-distribution for standard error*: if we were fitting a model in which ω_s was estimated independently for each site s, then the predictive variance for each site would be c_i. Similarly, because the sum of Poisson distributions is itself Poisson-distributed, and defining the sum $c_{total} = \sum_s^{n_s} c_s$, then the area-weighted total abundance should have predictive variance of approximately c_{total}. We therefore hypothesize that the standard error for predictive total abundance will be approximately $\sqrt{c_{total}}$ when the model does not account for spatial autocorrelation.

A single replicate of the simulation is consistent with both assumptions, e.g., where the epsilon bias-corrected estimate is within one standard error of the true value, and the standard error is very close to the square-root of true total abundance.

```
1 > summary(opt$SD, 'report')
2           Estimate Std. Error Est. (bias.correct) Std. (bias.correct)
3 sumlambda 898.4969    30.45608                 919                  NA
4 > sum(lambda_s)
5 [1]  953.4628
6 > sqrt(sum(c_i))
7 [1]  30.31501
```

We next move on to more complicated properties of this model. Specifically, if we only have samples available for a small subset of grid-cells, then estimating the total abundance will involve interpolation and standard errors will be substantially larger than in the case where samples are available for all grid cells. However, how will including one or both covariates affect the standard error of this estimate?

To explore this, we simulate 100 replicates of an experiment. For each replicate, we simulate the same model, but with $\mathbf{x}_1$ explaining a larger portion of the variance than $\mathbf{x}_2$, i.e., $\sigma_1^2 = 0.6^2$ and $\sigma_2^2 = 0.3^2$, and we also only simulate 100 grid-cells being sampled. For each replicate, we then fit two models model, either (1) not using covariates at all, or (2) using $\mathbf{x}_1$ as a covariate. For each replicate, we then extract the relative error in estimated total abundance, and the estimated standard error for estimated total abundance. Results confirm that the epsilon bias-corrected estimate of total abundance is approximately unbiased either with or without including $\mathbf{x}_1$ as covariate. However, including $\mathbf{x}_1$ as a covariate decreases the interquartile range for relative error and also decreases the average standard error for total abundance by approximately 20% (Fig. 7.5).

From these two exercises involving a log-Gaussian Cox process, we therefore conclude that we can obtain an approximately unbiased estimate of total abundance with either complete or partial sampling of a spatial domain, and that including an informative covariate can decrease the standard error in an estimate of total abundance.

7.4   Density and Detectability Covariates

Finally, we introduce a basic distinction between density and detectability covariates:

1. *Density covariates* are measurements that can be fitted as a predictor variable in a regression, where a change in the covariate is associated with a corresponding change in the target variable (e.g., population density in a species distribution model). As a consequence, predicting the expected value for the target variable at location s should be done conditional upon the values of density covariates at that location;

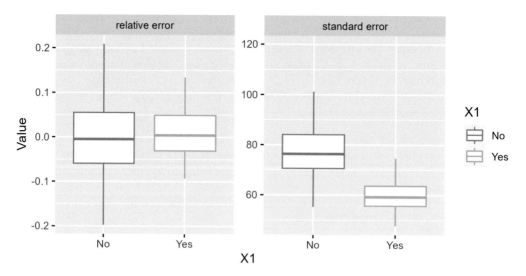

FIGURE 7.5: Relative error in epsilon bias-corrected estimate of total abundance (left panel) and estimated standard error (right panel) when either including $\mathbf{x}_1$ as a covariate or not (blue and red, respectively), showing a boxplot that summarizes 100 simulation replicates for each.

2. *Detectability covariates* are also measurements that can be fitted as predictor variable, but where changes in the covariate are not associated with changes in the target variable. Instead, these detectability covariates are associated with changes in the expected measurement, but believed to not indicate changes in the target variable itself. These detectability covariates typically include categorical variables indicating different sampling methods and other characteristics of sampling.

This distinction may seem obvious. However, different research communities often implicitly assume that all covariates are either habitat or detectability, and results will differ greatly depending upon this assumption. For example, *index standardization* in fisheries science has historically included covariates to control for differences in sampling and has conventionally involved interpreting estimates of an annual intercept as representing changes in population density [11]. Therefore, fitting covariates explains variance that might otherwise be attributed to the annual intercept. This treatment implicitly assumes that all covariates affect detectability, including variables like ocean temperature that in reality likely drive changes in population density itself. By contrast, species distribution models typically use the same set of variables when fitting the model and when predicting density at new locations. This treatment involves the implicit assumption that covariates affect population density rather than detectability, despite the fact that density covariates (e.g., tree cover) might affect sampling efficiency.

The simplest form of detectability covariate is an *offset variable*. For example, an analyst might model population density using a log-linked generalized linear model but where each measurement C_i resulted from sampling density over a spatial area $\mathcal{A}_i$ with area $a_i = |\mathcal{A}_i|$. In this case, expected density $\mathbb{E}(C_i) \propto a_i \lambda_i$, or using GLM notation:

$$\log\left(\mathbb{E}(C_i)\right) = \beta_0 + \omega_i + \log(a_i) \tag{7.12}$$

where e^{β_0} is the proportionality constant (i.e., detectability from Section 6.5.2), ω is other spatial terms, and $\log(a)$ represents the assumed linear scaling of the sample expectation with sample area.

To further demonstrate this distinction between density and detectability covariates, we also introduce the concept of an *integrated model*. We define an integrated model as one that is fitted to multiple data types and/or sources, which all measure an ecological variable but also differ in other respects. In this case, we estimate biomass density $D = NW$ (with units kilograms per area), defined as the product of numerical density D (with units numbers per area) and average weight W (with units kilograms per number). Samples of biomass B directly measure D, while counts C directly measure N. We also introduce the the *complementary log-log link function* (see Section B.2). Under the assumption that C individuals are present in the vicinity of sampling and that this count follows a Poisson distribution $C \sim \text{Poisson}(N)$, the probability of having at least one individual present, $\Pr(C > 0) = \pi = 1 - e^{-N}$. Assuming that we directly model log-abundance, this then results in a cloglog link function, which can transform encounter probability π to a linear predictor p:

$$\text{cloglog}(\pi) = \log(-\log(1 - \pi)) = p \tag{7.13}$$

where $p = \log(N)$ is the modeled linear predictor for log-abundance, or equivalently the inverse-cloglog function:

$$\text{cloglog}^{-1}(p) = 1 - e^{-e^p} = \pi \tag{7.14}$$

where this inverse-link transforms a linear predictor p to the expected encounter probability π.

Given this brief derivation of the cloglog link function, we next define a specific distribution for each of three types of data (see Section B.2)

- *Biomass*: we specify that biomass-sampling data follows a Tweedie distribution:

$$B \sim \text{Tweedie}(D, \phi, \psi) \tag{7.15}$$

where ϕ and ψ are parameters that represent the variance of samples, $\text{var}(Y) = \phi D^\psi$. This Tweedie describes a compound Poisson-gamma process whenever $1 < \psi < 2$ (recalling Section 2.1.1), i.e., where we sample a count from a Poisson distribution, and then each individual has weight that follows a gamma distribution;

- *Counts*: similarly, we specify that count data follow a Poisson distribution:

$$C \sim \text{Poisson}(\eta_1 D) \tag{7.16}$$

where η_1 is the log-ratio of expected counts and expected biomass samples at the same place and time, which captures different sampling efficiency as well as the conversion W from weight to numbers;

- *Encounters*: finally, we specify that binary encounter/non-encounter data follow a Bernoulli distribution using a complementary log-log link function:

$$E \sim \text{Bernoulli}\left(1 - e^{-\eta_2 D}\right) \tag{7.17}$$

where E is a binary 0/1 indicator indicating whether or not the species was encountered in a sample, and η_2 similarly converts from biomass to encounter-rate data.

This model has been explored previously [74], and we follow that paper in fitting it to samples of red snapper (*Lutjanus campechanus*) in the United States waters of the Gulf of Mexico. We specifically fit biomass samples from the a bottom trawl survey (the SEAMAP Groundfish Trawl Survey), count data from an accoustic and midwater trawl survey (the National Marine Fisheries Service Pelagic Acoustic Trawl Survey), and encounter/non-encounter data from a longline survey (the NMFS Red Snapper/Shark Bottom Longline Survey)[2]. We also obtain data regarding the depth of the seafloor below the ocean surface (termed *bathymetric depth*), using the ETOPO 2022 data [164] downloaded using R-package `marmap` [170]. We treat calendar year and bathmetric depth as density covariates, and specify a quadratic basis expansion to bathymetric depth so that we can estimate any optimum in the estimated bathymetric response function (see Section 5.2).

Expressing this integrated model in full, we arrive at the following specification:

$$p_i = \omega_{s_i} + \mathbf{x}_i\gamma + \mathbf{q}_i\eta$$
$$\omega \sim \mathrm{MVN}(\mathbf{0}, \sigma^2\mathbf{Q}^{-1})$$
$$y_i \sim \begin{cases} \mathrm{Tweedie}(\log^{-1}(p_i), \phi, \psi) & \text{if } y_i \text{ samples biomass} \\ \mathrm{Poisson}(\log^{-1}(p_i)) & \text{if } y_i \text{ samples counts} \\ \mathrm{Bernoulli}(\mathrm{cloglog}^{-1}(p_i)) & \text{if } y_i \text{ samples encounter/non-encounter} \end{cases} \tag{7.18}$$

where $\mathbf{x}_i$ is the vector of density covariates co-located with sample i, and $\mathbf{q}_i$ is the vector of detectability variables for sample i. In practice, this model can be implemented in TMB while using input vector `e_i` to indicate whether a given sample arises from a biomass, count, or encounter/non-encounter sample (Code 7.2). We use the SPDE method to assemble a sparse precision matrix from sparse matrices `M0`, `M1`, and `M2`, and again using the `density::GMRF` function to compute the negative log-likelihood from that distribution (as introduced in Section 5.5.2).

The main novelty in this code is the use of the TMB function `logspace_sub` when applying the inverse-cloglog link function to compute the log-likelihood for Bernoulli-distributed samples. This inverse-link function $1 - e^{-e^p}$ involves exponentiating the linear predictor and then exponentiating its negative value, and any sequence of exponential functions easily results in numerical under or overflow. However, we want to calculate the log-density of the Bernoulli distribution resulting from computing the log of this inverse-link function. We therefore use TMB code that evaluates $\log(e^X - e^Y)$ without actually computing either e^X or e^Y. To do so, we define $X = 0$ and $Y = -e^p$, and pass each to function `logspace_sub`. By doing so, we calculate the log of the inverse cloglog function while replacing two calls to an exponential function with only one.

CODE 7.2: TMB code for an integrated species distribution model that is fitted to biomass counts and encounter/non-encounter data.

```
1  #include <TMB.hpp>
2  template<class Type>
3  Type objective_function<Type>::operator() ()
4  {
5    using namespace density;
6
7    // Data
8    DATA_VECTOR( c_i );   // counts for observation i
9    DATA_IVECTOR( e_i );  // 0: Encounter;  1: Count;  2=Biomass
```

[2]Data are publicly available as file "multimodal_red_snapper_example.rda" downloaded with R-package VAST [221], and thank Arnaud Gruss for compiling these data in support of a previous publication [74].

```
10   DATA_MATRIX( X_ik );   // Habitat covariates for samples
11   DATA_MATRIX( X_gk );   // Habitat covariates for integration points
12   DATA_MATRIX( Q_ij );   // Detectability covariates for samples
13   DATA_SPARSE_MATRIX(M0); // SPDE matrix-1
14   DATA_SPARSE_MATRIX(M1); // SPDE matrix-2
15   DATA_SPARSE_MATRIX(M2); // SPDE matrix-3
16   DATA_SPARSE_MATRIX(A_is); // Project vertices to samples
17   DATA_SPARSE_MATRIX(A_gs); // Project vertices to integration points
18
19   // Parameters
20   PARAMETER( ln_tau );
21   PARAMETER( ln_kappa );
22   PARAMETER( ln_phi );
23   PARAMETER( finv_power );
24   PARAMETER_VECTOR( gamma_k );
25   PARAMETER_VECTOR( eta_j );
26   PARAMETER_VECTOR( omega_s );
27
28   // Global variables
29   Type jnll = 0;
30   Type phi = exp(ln_phi);
31   Type power = Type(1.0) + invlogit(finv_power);
32   vector<Type> logmu_g = A_gs*omega_s + X_gk*gamma_k;
33   vector<Type> logmu_i = A_is*omega_s + Q_ij*eta_j + X_ik*gamma_k;
34
35   // Probability of random effects
36   Eigen::SparseMatrix<Type> Q = (exp(4*ln_kappa)*M0 + Type(2.0)*exp(2*
       ln_kappa)*M1 + M2) * exp(2*ln_tau);
37   jnll += GMRF(Q)( omega_s );
38
39   // Likelihood of data
40   for( int i=0; i<c_i.size(); i++){
41     // Bernoulli
42     if(e_i(i)==0){
43       if( c_i(i) > 0 ){
44         jnll -= logspace_sub( Type(log(1.0)), -1*exp(logmu_i(i)) );
45       }else{
46         jnll -= -1*exp(logmu_i(i));
47       }
48     }
49     // Poisson
50     if(e_i(i)==1){
51       jnll -= dpois( c_i(i), exp(logmu_i(i)), true );
52     }
53     // Tweedie
54     if(e_i(i)==2){
55       jnll -= dtweedie( c_i(i), exp(logmu_i(i)), phi, power, true );
56     }
57   }
58
59   // Reporting
60   REPORT( logmu_g );
61   return jnll;
62 }
```

To model detectability in this case study, we obtain an indicator variable (a.k.a. factor) that lists the sampling method for each sample, and convert this to an indicator matrix. We then drop one column of this matrix to avoid specifying two intercepts (the first in density covariates **X** and the second in catchability covariates **Q**), which would then be confounded. For example, this might result in a detectability matrix for three hypothetical samples (rows):

$$\mathbf{Q} = \begin{bmatrix} 0 & 0 \\ 1 & 0 \\ 0 & 1 \end{bmatrix} \tag{7.19}$$

where the first row corresponds to a biomass sample, the second row to a count, and the third row to a sample of encounter/non-encounter data, such that η_1 is the detectability coefficient for counts relative to biomass samples, and η_2 is the detectability coefficient for encounter/non-encounter data relative to biomass samples.

Fitting this integrated model to these three types of data, we estimate that red snapper has highest densities in inshore areas from Texas through immediately east of the Mississippi delta, with lower densities offshore and particularly in the Western Florida shelf (Fig. 7.6). Predictions have a coefficient of variation of approximately 6–10% throughout much of this area, with higher uncertainty in areas with lower density. This relationship (where standard error decreases where densities are higher) presumably results from the specified distributions for sampling, where the standard deviation of sampling decreases with increased mean for the Poisson and Tweedie distributions. Similarly, we estimate that densities decline rapidly with increasing bathymetry, and identify higher median density in some years (e.g., 2012) than others (e.g., 2010), although there is little evidence of an overall trend in the estimated intercept values (Fig. 7.7).

7.5   Chapter Summary

In summary, we have showed that:

1. Covariates are included in spatio-temporal models for many different purposes, including (1) mitigating bias that would otherwise arise from ignoring differences in sampling methods or gears, (2) predicting densities in areas that resemble those that are sampled (termed interpolation), (3) attributing observed patterns to different covariates, and (4) predicting densities in conditions that are different than those that are sampled (termed counter-factual prediction, and generally measuring transferability);

2. Attribution requires some inference about causality, and causality can be formally expressed using do-calculus, which involves predicting a change in one or more variables based explicitly on either endogenous mechanisms (i.e., the effect of a change in some other modeled variable) or exogenous mechanisms (i.e., an experimental treatment or novel condition). Do-calculus involves defining a causal map, and if all mechanisms are linear then parameters can be estimated by fitting a structural equation model. Replacing an explicit path diagram with a standard regression model (where all predictor variables have an immediate effect on a specified response variable) can still result in good predictive performance when interpolating density, but often will degrade performance when extrapolating density;

3. Covariates can be broadly classified as either representing habitat (i.e., variables that are associated with local population density) or detectability (i.e., variables that measure a difference in sampling performance independently of local density). Using point-count data, it is often necessary to determine whether a given

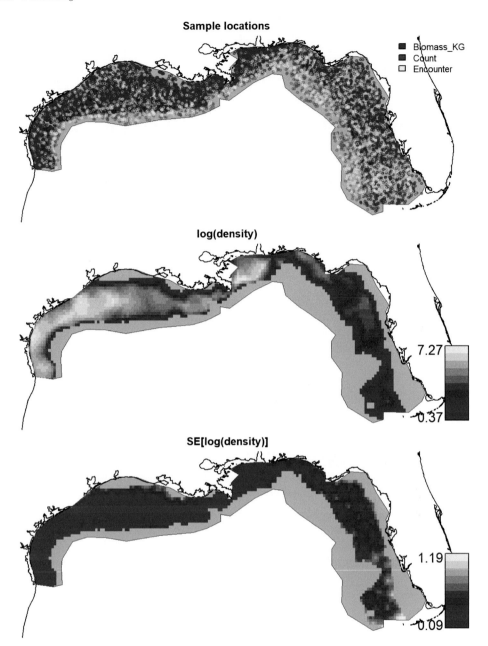

FIGURE 7.6: The spatial distribution of biomass, count, and encounter/non-encounter samples (top panel) fitted by an integrated model for red snapper in the Gulf of Mexico overlaid on the modeled spatial domain (grey polygon), as well as predicted log-biomass density (middle panel), and the standard error in log-biomass density (bottom panel) obtained from sampling the joint precision of fixed and random effects (Section 6.2.1), where the latter two only show log-density or standard errors for locations with density greater than 0.1% of the maximum value (and other areas are displayed as grey) to allow greater visual contrast for areas that are plotted.

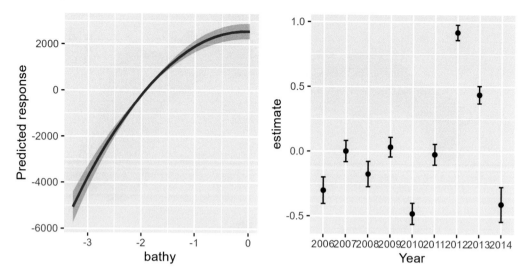

FIGURE 7.7: The estimated effect of bathymetry on biomass density for red snapper (left panel) when fitting a quadratic response as a density covariate, and the estimate of the annually varying intercept for each year with available data (right panel), where each shows a 95% confidence interval.

    variable affects density or detectability based on ecological and system knowledge. When including both density and detectability variables, it is necessary to exclude an intercept from one or the other to avoid confounding;

4. Including density covariates can explain variance that would otherwise be attributed to a spatially correlated residual term. By doing so, density covariates can substantially decrease the standard error for predictions of local density or spatially integrated abundance;

5. Integrated models typically involve fitting more than one type of data, where these data types are all informative about some variables and/or parameters but differ systematically in other respects. For example, it is possible to fit biomass, counts, and encounter/non-encounter data jointly, particularly when using a complementary log-log link function to derive the encounter probability from numerical density.

7.6   Exercises

1. In Eq. 7.13, we introduce the complementary log-log link function and claim that the inverse cloglog can take as input the log-linked linear predictor p for a Poisson distribution $C \sim \text{Poisson}(e^p)$ and return the probability of an encounter, $\Pr(C > 0)$. Confirm that this claim is true, either analytically from the probability mass function for the Poisson distribution or numerically by simulating replicates from a Poisson distribution and calculating the resulting encounter frequency for different values of p.

2. In Eq. 7.1, we illustrated the challenges with extrapolation or counterfactual prediction when fitting a generalized linear model (GLM) to covariates that are not independent. We further showed that a structural equation model (SEM) can incorporate information regarding the relationship among covariates to address these challenges. However, we did not compare GLM and SEM performance in the case when GLMs are expected to perform well. Change the data-generating process to incorporate independent covariates:

$$X \sim \text{Normal}(0, 1)$$
$$Y \sim \text{Normal}(0, 0.5) \qquad (7.20)$$
$$Z \sim \text{Poisson}\left(e^{X+Y}\right)$$

We will continue to specify the SEM using the original specification (Eq. 7.1), which no longer matches the true data-generating process. Now repeat the simulation experiment using this alternative data-generating process, and compute extrapolation and counterfactual performance for both the GLM and SEM. Does the GLM continue to have different performance for extrapolation and counterfactual prediction? How does the SEM perform given that the structural assumptions are now mis-specified?

Part III

Advanced

8

Spatio-Temporal Models with Seasonal or Multi-year Dynamics

8.1 Monitoring Trends in Abundance and Distribution

Scientists monitor changes in natural resources, species, or communities over time, and this information is then used by local communities, businesses, and the general public to debate changes in environmental policy and management. As a few examples:

- Voters and legislatures in the United States have defined various state and federal Endangered Species Acts [66], which outline processes to list plant and animal species as endangered or threatened, and mandate legal protections for those listed species. Criteria for listing include (1) that the species is declining, or (2) that its habitat or range is being damaged or curtailed;

- The International Union for Concerned Scientists (IUCN) maintains the IUCN Red List, which identifies which taxa are data-deficient and then assigns other taxa to categories of increasing risk from Least Concern to Critically Endangered, or to different categories for Extinct taxa. Criteria are evaluated based on evidence of (Criterion A) a decline in population abundance or (Criterion B) a change in area of occupancy [102], often based on time-series or spatio-temporal models that infer abundance or habitat utilization [205];

- Markets exist or are being developed for many types of *ecological services*. For example, carbon offset markets seek to attract financial capital for efforts to store carbon and other emissions that would otherwise contribute to global climate change. The Warsaw Framework for REDD+ established a process to provide global financing for forest management that results in reduced emissions, while also mandating a process to monitor and verify progress and associated uncertainties [177];

- Non-governmental organizations such as the Marine Stewardship Council establish a process by which commercial products can be certified as sustainable, thereby providing a signal to consumers that can affect product access or price. These certifications typically involve a process to monitor changes over time, and the success of the certifying standards can itself be assessed by comparing trends for certified vs. non-certified organizations [76].

In general, these examples involve monitoring changes in the total or average value for a system variable over time, and calculating system variables will typically require spatial integration (Section 6.2). However, it also entails two further challenges. First, it requires extending our previous spatial models to include spatial variables that change over time. It also requires defining new system measurements, corresponding to area of occupancy, range edges and other measurements that characterize spatial distribution and dynamics.

Complicating matters further, ecological variables have a spatial pattern that varies at a variety of different time-scales. Important time-scales in ecological systems often include:

DOI: 10.1201/9781003410294-8

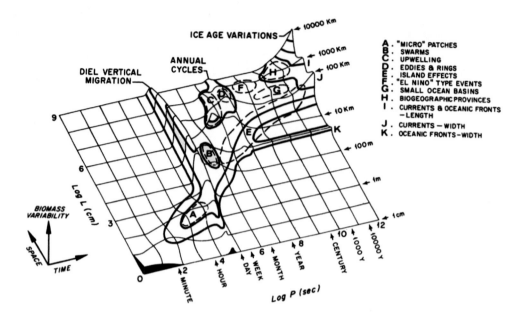

FIGURE 8.1: Stommel diagram showing the hypothesized variance (vertical axis) arising at a given spatial (y-axis) and temporal (x-axis) scale for an idealized marine plankton community, with major biophysical features indicated as letters and labeled on the right-hand side. Reprinted from [88] with permission from SNCSC and available at https://doi.org/10.1007/978-1-4899-2195-6_12.

- *Daily*: individuals often have daily cycles of higher and lower activity-levels, such that individual tracks will show daily movement between resting and foraging areas. Similarly, point-count samples of species density will typically show differences in habitat utilization between day and night, including birds or bats returning to rookeries to sleep and the diel vertical migration of zooplankton;

- *Seasonal*: similarly, populations often show predictable cycles in abundance and distribution on seasonal time-scales. This includes increased vegetation densities in spring and summer, seasonal migrations of mobile consumers, and insects that produce multiple generations per year. Therefore, estimates of habitat utilization may differ between seasons;

- *Interannual*: finally, populations often have fluctuations in spatial distribution or total abundance arising over many years. This includes population cycles for cicadas, years with greater or lesser reproductive output (termed *masting*) for fruiting trees, and population cycles arising from age-structured population and community dynamics. Therefore, ecological inference about abundance and habitat utilization may differ substantially among years.

The magnitude of variation occurring at different spatial and temporal scales is sometimes conceptualized using a *Stommel diagram* [216], as popularized in the famous paper *The problem of scale in ecology* [129]. For example, inspecting the Stommel diagram for an idealized community of marine plankton [88] suggests that daily time-scales have substantial variance across a wide range of spatial scales ranging from 100 m to 100 km (the *Diel vertical migration*), while annual cycles have similar variance on the larger end of this spatial range (Fig. 8.1). This approach has been used to identify spatial and temporal scales that are expected to have substantial variance for marine ecosystems in general [89], and could similarly be adapted for use in other ecosystems.

Constructing a spatio-temporal model therefore requires some careful thought regarding how different temporal scales are treated. For example, a model might define a reference level for one time-scale (e.g., estimating summertime habitat utilization and only fitting to summertime samples), integrate across another variable (e.g., including samplings from dawn to dusk and accepting some extra uncertainty arising from differences in distribution between dawn and mid-day behavior) and explicitly estimate differences in another time-scale (i.e., estimating densities occurring in each year).

8.2 Infill and Sprawl Asymptotics

Despite these different time-scales appearing interchangeably in a Stommel diagram, they have different implications when viewed statistically[1] [36]. To explain, we first note the distinction between *sprawl and infill asymptotics*. This distinction applies to designs occurring over space and time, and we provide examples while outlining both below:

- *Sprawl asymptotics* refers to a design where we can increase the spatial or temporal extent but have a fixed density of samples within that extent. In a time-series design, for example, we may be able to obtain only a single measurement of population abundance in each year. In this case, we can only increase the amount of data available for statistical inference by extending a study over many years. As we asymptotically increase the number of years with available data, we will accumulate information about the processes governing changes in abundance over time (parameters representing population dynamics). However, we will never accumulate additional data regarding abundance in any single time. Intuitively, we might hope to eventually obtain perfect information about parameters but not about state variables;

- *Infill asymptotics* refers to the opposite case, e.g., when we have a fixed spatial or temporal domain, and can obtain additional samples within that fixed domain but can never sample outside that domain. In this case, the density of sampling (number of samples per area and/or time) increases asymptotically, and we could theoretically get perfect inference about a state-variable (e.g., population abundance) within that spatial domain. By contrast, we cannot obtain samples from locations outside that domain; this then limits our ability to accumulate information about the underlying relationship between ecological drivers and our target variable (i.e., parameters representing spatial or temporal dynamics).

We highlight this distinction because ecologists can often obtain multiple samples at daily and seasonal time-scales (where cyclic drivers occur over a fixed interval), such that infill asymptotics apply at those temporal scales. However, ecologists who want more information about the relationship between variables at an annual scale must often wait for additional years, such that sprawl asymptotics generally apply at interannual time-scales.

We further note that spatially correlated random effects can be defined anywhere within a modeled domain. We can therefore view a sample from a spatial random effect as a realization of a Gaussian process (GP) (see discussion in Section 3.6), and this GP is sometimes called a *Gaussian random field* when it is defined over two or more dimensions. When we

---

[1]See https://github.com/james-thorson/Spatio-temporal-models-for-ecologists/Chap_8 for code associated with this chapter.

can obtain progressively more samples within a fixed domain (such that infill asymptotics apply), we get progressively more and more information about the value of the spatial random effect, and in many cases, we can expect that the estimate will converge on the true value regardless of whether the specified semi-variogram is a good or poor approximation to the true data-generating process. To make sense of this result, we can invoke a *representation theorem for Gaussian processes* [118], and representation theorems such as this are useful for explaining the asymptotic consistency of estimated mixed-effects models under infill asymptotics.

As one concrete example, this distinction affects our capacity to estimate parameters governing spatial and temporal autocorrelation. For example, consider a spatial model fitted to 50 random samples within a 1 square-kilometer spatial domain, in which the true data-generating process follows an exponential correlation function with no measurement error, and we fit the same exponential function to those samples. We simulate 100 replicates of this experiment, and for each replicate, we use the geoR package [188] to first simulate the spatial field, then compute the pairwise difference between each pair of samples and calculate the average difference for different binned distances (i.e., construct a variogram), fit an exponential semi-variance function to these pairwise differences, and compare the estimated semi-variance function against the true (known) semi-variance (see Code 8.1). We specifically compare performance using this original design with two alternative designs, where we either increase sampling density in the same spatial domain (i.e., moving toward performance under infill asymptotics), or a design that increases the spatial area but using the same sampling density per area (i.e., moving toward performance under sprawl asymptotics).

CODE 8.1: R code specifying a simulation experiment that compares performance for estimating an exponential semivariance function with either increased sample density or increased sample area.

```
 1  library(geoR)
 2  # Simulation settings
 3  area_c = c("original"=1, "infill"=1, "sprawl"=10)
 4  nsamp_c = c("original"=50, "infill"=500, "sprawl"=500)
 5  nrep = 100
 6  cov.pars = c("sigma2"=1, "range"=0.4)
 7  results_rcz = array(NA, dim=c(nrep,length(area_c),length(cov.pars)),
 8                      dimnames=list(NULL,names(area_c),names(cov.pars)) )
 9
10  # Run experiment
11  for( config in seq_along(area_c) ){
12  for( rep in 1:nrep ){
13    loc = matrix( runif(n=2*nsamp_c[config], max=sqrt(area_c[config])), ncol=2)
14    # Simulate a spatial field
15    field = grf( grid = loc,
16                 cov.pars = cov.pars,
17                 cov.model= "exponential" )
18    # Construct variogram
19    var = variog( field )
20    # Fit variogram model
21    varhat = variofit( var,
22                       cov.model= "exponential",
23                       fix.nugget = TRUE,
24                       limits = geoR::pars.limits("phi"=c(lower=0.01, upper=Inf)),
25                       ini.cov.pars = cov.pars )
26    # Compile results
27    results_rcz[rep,config,] = varhat$cov.pars
28  }}
```

Results from this experiment (Fig. 8.2) show that 50 samples over the original domain results in many replicates that cannot identify the variance that occurs as distance increases

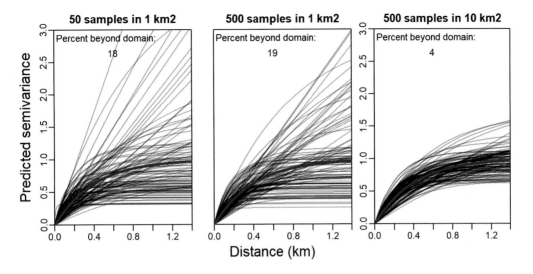

FIGURE 8.2: Illustrating performance when estimating an exponential semi-variance in 100 simulation replicates (black lines) compared with the true semi-variance function (blue line), in a design with 50 random samples within a square 1 km^2 spatial domain (left column), or ten times more samples in the same domain along the lines of *infill asymptotics* (500 samples in 1 km^2; middle panel), or ten times more samples over ten times larger area, hence maintaining a constant sampling density along the lines of *sprawl asymptotics* (500 samples in 10 km^2; right panel). For each design, we also list the percent of replicates where the estimated semi-variance has a geostatistical range beyond the maximum pairwise distance in the original domain, i.e., when the model cannot usefully identify the sill of the semivariance function.

asymptotically (the *sill* of the semi-variance function, see Section 3.3). These 18 replicates are associated with estimates of the semi-variance function that are essentially straight lines, without any obvious decline in slope for increasing distance. Perhaps surprisingly, increasing the number of samples by 10-fold within that same domain still results in almost 20% of replicates that cannot identify the asymptote of the semi-variance function. However, the same increase in sample size over a larger area (i.e., increasing the area 10-fold while maintaining the original sampling density) results in a large decrease in the replicates that are unable to identify this asymptote. By contrast, both designs involving 500 samples appear to perform similarly in terms of their ability to estimate the slope at the origin, which corresponds to the predicted rate that variance accumulates per distance for two nearby points.

This experiment illustrates a few patterns that also typically hold in other circumstances:

1. The benefits of increased sampling when estimating parameters for a spatial or temporal model depend greatly on whether the increase arises from increasing sample density (i.e., infill asymptotics) or an increasing domain (i.e., sprawl asymptotics);

2. To improve performance when estimating parameters (i.e., the correlation across space or time), it is often more helpful to increase the domain being sampled than to increase the sampling density within the existing domain. In fact, given a fixed domain and infill asymptotics, we might only be able to accurately estimate

the rate that variance accumulates per distance (the slope at the origin of the semi-variance function), and not ever have a well-performing estimator of the sill.

And what does this have to do with seasonal vs. inter-annual variation in a spatio-temporal model? We note that ecologists are often able to wait a year to get a chance to resample habitat utilization within a given season. Therefore, spatio-temporal variation arising at a seasonal time-scale is likely to have infill asymptotics, and we might hope to eventually accumulate perfect information about seasonal habitat use all else being equal. However, we can never go back and obtain new samples from the past. Ecologists must therefore accept that we will often have *ecological surprise* [44] when estimating dynamics over spatial and temporal scales where sprawl asymptotics apply.

8.3   Seasonal Adjustment and Spatially Varying Coefficient Models

We next demonstrate different approaches to extend spatial models to include changes in spatial variables over time. To do so, we extend the analysis from Section 6.2, where we fitted spatial models to ozone concentrations measured on July 1, 2019. We specifically illustrate how to extend spatial models to account for the daily or seasonal timing of samples by fitting daily ozone concentrations arising from Jan. 1 to Dec. 31, 2019. But to do so, we first take a detour and discuss seasonal adjustment and spatially varying coefficient models.

8.3.1   Seasonal Adjustment

To address seasonal time-scales, we first introduce the seasonal adjustment for time-series data that is commonly used in econometrics. For example, an economist might have a time-series of Gross Domestic Product (GDP) $x_{u,y}$ measured in each quarter $u \in \{1, 2, 3, 4\}$ and year y. We use two subscripts for time to highlight the different time-scales involved, but a vector of GPD over time can be extracted as vec($\mathbf{X}$).

GPD has a typical fluctuation during each year, and recessions are typically defined as when GPD shrinks for two consecutive quarters. However, defining a recession in this way requires first accounting for seasonal fluctuations in GPD, to avoid confusing a recession with a predictable seasonal downturn. In this case, we can calculate the average difference for GPD in a season relative to GPD in a reference-season, and subtract this difference for each season relative to the reference season u^*:

$$\bar{x}_u = \frac{1}{n_y} \sum_{y=1}^{n_y} x_{u,y}$$

$$x_{u,y}^* = x_{u,y} - \bar{x}_u + \bar{x}_{u^*}$$

(8.1)

where n_u is the number of seasons (i.e., $n_i = 4$ using quarterly data), n_y is the number of years analyzed, and vec($\mathbf{X}^*$) is an adjusted measurement of GPD that has removed the average seasonal fluctuations.

We generalize this to define *spatial seasonal adjustment*. In doing so, we have a few requirements that such an approach must satisfy:

1. *Continuous or discrete seasonal timing*: in Eq. 8.1, we divide the calendar year into a set of discrete seasons (typically quarters or months). However, ecological data are often collected over a series of weeks, and movement and growth may be

substantial during this interval. We therefore want a generalization of seasonal adjustment that can treat seasonal timing either using discrete seasons, or using a continuous-time process;

2. *Seasonal correction varies spatially*: similarly, Eq. 8.1 does not include any indexing for location s. However, seasonal patterns often vary spatially, e.g., where the onset of spring typically occurs later at poleward locations. We therefore seek a seasonal adjustment method that applies a seasonal-adjustment term that can differ by location.

Both of these requirements are satisfied using a spatially varying coefficient model involving a cyclic basis spline, which we introduce in subsequent subsections.

8.3.2   Spatially Varying Coefficients

So far, we have discussed generalized linear mixed models with the following form:

$$g(\mu_i) = \sum_{j=1}^{n_j} x_{i,j}\beta_j + \omega_i \tag{8.2}$$

where $\mathbf{X}$ is the design matrix for j covariates which includes a column of 1s representing the intercept, each covariate has estimated response β_j that is constant across space, and $\omega_i = \mathbf{A}_i\omega$ is the value of the residual spatial variable at s_i given random effects ω and projection matrix $\mathbf{A}$. In this model, the spatial variable ω can be interpreted as a *spatially varying intercept*, in the sense that it controls the value of μ_i in the absence of covariate effects $\mathbf{x}_i = \mathbf{0}$.

As alternative, we next generalize this by introducing *spatially varying coefficients*, where we include both a spatially varying intercept ω but also spatially varying slopes ξ:

$$g(\mu_i) = \sum_{j=1}^{n_j} x_{i,j}\beta_j + \sum_{k=1}^{n_k} z_{i,k}\xi_{i,k} + \omega_i \tag{8.3}$$

$$\xi_k \sim \text{MVN}(\mathbf{0}, \mathbf{Q}^{-1})$$

where $\mathbf{Z}$ is the design matrix with zero-centered spatially varying slopes $\xi_{i,k} = \mathbf{A}_i\xi_k$. When the same design matrix is supplied for both $\mathbf{X}$ and $\mathbf{Z}$ then this can be written instead as:

$$g(\mu_i) = \sum_{j=1}^{n_j} x_{i,j}(\beta_j + \xi_{i,k}) + \omega_i \tag{8.4}$$

where it is more obvious that β_j is the spatially averaged response and $\xi_{i,k}$ the is the local difference relative to this average for covariate $x_{i,j}$ at location s_i. Alternatively, there is no reason to require the same covariate matrix for spatially constant and spatially varying slopes.

Spatially varying coefficient models have been widely used in statistics [63, 87], but despite a few exceptions [56, 222] have seen relatively little use in ecology. Despite this limited use in ecological studies, they could be used to address many common questions including:

1. *Local response to regional environmental conditions*: in many cases, an annual index of atmospheric or oceanographic conditions can result in simultaneous responses for species at geographically distant locations (termed an *ecological teleconnection*). For example, the wintertime production of sea ice in the Bering Sea

drives the spatial extent of cold near-bottom waters in the subsequent spring and summer, and this in turn drives the seasonal migration of demersal fishes. In this case, it is helpful to include a regional index (summer cold-pool extent) as a covariate with a spatially varying response [222];

2. *Density-dependent habitat selection*: as a special case of local responses to regional conditions, it may be helpful to specify total population abundance as a covariate and estimate a spatially varying response to this total. The estimated map of responses then identifies habitats that have a faster- or slower-than-proportional response to changes in total abundance [220]. This response-map then approximates the action of density-dependent habitat selection;

3. *Context-dependent habitat responses*: alternatively, an analyst might know a priori that two variables have an interactive effect on local densities, but only be able to measure one of the two variables. For example, the importance of predator refuges will likely depend upon the density of predators, which might itself be difficult to measure directly. In this case, the response to the measured variable (predator refuges) is likely to vary among habitats due to the effect of the other missing variable (predator densities). One short-term response would then be to include predator-refuge as a covariate, and explore the impact of estimating a spatially varying response to this covariate. Estimates could then be corroborated or refuted by future studies that directly measure the missing covariate;

4. *Spatially varying detectability*: finally, an ecologist might include sampling gear as an indicator matrix. For example, if data are available for two sampling gears, then covariate $z_{i,k}$ would be 0 for the primary gear and 1 for the alternative gear (see Section 7.4). An analyst might then estimate a spatially varying response to this indicator variable. The spatial response $\beta_j + \xi_{i,j}$ would then represent the difference in detectability between these two gears at location i.

Other ecological uses for spatially varying coefficients are reviewed in detail elsewhere [239].

8.3.3   Seasonal Spatial Adjustment

We also previously discussed spline basis-expansion (Section 5.2), and how this basis expansion is similar to the basis functions implied by a given spatial correlation function (Section 5.4). We now extend this by introducing *periodic splines* (Fig. 8.3). A periodic spline specifies that the resulting response function $f(x)$ has the same value and derivatives at some specified period T, i.e., $f(x + T) = f(x)$. For use in seasonal modeling, we can specify a period $T = 365.24$ days, such that the response function represents seasonal variation that is repeated every year. Alternatively, in daily models using high-frequency data, we might specify $T = 24$ hours, such that estimated dynamics recur daily.

To address seasonal spatial variation, we then include the basis functions resulting from a periodic spline as a covariate in the spatially varying coefficient model (Eq. 8.4). To demonstrate, we specifically fit to 87,520 ozone measurements arising throughout 2019, use six periodic splines with knots distributed evenly throughout the year, and then predict concentrations at the beginning of each month. Predictions from this seasonal spatial model for July 1 (Fig. 8.4) are generally similar to estimates fitted only to data from July 1 (Fig. 6.1) in terms of showing lower concentrations in Florida and northern New York and also identifying hotspots along the border between Georgia and South Carolina. However, they also differ in some cases, e.g., where the seasonal model shows lower concentrations near Atlanta in northern Georgia.

Importantly, the seasonal spatial model (Fig. 8.4) includes only six basis functions with evenly spaced knots but can be mapped for twelve months or at higher temporal frequency.

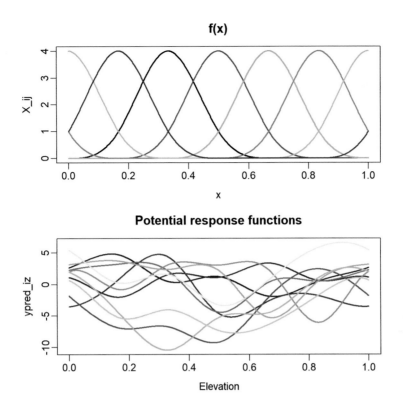

FIGURE 8.3: Illustration of the basis functions resulting from a periodic spline with six degrees of freedom and a period $T = 1$ defined such that the response function has the same value and derivatives at the boundaries $x = 0$ and $x = 1$ (top panel), and ten random simulations of the response function that could arise from using this basis (bottom panel).

Estimates at high temporal frequency are then interpolated from measurements that are nearby in space and time, e.g., predicted ozone densities on Feb. 1 are intermediate between those estimated for Jan. and March. Because we use a cyclic spline, this similarity also wraps around the seasonal domain, i.e., where the estimated ozone density in December is similar to Nov. and January. Infill asymptotics then apply to the estimated seasonal pattern, because we will never need to estimate densities for any season beyond this modeled range.

8.4 Interannual Dynamics

We next outline how to incorporate interannual variation within a spatio-temporal model. This differs from seasonal variation because interannual variation is likely subject to sprawl asymptotics, while seasonal variation is likely subject to infill asymptotics. We also outline several different model structures, and which is suitable likely depends upon how the results are being interpreted or used.

In particular, we introduce different forms of spatio-temporal variation by referring to *main effects and interactions*, to link our discussion to the vocabulary that ecologists use when analyzing experimental data [275]. We restrict our presentation to separable models,

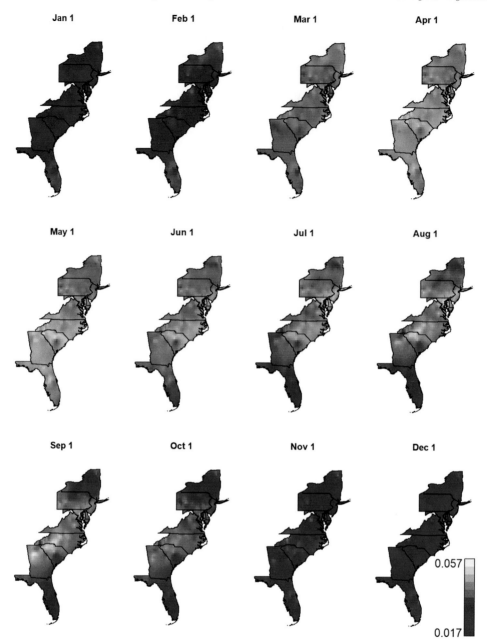

FIGURE 8.4: Predicted ozone concentrations for each month (with a shared color legend in the bottom-right panel), using a spatially varying response to the periodic spline with six degrees of freedom and a period of 1 year.

such that the spatio-temporal covariance $\mathbf{Q}_{full}^{-1} = \mathbf{Q}_{spatial}^{-1} \otimes \mathbf{Q}_{time}^{-1}$ is formed as the Kronecker product of a spatial covariance $\mathbf{Q}_{spatial}^{-1}$ and a temporal precision $\mathbf{Q}_{time}^{-1}$ (see Section 5.4). We note that there is a large and growing literature regarding non-separable models, where the covariance cannot be constructed as the Kronecker product of smaller processes. For example, research has developed computationally efficient approaches to construct the spatio-temporal precision resulting from advection and diffusion in atmospheric dynamics [28]. Some types of non-separable model are simple to specify in TMB, where e.g. a non-separable process can be computed by specifying a first-order Markov process:

$$\mathbf{n}_{t+1} = f(\mathbf{n}_t) \times e^{\epsilon_t}$$
$$\epsilon_t \sim \mathrm{MVN}(\mathbf{0}, \mathbf{Q}^{-1}) \tag{8.5}$$

where $\mathbf{n}_t$ is a vector of abundance at a set of locations in each time t, $f(.)$ is a production function that includes nonlinear effects among locations, and ϵ_t is a spatially correlated process error that represents the net effect of exogenous and endogenous drivers that are not included in function f representing system dynamics. Specified this way, dynamics f might include terms representing animal movement, density dependence, and other population-dynamics processes. This specification extends state-space time-series models (Section 3.2) by replacing a scalar n_t representing population abundance with a vector $\mathbf{n}_t$ representing abundance in different areas. As we saw in Chapters 3 and 5, some (but not all) functions f could in fact be re-parameterized as a separable model. Specifying first-order Markov dynamics (Eq. 8.5) simply requires computing the probability in each time conditional upon the previous, and this can be computed efficiently even when the resulting dynamics cannot be represented by a separable precision for ϵ across all years.

However, specifying non-separable spatio-temporal models (e.g., applying Eq. 8.5 directly in TMB code) also has several important drawbacks:

- *Slower inner optimizer*: separable models can often be parameterized such that the inner optimizer can rapidly optimize random effects given the specified value of fixed effects. By contrast, non-separable models are often harder to parameterize such that the inner optimizer is efficient;

- *Decreased sparsity*: separable models typically maintain any sparsity that is specified for spatial, temporal, or other model dimensions (see Section 5.4). Without this sparsity, computing the Laplace approximation for non-separable models becomes much slower;

- *Interpretive difficulties*: separable models can often be explained using a small number of simple concepts. While this is also true for some non-separable models, it is also possible to specify and fit a non-separable model that is then very difficult to explain or derive from ecological principles. This lack of simple explanation then results in to difficulties when interpreting outputs, identifying bugs, or comparing results with past research.

We therefore proceed instead by introducing spatio-temporal models using a smaller range of separable specifications, and suggest that these are often worth fitting before exploring nonseparable versions. In later chapters, we then introduce concepts (e.g., animal movement) that could be useful to include in non-separable spatio-temporal models [234].

8.4.1 Main Effects for Space and Year

The simplest model for interannual variation involves calculating a variable as the combination of separate effects for year and location. For example, we might calculate the density $d_{s,t}$ using a log-linked linear predictor:

$$\log(d_{s,t}) = \beta_t + \omega_s$$
$$\omega \sim \mathrm{MVN}(\mathbf{0}, \boldsymbol{\Sigma}) \tag{8.6}$$

where β_t defines the median density in year t and ω_s defines the density for location s relative to the median location that is typical across years. This model has several useful properties including:

1. *Proportional changes over time*: for any two years t_1 and t_2 the difference $\beta_{t_2} - \beta_{t_1}$ defines the log-ratio for density at every possible site;

2. *Rank-order of habitat quality is constant over time*: the spatial term ω_s entirely defines the rank-order of densities across space for a given species. So for any two sites s_1 and s_2, if $\omega_{s_2} > \omega_{s_1}$ then density will be higher at site s_2 than s_1 regardless of the year being considered;

3. *Predictive variance can be low even in areas with no data*: when predicting density at a location s and time t, the variance of this prediction only depends upon uncertainty about β_t and ω_s. If there is extensive data in a single year t_1 then it is possible to precisely estimate ω_s. Then, when predicting density in a new year t_2, all samples are equally informative about β_{t_2} regardless of where those samples occur. Therefore, predicted density might have a low variance even in an area that is geographically distant from all available data in that year.

These characteristics cause this *main effects* model to represent a useful null model for thinking about interannual variation in spatial densities. However, the 3rd property also limits its usefulness in real-world contexts.

8.4.2 Independent Interaction of Space and Time

Adding complexity progressively, we extend this "main effects" model for interannual spatial dynamics by using a log-linked linear predictor that includes both spatial and spatio-temporal variation:

$$\log(d_{s,t}) = \beta_t + \omega_s + \epsilon_{s,t}$$
$$\omega \sim \text{MVN}(\mathbf{0}, \sigma_\omega^2 \mathbf{R}_\omega) \tag{8.7}$$
$$\epsilon_t \sim \text{MVN}(\mathbf{0}, \sigma_\epsilon^2 \mathbf{R}_\epsilon)$$

where we have now added a separate spatio-temporal variable ϵ_t for every year, with spatial correlation $\mathbf{R}_\epsilon$ and spatio-temporal variance σ_ϵ^2, and these spatial correlation and variance are also specified for the spatial variable ω. This results in the following properties:

1. *Spatial differences in habitat quality tend to be preserved*: both ω_s and $\epsilon_{s,t}$ control the relative density for any pair of sites in a given year. When spatio-temporal variance is smaller than spatial variation, $\sigma_\epsilon^2 << \sigma_\omega^2$, then the relative ranking of densities among sites tends to be preserved over time. Alternatively, when spatio-temporal variance is bigger $\sigma_\epsilon^2 >> \sigma_\omega^2$, then the relative ranking of densities tends to be scrambled for each pair of years;

2. *Areas with no samples in a year tend to have higher predictive variance*: when predicting densities at a new location using this model, it is necessary to predict the value of both temporal variation β, spatial variation ω and spatio-temporal variation ϵ. To precisely estimate $\epsilon_{s,t}$ it is then necessary to have data near location s in that exact year. Therefore, predictive variance tends to reflect the density of data in that exact time and location.

For completeness, we note that it is possible to combine these model terms, $\epsilon_{s,t}^* = \omega_s + \epsilon_{s,t}$. We can instead form the covariance of that combined term directly as:

$$\text{vec}(\mathbf{E}^*) \sim \text{MVN}(\mathbf{0}, \boldsymbol{\Sigma}_\omega \otimes \mathbf{1} + \boldsymbol{\Sigma}_\epsilon \otimes \mathbf{I}) \tag{8.8}$$

where $\mathbf{1}$ is a matrix composed of 1s, and $\mathbf{I}$ is a diagonal matrix. We can therefore specify $\epsilon_{s,t}^*$ using a separable covariance (recalling the definition from 5.4).

8.4.3 Autocorrelated Interaction of Space and Time

Next, we extend the model that includes an interaction of space and time by introducing temporal autocorrelation. This can include either:

- autocorrelation in the temporal main effect, β_t; and/or

- autocorrelation in the spatio-temporal term $\epsilon_{s,t}$.

Specifically, these involve modifying Eq. 8.7 as:

$$
\begin{aligned}
\log(d_{s,t}) &= \beta_t + \omega_s + \epsilon_{s,t} \\
\omega &\sim \text{MVN}(\mathbf{0}, \boldsymbol{\Sigma}_\omega) \\
\epsilon_t &\sim \begin{cases} \text{MVN}\left(\mathbf{0}, \frac{1}{1-\rho_\epsilon^2}\boldsymbol{\Sigma}_\epsilon\right) & \text{if } t = 1 \\ \text{MVN}(\rho_\epsilon \epsilon_{t-1}, \boldsymbol{\Sigma}_\epsilon) & \text{if } t > 1 \end{cases}
\end{aligned}
\tag{8.9}
$$

where ρ_ϵ is the magnitude for first-order autocorrelation in spatio-temporal variation, and:

$$
\beta_t \sim \begin{cases} \text{N}\left(0, \frac{1}{1-\rho_\beta^2}\sigma_\beta^2\right) & \text{if } t = 1 \\ \text{N}(\rho_\beta \beta_{t-1}, \sigma_\beta^2) & \text{if } t > 1 \end{cases}
\tag{8.10}
$$

where ρ_β is the magnitude of autocorrelated variation in median density. In both of these autocorrelated terms, we specify that the variance in the initial year is equal to the stationary variance, which involves correction term $\frac{1}{1-\rho^2}$ (see below Eq. 3.8). However, these autocorrelated terms do not have a stationary distribution when $|\rho| \geq 1$, so we would drop the term $\frac{1}{1-\rho^2}$ if the process approaches a random walk.

Similar to the previous simplified models, this autocorrelated spatio-temporal model has several useful properties:

1. *Density hotspots propagate through time*: when the spatio-temporal term is positively autocorrelated (i.e., $\rho_\epsilon > 0$), hotspots in density are likely to persist over time. This then allows for useful interpolation of spatial distribution for years when sampling is not otherwise available [166];

2. *Forecasts converge on a stationary distribution*: similarly, when autocorrelation is within plausible bounds (i.e., $|\rho_\epsilon| < 1$ and $|\rho_\beta| < 1$), then both β and ϵ are stationary and converge on a finite variance (recalling Section 3.3). In this case, forecasting density from the spatio-temporal model converges on a stationary distribution, which presumably represents the landscape-level variation that is expected over multi-decadal time scales.

When the autocorrelation terms are set equal, $\rho = \rho_\beta = \rho_\epsilon$, this model has been called a *spatial Gompertz* process [240]. In this case, the value $1 - \rho$ then measures the strength of density dependence, similar to its role in the conventional Gompertz model (see Section 3.1).

8.5 Measuring Changes in Spatial Distribution

Given this set of models, we can predict density $d_{s,t}$ across space and among years. To illustrate, we introduce a new case-study, involving walleye pollock (*Gadus chalcogrammus*) in the eastern and northern Bering Sea. Walleye pollock supports a commercial fishery in the Bering Sea that has a commercial value exceeding 1 billion US dollars annually [4]. The fishery is operated by a combination of shore-based and at-sea processing facilities, and the catch is frozen to provide low-cost fish worldwide. The pollock population in the eastern Bering Sea remains one of the most intensively studied fishes worldwide, with a summertime bottom trawl survey in the eastern Bering Sea using an annual fixed station design with approximately 370 stations annually from 1982 to the present day. In recent years, however, the eastern Bering Sea population has expanded northward into areas in the northern Bering Sea, colonizing habitats that were previously too cold for pollock. The barrier to northward summertime movement is thought to be a body of near-freezing waters near the seafloor, and this *cold pool* is created by ice production near St. Lawrence Island during the preceding winter and spring. Therefore, a summertime index of "cold pool extent" is often used to measure bottom-up oceanographic constraints on summertime distribution for this and other species [75]. As pollock have moved northward, there has been increased survey effort in the northern Bering Sea. This ecosystem was previously dominated by an alternative community of Boreal species, and was surveyed following a bottom trawl survey using varied sampling designs (but a consistent sampling gear) in 1982, 1985, 1988, and 1991, and then again in 2010, 2017, 2018, and 2019.

We fit these data using a Tweedie distribution for survey catches (Eq. 8.7) and the SPDE methods for spatial correlations. We specify a spatio-temporal model that includes an autocorrelated interaction of space and time (i.e., using Eq. 8.7) but treating β_t as a vector of fixed effects instead of specifying that it follows an autoregressive process (Eq. 8.10). After fitting, we sample from the joint precision matrix (see Eq. 2.28) to visualize predictive uncertainty for density estimates. Results confirm that predictive standard deviations are low (i.e., 0.2) where data are available in a given year, and are extremely large (i.e., >2) in areas without data in a given year. Similarly, standard errors in the northern Bering Sea in 2016 (i.e., the year before data were available) are lower than in previous years (2014 and 2015) as the variance bridges toward the high precision in 2017 resulting from sampling data in that region (Fig. 8.5).

8.5.1 Indices of Spatio-Temporal Dynamics

We next explore how these predicted densities can be summarized to yield insights about changes in distribution and occupied habitat over time. As we will see, these methods provide a complementary picture to the estimates of total abundance introduced in Section 6.2.2. In addition to the total biomass over the survey domain, we calculate the following metrics:

- *Center of gravity*: for each extrapolation point g we define some measurement of its spatial position z_g along some axis. This measurement could represent geographic location, altitude/depth, or other way of ordering locations along some axis. We then use this as a covariate in a covariate-weighted average (Section 6.2.2). In this example, we define location z_g as the distance of each point from the equator, and use this to measure poleward shifts in the centroid of the spatial distribution;

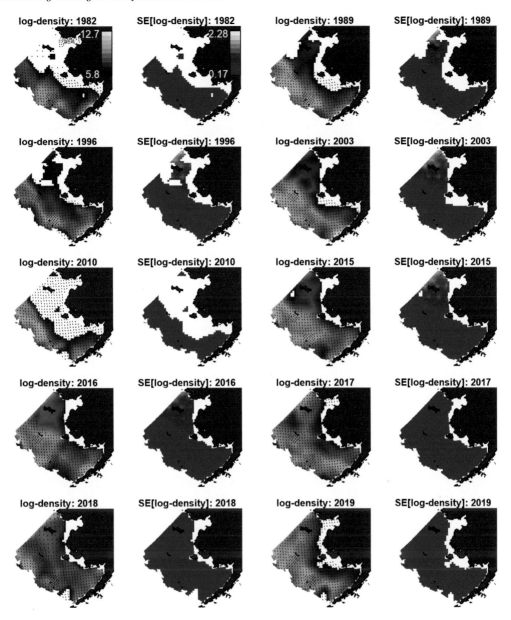

FIGURE 8.5: Predicted biomass density (on log-scale; 1st and 3rd columns) and predictive standard deviation for log-density (2nd and 4th columns) in selected years, fitted to annual data for the eastern Bering Sea but only having data in the northern Bering Sea in 1982/1985/1988/1991/2010/2017/2018/2019 (with sample locations shown as bullets on density plots). To improve contrast in plots, we do not plot densities or standard errors (and instead leave those areas white) for any modeled location that has negligible density (defined as $< 0.1\%$ of the maximum estimated density).

- *Equivalent area*: alternatively, we may seek to measure range expansion/contraction. This can then be used, e.g., to evaluate whether range size changes during population declines [232]. To do so, we define the *equivalent area* (a.k.a. effective area occupied) as the area necessary to contain the population at its average density [264]. This is then calculated as:

$$B_t = \sum_{g=1}^{n_g} a_g d_{g,t}$$

$$D_t = \sum_{g=1}^{n_g} d_{g,t} \frac{a_g d_{g,t}}{B_t} \tag{8.11}$$

$$A_t = \frac{B_t}{D_t}$$

where B_t is total abundance, D_t is average density, and equivalent area A_t is their ratio. There are many other ways to define area occupied, but this has the benefit that it (1) does not require defining any threshold value to convert estimated densities to some binary presence/absence metric, and (2) it will result in the same value if the spatial configuration of habitats is changed (i.e., it does not depend on occupied habitats being contiguous);

- *Range edge*: finally, we may seek to measure shifts in range edges as an ecological definition of colonizing new habitats [60]. To do so, we order extrapolation points based on their location z_g along some coordinate system, calculate the cumulative sum of abundance along these ordered locations, and identify the location where this cumulative distribution crosses some specified quantile. For example, we here calculate the poleward range edge as the location where 95% of population biomass is distributed to the south.

We note that many other metrics can also be calculated where, e.g., the *spreading area* measures range expansion/contraction in a way that is related (but not identical) to the equivalent area presented here [264]. We calculate each of these metrics for each of 500 samples from the joint precision matrix and inspect the resulting time series to infer changes in spatial distribution and population dynamics.

Results for pollock (Fig. 8.6) shows that the bottom trawl survey measured available population biomass of 4-8 billion kg from 1982 to 2019, with lowest abundance in 1996–1999 and again 2004–2010. Importantly, estimated biomass is most imprecise in 2013-2016. This occurs because the model estimates a substantial density in the northern Bering Sea in 2017, and therefore also expects increasing densities between 2010 and 2017. However, there is no data to measure this proportion from 2011-2016, so the model compensates by having large standard errors for total biomass in those years. The estimated center of gravity in 2017–2019 was northward of any previous year, and shifted nearly 150 km from 2015 to 2019. More dramatically, however, the northward range edge shifted over 400 km north from 2010 (a well-sampled year when the northern range edge was precisely measured) to 2017 (the next year with complete data and therefore a precise estimate of the northward range edge). This northward range expansion is associated with a dramatic increase in the effective-area occupied from a historical low in 2010 to a high in 2017. Collectively, these time-series estimates paint a comprehensive picture of a commercially important fish shifting rapidly northwards over time.

8.5.2 Estimating Local Trends

So far we have shown how to convert output of a spatio-temporal model to time-series that measure changes in spatial attributes for a given population. However, these do not provide much context for interpreting fine-scale differences in spatial distribution. We seek a method that summarizes distribution shifts with a level of detail that is intermediate

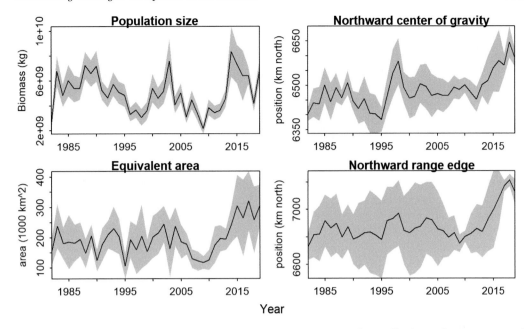

FIGURE 8.6: Measurements of spatio-temporal dynamics for pollock in the eastern and northern Bering Sea from 1982-2019, including total biomass (top-left panel), northward center of gravity (top-right), effective area occupied (bottom-left panel), and northward range edge (bottom-right).

between inspecting density maps for each individual year and reducing these down to individual time-series.

We therefore introduce a method that measures the distance a fish would need to move to maintain a constant population density. This metric is calculated identically to *climate velocity* [20, 21]. However, climate-velocity is typically applied to a measurement of ocean temperature, and measures the distance that must be moved to keep pace with climate by maintaining a constant temperature. Our metric is instead intended as a high-level summary of local shifts in distribution, and we therefore call it *local shifts*.

To calculate local shifts, let's imagine that we have estimated densities $D(s, t)$ continuously across space and time. We then define the spatial gradient $\nabla D(s, t)$, defined as the derivative of the density function with respect to spatial coordinates. Evaluated at a location s, this results in a vector pointing in the direction of greatest increase in density, with vector length corresponding to the magnitude of that slope, and having units density per distance. Similarly, we define the temporal gradient $\frac{d}{dt} D(s, t)$, which indicates whether a location has an increasing or decreasing trend over time with unit density per time. Finally, we define the local trend as this ratio:

$$\mathbf{v}(s, t) = \frac{\frac{d}{dt} D(s, t)}{-\nabla D(s, t)} \tag{8.12}$$

where this results in a vector $\mathbf{v}(s, t)$ with units distance per time that points in the direction of movement that would be necessary to maintain a constant population density.

However, we do not actually have a continuous estimate of population density $D(s, t)$, and instead have discretized this across space and time $d_{s,t}$. We therefore replace derivatives with difference operators. Specifically, we convert densities at our extrapolation points to values on a grid using the `terra` package. We then use the `terra::terrain` function to calculate the spatial slope (i.e., a discretized approximation to the negative spatial gradient $-\nabla D(s,t)$), and a series of linear models to calculate a temporal trend (i.e., the discretized approximation to $\frac{d}{dt}D(s,t)$) over one or more years.

To illustrate the potential for this local-trend calculation, we apply it to estimated log-densities in 2010 and 2019. These years bookend the rapid shifts northward in both center of gravity and northward range edges, as well as a huge increase in effective area occupied (Fig. 8.6). Inspecting results, we see that locations in the middle and eastern portion of the eastern Bering Sea have trends that require a northeastward movement to maintain a constant density (Fig. 8.7). However, local trends are largest and generally point northward in the area between the eastern and northern Bering Seas. Comparing this local-trends plot with densities in 2010 and 2019 (Fig. 8.5), we confirm that the local-shift metric is a useful depiction of the northward shift in the northern range edge.

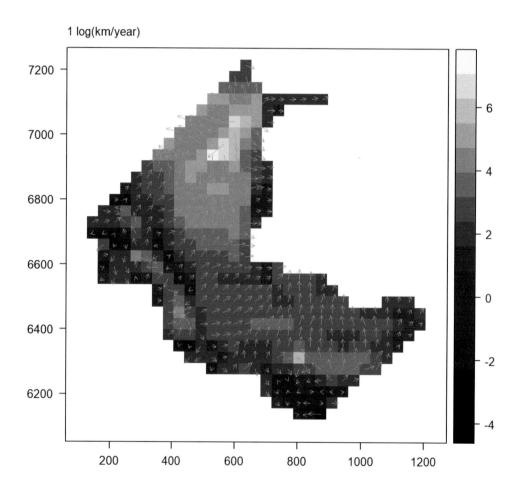

FIGURE 8.7: Visualizing the distance (log kilometers per year) required to maintain a constant population density between 2010 (the year with the smallest equivalent area) and 2019 (after a large expansion northward).

8.6   Chapter Summary

In summary, we have showed that:

1. Spatio-temporal dynamics involves variance at a variety of spatial and temporal scales, and dominant scales can be visualized and summarized using a Stommel diagram. Defining a spatio-temporal model typically involves conditioning results upon a fixed value of some temporal scales (e.g., only using daytime samples to estimate daytime distribution), implicitly averaging across other scales, and explicitly modelling spatio-temporal variation in a selected set of scales;

2. The consequences of increasing sample size depend on whether sampling densities increase within a fixed spatial and temporal domain (i.e., moving toward infill asymptotics) or sampling densities are constant but with increasing spatial or temporal domain (i.e., moving toward sprawl asymptotics). In particular, sprawl asymptotics is typically necessary to obtain accurate estimates of parameters, while infill asymptotics can improve estimates of state-variables;

3. Spatially varying coefficient models involve estimating a covariate slope that varies across space. Infill asymptotics typically apply to estimates of seasonal spatial patterns, and these can be estimated by including a spatially varying coefficient for a periodic spline with an annual period. The resulting model continuously interpolates the response variable for any season;

4. Modeling interannual dynamics in a spatio-temporal model involves some minimal choices about whether to include just the main effect of space and year, or also a space-by-year interaction. This interaction, in turn, can be independent among years, or autocorrelated to ensure that hotspots are propagated through time;

5. Spatio-temporal models can be summarized to extract many measurements of shifting habitat utilization, including center-of-gravity, effective area occupied, and range edges. In instances where data are spatially unbalanced, a spatio-temporal model with an autocorrelated space-by-year interaction can appropriately identify an increase in predictive variance for years with spatially restricted sampling. Spatio-temporal patterns can also be summarized using a metric of local shifts, defined as the direction and distance that an animal would move to maintain a constant population density. This can then be calculated as the ratio of temporal trend and spatial gradient, and results provide a summary of distribution shifts that is complementary to raw density estimates or high-level time-series summaries.

8.7   Exercise

1. In Section 8.5, we introduced center-of-gravity and range-edge indices as ways to summarize changes in spatial distribution over time. We then illustrated these by fitting to data for Alaska pollock using an autocorrelated interaction of space and time (Eq. 8.9). Please refit this model, but instead using the independent interaction of space and time (Eq. 8.7) and compare the estimates of spatial indices with those presented in Fig. 8.6. How does it affect the estimates for years with spatially unbalanced sampling (i.e., 2011–2016)?

2. Continuing this first exercise, please fit the autocorrelated interaction of space and time and add TMB code (e.g., using SIMULATE) to simulate new samples conditional upon the estimates of spatial variables reflecting a progressive northward movement from 2010-2017. Then refit both the independent (Eq. 8.7) and autocorrelated (Eq. 8.9) models. Do the confidence intervals typically include the true value for spatial indices using both models? When might you prefer to fit one model or the other?

9

Ecological Teleconnections

9.1 Nonlocal Drivers of Ecological Dynamics

The most spectacular events in the natural world often arise when different individuals change their behavior simultaneously across a large landscape. In eastern North America, for example, deciduous trees will rapidly change from summer greens to fall colors over the course of several weeks. These fall colors then indicate the passing of another summer. Other famous examples include:

- Periodical cicadas that emerge in the northeastern US every 13 or 17 years to reproduce and then die, strewing exoskeletons across the east coast of North America;

- Mass coral spawning, where immobile corals across a large seascape synchronously release gametes for broadcast spawning, likely due to shared responses to the timing of moonrise relative to sunset [132];

- Lemming outbreaks in Norwegian grasslands that support the survival of juvenile predators in Arctic ecosystems [101].

These are all examples of *spatial synchrony*, where population densities and/or demographics simultaneously undergo some rapid change at geographically distant locations. Examples such as these have interested naturalists for thousands of years. They also have important implications for ecosystem function, e.g., for how bears can follow a *red wave* of salmon that spawn at different times across their foraging landscape [40].

These four examples of population synchrony result from many different mechanisms, including synchronous environmental conditions (*exogenous synchrony*) as well as evolutionary pressures to maximize fitness by synchronizing reproductive or other demographic processes (*endogenous synchrony*). In each case, an ecologist might then seek to describe the spatial scale at which dynamics are synchronized, the spatial and temporal scale over which synchrony is predictable, the relative importance of various endogenous and exogenous drivers, as well as how this synchrony affects community structure and ecosystem function.

Given the ubiquity and importance of population synchrony, we introduce techniques to describe broad-scale synchrony in biological systems. To do so, we first introduce a toolkit that is borrowed from physical scientists. In particular, oceanographers and atmospheric scientists often study processes that are synchronized even at geographically distant locations, (termed atmospheric, oceanographic, or geophysical *teleconnections*). As we have already discussed, synchrony at geographically distant locations also occurs in ecological systems, and we will use the term *ecological teleconnections* for these specific types of synchrony.

DOI: 10.1201/9781003410294-9

9.2 Exploratory Factor Analysis and Empirical Orthogonal Functions

Physical systems often exhibit predictable oscillations that are driven by daily and seasonal cycles of solar radiation (see Fig. 8.1)[1]. However, they often also have other modes of variation that repeat with different frequencies and are not associated with solar forcing. In temperate freshwater lakes, for example, an internal layer arises during spring warming that separates warm surface water from denser cold waters below, and gravity favors an outcome where this *thermocline* occurs at a constant depth across the entire lake. However, strong winds from summer storms can then push surface waters downwind, such that the thermocline is deeper downwind than upwind. When the first summer storm passes and winds stop, the thermocline may start rocking back and forth like the surface of the water in a bathtub. The depth of this thermocline relative to the lake surface then has a periodic oscillation during summer months, with frequency determined by the size and shape of the lake [215]. On a larger scale, sea surface temperature in the Atlantic Ocean exhibits cyclic variation that typically occurs with two cycles per year [13], while the El Nino-Southern Oscillation (ENSO) results in cyclic variation in sea surface temperature in the equatorial Pacific Ocean occurring every 2–7 years. In both of these cases, the frequency of physical oscillations arises from the speed at which physical dynamics propagate through ocean waters in relation to the size of physical boundaries. These endogenous physical cycles can then propagate up to cyclic dynamics in biological systems. Importantly, these types of oscillation result in localized impacts on physical measurements (e.g., atmospheric/oceanographic temperature and pressure) in areas that are separated by large geographic distances. For example, the ENSO cycle has characteristic consequences for oceanography and weather both in South America and the Western Pacific island nations.

Cycles in physical systems such as these have been studied using *empirical orthogonal functions* (EOF) for over 50 years [73, 117]. Early studies would implement EOF by:

1. Assembling a complete matrix $\mathbf{Y}$ of measurements $y_{s,t}$ for each site s and time t, e.g., measuring daily sea surface temperature at different oceanographic stations in the Pacific Ocean;

2. Centering these data for each site, such that the mean across times $t \in \{1, 2, ..., T\}$ of centered matrix $y_{s,t}^*$ is zero for each location s;

3. Calculating the $T \times T$ covariance $\mathbf{\Sigma}$ in $y_{s,t}^*$ for each pair of times t_1 and t_2 and assembling a resulting covariance matrix;

4. Applying principal components analysis to covariance $\mathbf{\Sigma}$, i.e., computing the eigendecomposition (see Section B.4.1), extracting one or more eigenvectors $\mathbf{v}_l$ each with length T, and recording these as *modes of variability*;

5. For each mode of variability $\mathbf{v}_l$, calculating the correlation between it and the value $\mathbf{y}_s$ at each location, then recording this map of correlations as the *response map* associated with that mode of variability.

This procedure for calculating an EOF has then resulted in insights about atmospheric and oceanographic teleconnections over the past decades. In particular, modes of variability are often then correlated with other time-series (e.g., salmon returns in rivers across North

---

[1]See https://github.com/james-thorson/Spatio-temporal-models-for-ecologists/Chap_9 for code associated with this chapter.

America, [144]) to generate and support hypotheses about demographic drivers for highly mobile animals.

However, this PCA approach to fitting an EOF model has several drawbacks:

A *Complete data*: it requires having no missing entries in data matrix $\mathbf{Y}$, and it is unclear how to generalize this approach to ecological surveys, e.g., point-count data that follow a random design and are not already gridded;

B *Normally distributed data*: calculating a covariance matrix can be statistically inefficient for data that do not follow a normal distribution, e.g., the "dust-bunny" (i.e., zero-inflated and right-skewed) distribution that is common when sampling species densities in ecological communities [149];

C *Failure to propagate variance*: it is unclear how to estimate the total uncertainty resulting from the sequence of computational steps in the algorithm;

D *Inability to partition variance*: similarly, centering the data prior to analysis makes it difficult to compare the importance of EOF modes with the long-term spatial average of the data.

Due to these limitations, we instead present a spatio-temporal model that has similar properties as this conventional EOF analysis [224, 261], which we will call an *EOF-GLLVM* for reasons explained later. In particular, we specify:

$$
Y_t(s) \sim f(\mu_t(s), \theta)
$$
$$
g(\mu_t(s)) = \beta_0 + \omega_s + \sum_{l=1}^{n_{factor}} \lambda_{l,t} \epsilon_l(s) \tag{9.1}
$$

where f is a measurement distribution with variance θ, g is a link-function, β_0 is an intercept, ω_s is long-term average relative at location s relative to β_0, $\lambda_{l,t}$ is the association between time t and factor l, and $\epsilon_l(s)$ is the response map indicating the response at location s to factor l. In Section 8.4, we would call ω_s the spatial main effect and $\sum_{l=1}^{n_{factor}} \lambda_{l,t} \epsilon_l(s)$ the spatio-temporal interaction. We estimate temporal indices λ_l as fixed effects and response maps ϵ_l as random effects, and the model is completed by specifying a distribution for spatial variables:

$$
\epsilon_l \sim \text{MVN}(\mathbf{0}, \mathbf{\Sigma}_\epsilon)
$$
$$
\omega \sim \text{MVN}(\mathbf{0}, \mathbf{\Sigma}_\omega) \tag{9.2}
$$

where $\mathbf{\Sigma}_\epsilon$ is the spatial covariance for spatial responses to EOF indices, and $\mathbf{\Sigma}_\omega$ is the spatial covariance for average spatial patterns. Each of these covariances could in turn be specified by constructing a sparse precision matrix using a CAR, SAR, or SPDE method (see Chapter 5).

We can interpret Eq. 9.2 as a *generalized linear latent variable model* (GLLVM) [157] as reviewed for ecologists elsewhere [256]. A GLLVM is a type of generalized linear model (GLM) that involves estimating one or more latent variables E and multiplying these by a loadings matrix. The EOF-GLLVM specifically estimates one or more spatially correlated latent variables as well as spatially constant factor loadings $\mathbf{\Lambda}$. Comparing model components with the EOF algorithm above, we can also see that $\beta_0 + \omega_s$ accounts for the process of centering data in Step-2, λ_l is a temporal index similar to that calculated in Step-3, and ϵ_l is similar to the response map calculated in Step-5 for each mode of variability l.

This EOF-GLLVM then addresses the drawbacks to the conventional EOF listed previously, i.e., it (A) can be fitted to incomplete data, (B) uses a link function and specified response for non-normal and zero-inflated data, (C) allows us to apply the generalized delta method to propagate variance for all steps jointly, and (D) has an explicit variance for spatial effects ω vs. indices λ_f and spatial responses ϵ_f. As we will demonstrate later, it also provides inference regarding ecological and physical teleconnections.

CODE 9.1: R code showing how to rotate a loadings matrix $\mathbf{L\_tf}$ and spatial variable $\mathbf{x\_sf}$ using a PCA rotation.

```
 1  rotate_pca <-
 2  function( L_tf,
 3             x_sf,
 4             order = NULL ){
 5
 6    # Eigen-decomposition
 7    Cov_tmp = L_tf %*% t(L_tf)
 8    Cov_tmp = 0.5*Cov_tmp + 0.5*t(Cov_tmp) # Ensure symmetric
 9    Eigen = eigen(Cov_tmp)
10
11    # Desired loadings matrix
12    L_tf_rot = (Eigen$vectors%*%diag(sqrt(Eigen$values)))[,1:ncol(L_tf),drop=
         FALSE]
13
14    # My new factors
15    require(corpcor)
16    H = pseudoinverse(L_tf_rot) %*% L_tf
17    x_sf = t(H %*% t(x_sf))
18
19    # Get all loadings matrices to be increasing or decreasing
20    if( !is.null(order) ){
21      for( f in 1:ncol(L_tf) ){
22        Lm = lm( L_tf_rot[,f] ~ 1 + I(1:nrow(L_tf)) )
23        Sign = sign(Lm$coef[2]) * ifelse(order=="decreasing", -1, 1)
24        L_tf_rot[,f] = L_tf_rot[,f] * Sign
25        x_sf[,f] = x_sf[,f] * Sign
26      }
27    }
28
29    # return
30    out = list( "L_tf"=L_tf_rot, "x_sf"=x_sf, "H"=H)
31    return(out)
32  }
```

Despite these advantages, however, we also note several complexities that result from this model:

- *Index rotations*: the loadings matrix Λ represents the association $\lambda_{l,t}$ of each year t and each EOF index l. In Section 4.4.1, we noted that Λ must have zeros above the diagonal to ensures that parameters are uniquely identified. We also showed that the estimated matrix of factor loadings can be multiplied by a *rotation matrix* (and the estimated response map multiplied by the inverse of this rotation matrix) without otherwise affecting results (see Eq. 4.21). Finally, we noted that ecologists often apply a *varimax rotation*, which ensures that each factor l has a strong association with a minimal number of years and a weak association with other years.

However, the EOF algorithm instead calculates indices using PCA, and therefore defines indices such that the 1st eigenvector explains as much spatio-temporal variance as possible, the 2nd explains as much as possible conditional upon the 1st, and so on. To mimic this behavior, we therefore introduce the *PCA rotation* (Code 9.1) which rotates Λ in Eq. 9.2 to be similar to the PCA algorithm. This PCA rotation involves extracting the estimated covariance $\mathbf{V} = \Lambda\Lambda^T$, calculating its eigendecomposition, then calculating the rotation that transforms Λ to the first n_f eigenvectors of $\mathbf{V}$, and then applying the pseudoinverse of that rotation to response maps E. This rotation then ensures that the first rotated index (and associated response map) explains the largest possible component of spatio-temporal variance, the second explains the most possible conditional upon that 1st, etc., and hence has a similar interpretation as the original EOF algorithm;

- *Label switching*: similar to the potential for rotations to not affect results, it is possible to switch the sign of a given index (and its associated response-map) via multiplication with a reflection matrix. Technically the marginal likelihood resulting from 9.1 is multimodal, where each mode involves a switch in the sign of each index and also the associated response map (termed *label switching*). In practice, this multi-modality is trivial, because we can apply some algorithm to switch the labels, i.e., by ensuring that indices are transformed to have a positive slope or applying some other heuristic rule.

We note that this EOF-GLLVM then reduces to the spatio-temporal index model (Eq. 8.7 in Section 8.4.2) when specifying $\Lambda = \sigma\mathbf{I}$. However, the spatio-temporal index model estimates spatio-temporal variation as a random effect with as many columns as the number of modeled years. By contrast, the EOF-GLLVM replaces this by estimating response maps as random effects with as many columns as the number of estimated factors, while also estimating additional fixed effects in the loadings matrix Λ. We demonstrate the EOF-GLLVM using the SPDE method in TMB using Code 9.2.

CODE 9.2: TMB code to fit a generalization of empirical orthogonal function analysis as a spatial generalized linear mixed model.

```
1  #include <TMB.hpp>
2  template<class Type>
3  Type objective_function<Type>::operator() ()
4  {
5    using namespace density;
6
7    // Data
8    DATA_VECTOR( y_i );
9    DATA_IVECTOR( t_i );
10
11   // SPDE objects
12   DATA_SPARSE_MATRIX(M0);
13   DATA_SPARSE_MATRIX(M1);
14   DATA_SPARSE_MATRIX(M2);
15
16   // Projection matrices
17   DATA_SPARSE_MATRIX(A_is);
18   DATA_SPARSE_MATRIX(A_gs);
19   DATA_VECTOR( a_g );
20
21   // Parameters
22   PARAMETER( beta0 );
23   PARAMETER( ln_tau );
24   PARAMETER( ln_kappa );
25   PARAMETER( ln_sigma );
26   PARAMETER_MATRIX( L_ft );
27
28   // Random effects
29   PARAMETER_VECTOR( omega_s );
30   PARAMETER_MATRIX( epsilon_sf );
31
32   // Probability of random effects
33   Type jnll = 0;
34   Eigen::SparseMatrix<Type> Q = exp(2*ln_tau) * (exp(4*ln_kappa)*M0 + Type
     (2.0)*exp(2*ln_kappa)*M1 + M2);
```

```
35    jnll += GMRF(Q)( omega_s );
36    for( int j=0; j<epsilon_sf.cols(); j++ ){
37      jnll += GMRF(Q)( epsilon_sf.col(j) );
38    }
39
40    // Projection to data
41    matrix<Type> epsilon_it( y_i.size(), L_ft.cols() );
42    epsilon_it = A_is * epsilon_sf * L_ft;
43    vector<Type> omega_i( y_i.size() );
44    omega_i = A_is * omega_s.matrix();
45
46    // Probability of data conditional on random effects
47    for( int i=0; i<y_i.size(); i++){
48      jnll -= dnorm( y_i(i), beta0 + omega_i(i) + epsilon_it(i,t_i(i)), exp(
         ln_sigma), true );
49    }
50
51    // Extrapolation
52    matrix<Type> epsilon_gf( A_gs.rows(), epsilon_sf.cols() );
53    epsilon_gf = A_gs * epsilon_sf;
54    matrix<Type> epsilon_gt( A_gs.rows(), L_ft.cols() );
55    epsilon_gt = epsilon_gf * L_ft;
56    vector<Type> omega_g( A_gs.rows() );
57    omega_g = A_gs * omega_s.matrix();
58
59    // Prediction
60    matrix<Type> yhat_gt( epsilon_gt.rows(), epsilon_gt.cols() );
61    vector<Type> a_t( epsilon_gt.cols() );
62    a_t.setZero();
63    for( int t=0; t<epsilon_gt.cols(); t++ ){
64    for( int g=0; g<epsilon_gt.rows(); g++ ){
65      yhat_gt(g,t) = beta0 + omega_g(g) + epsilon_gt(g,t);
66      a_t(t) += yhat_gt(g,t) * a_g(g);
67    }}
68
69    // Reporting
70    REPORT( omega_g );
71    REPORT( epsilon_gf );
72    REPORT( yhat_gt );
73    REPORT( a_t );
74    return jnll;
75 }
```

We first illustrate the usefulness of this EOF-GLLVM using a physical system that is undergoing rapid dynamics due to directional climate change as well as climate oscillations. We specifically download sea-ice concentrations for the Arctic in September, which are available based on satellite observations [32]. We download data from 1997 to 2017 from PolarWatch[2], and then restrict data to within 3000 km of the north pole. For illustration, we first fit the spatio-temporal index model (i.e., $\Lambda = \sigma\mathbf{I}$) using an identity link and normal distribution, and present results to visualize spatial and temporal patterns in sea ice over the past two decades. Predicted summer sea-ice concentrations (Fig. 9.1) show notable declines in average fall extent from early years (1997–2000) to late years (2014–2017). For example, a northern passage (resulting from ice-free waters north of Russia) seems feasible from 2011 to 2015. However, the spatio-temporal index model does not itself estimate any parameters that can be directly interpreted to draw large-scale inference about dominant patterns; instead, it is helpful primarily to visualize the apparent decrease in sea-ice extent in the waters near Alaska and Russia.

---

[2]These data were provided by NOAA's Center for Satellite Applications & Research (STAR) and the CoastWatch program and distributed by NOAA/NMFS/SWFSC/ERD.

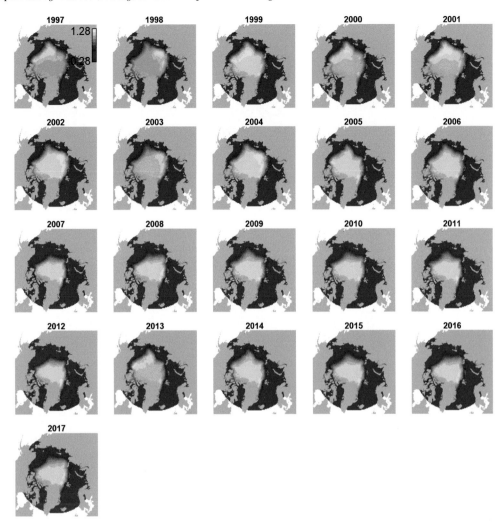

FIGURE 9.1: Predicted concentration of sea ice in September from 1997 to 2017 fitted using a spatio-temporal index model, based on satellite measurements and restricted to locations within 3000 km of the north pole.

We then contrast these predictions with those arising from an EOF-GLLVM while estimating two modes of variability. We specifically compare predictions of sea-ice area, calculated as the area-weighted sum of sea-ice concentrations from both the "index standardization" and the EOF-GLLVM models. This shows that both models estimate nearly identical trends and interannual variability, including estimates of low sea-ice extent in 2007 and 2012 (Fig. 9.2). We therefore conclude that the EOF-GLLVM is a suitable approximation for the full interannual dynamics of Arctic sea ice concentrations over this period, despite estimating considerably fewer random effects than the spatio-temporal index model.

Given that the EOF model is a suitable approximation to the full spatio-temporal dynamics, we next proceed to visualize and interpret individual model components (Fig. 9.3). As expected, the average spatial component ω captures the multidecadal average concentration of sea ice, including consistently high concentrations from the north pole to Greenland

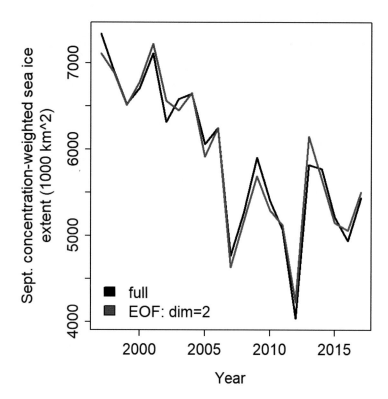

FIGURE 9.2: Predicted spatial extent of Sept. Arctic sea ice from the spatio-temporal index standardization model (Eq. 8.7) or the EOF model with two modes of variability (Eq. 9.1).

and west to the Canadian Arctic archipelago. By contrast, the first mode of variability is associated with elevated sea-ice concentrations north of the Pacific gateway to the Arctic (i.e., poleward of Alaska and eastern Russia), while the second mode is associated with elevated sea-ice poleward from the Kara and Laptev Seas north of central Russia. The first mode (purple line in bottom-right panel of Fig. 9.3) showed a precipitous decline 2006–2007 that persists over the following years (with a notable exception of 2013–2014), while the second mode (yellow line in bottom-right panel) showed more consistency over time with the exception of a decline and subsequent recovery from 2010 to 2016. From this dimension-reduction exercise, we therefore conclude that:

- Two dominant modes of physical variability can explain synchronous changes in summer sea-ice concentration through the Arctic;

- The primary mode of variability represents progressive warming in the Pacific gateway to the Arctic; and

- The secondary mode represents fluctuations and a slower decline in the Kara and Laptev seas that might enable a northern passage across Russian waters.

These dynamics are much more clearly visualized using EOF (Fig. 9.3) than inspecting maps of sea-ice concentration arising from the index-standardization model (Fig. 9.1), despite these two models resulting in essentially identical predictions of underlying dynamics.

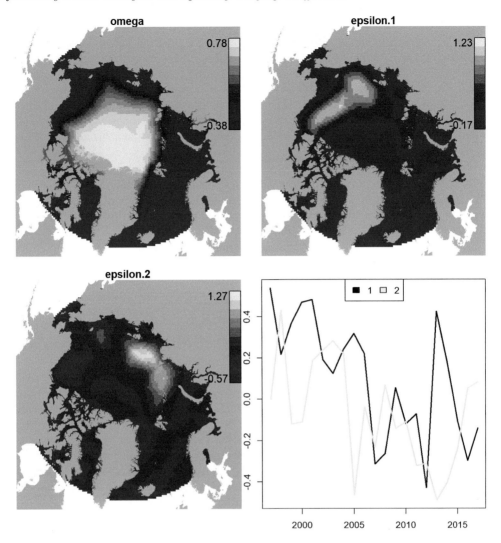

FIGURE 9.3: Empirical orthogonal function model components that explain Arctic Sea ice concentrations (see Section 9.1), including the average spatial component (top-left panel), the spatial response for both models (top-right and bottom-left), and the temporal index associated with each mode (bottom-right).

9.3 Confirmatory Factor Analysis and Spatially Varying Coefficients

In the preceding section, we summarized spatio-temporal dynamics using an EOF model. This then represented physical teleconnections (synchronous dynamics at geographically distant locations) using "modes of variability" that each include a time-series and response map. Notably, we imposed no structure on the loadings matrix Λ representing the association of each year with each mode. This unstructured loadings matrix also arose in Section 4.4.1, where it was called a "factor analysis" model. Using an unstructured loadings matrix for factor analysis is also sometimes called *exploratory factor analysis.*

As alternative to exploratory factor analysis, we next introduce *confirmatory factor analysis.* As their names imply, exploratory and confirmatory factor models can be used iteratively to identify and then attribute observed variation to hypothesized ecological processes [75, 82], including physical and ecological teleconnections. To do so, we here replace the unstructured loadings matrix (Eq. 4.16) with one or more columns containing known values or constraints. For example, we might specify that loadings matrix Λ has a single column containing a time-series that is hypothesized to drive synchronous variation in a target variable.

To demonstrate, we expand the case study for Alaska pollock that was previously used to introduce interannual variation in spatio-temporal models (Section 8.4). We specifically expand Eq. 8.7 by adding a spatially varying response to a time-series measuring the spatial extent of cold waters near the seafloor (termed "cold pool extent" and introduced in Section 8.5):

$$
\begin{aligned}
\log(d_{s,t}) &= \beta_t + \omega_s + \epsilon_{s,t} + x_t \xi_s \\
\omega &\sim \text{MVN}(\mathbf{0}, \boldsymbol{\Sigma}_\omega) \\
\epsilon_t &\sim \text{MVN}(\mathbf{0}, \boldsymbol{\Sigma}_\epsilon) \\
\xi &\sim \text{MVN}(\mathbf{0}, \boldsymbol{\Sigma}_\xi)
\end{aligned}
\tag{9.3}
$$

where x_t is now replacing a column of the exploratory factor-analysis model with a time series measuring an hypothesized mechanistic driver of spatio-temporal dynamics.

In contrast to exploratory factor analysis, we can use confirmatory factor analysis to attribute large-scale patterns to a known covariate that is assumed to be exogenous (see Section 7.1 for a discussion of exogeneity and attribution). In this case, we visualize the expected changes in population density that are attributed to geophysical processes that are associated with cold-pool extent. Examining the cold-pool response map for pollock (Fig. 9.4) shows that a larger-than-average cold-pool extent is associated with lower pollock densities in the waters south of St. Lawrence Island (i.e., essentially in the northwestern portion of the model domain), and elevated densities along the southwestern border of the survey area. Given the lower frequency of sampling in the northern Bering Sea, it is unsurprising that the response map in these northern areas is estimated with greater uncertainty (right panel of Fig. 9.4). In this specific application, decreased pollock densities south of St. Lawrence Island occurring in summers with a large cold pool is typically explained by noting that adult pollock migrate to avoid near freezing waters, and that the cold-pool arises mainly in those impacted areas.

An analyst may also wonder about the potential that a spatial response like this could arise by chance. To assess the "statistical significance" of this confirmatory factor analysis, we could use a variety of statistical comparisons and diagnostics including using a likelihood-ratio test or the Akaike Information Criterion [1] to compare models with or without the spatially varying response. However, for illustration, we here refit the same model while randomizing the order of the cold-pool index (i.e., sampling without replacement from x_t). This randomized index is expected to have no remaining information about the dynamics of pollock, so we expect that a well-performing model will tend to estimate that it has a negligible impact. As expected, this experiment results in an estimated standard deviation for the spatially varying response that approaches zero (Fig. 9.5). This therefore confirms that including an uninformative index in a confirmatory factor-analysis model will tend to be eliminated from the model due to the shrinkage that occurs when estimating the variance of random effects.

Beyond introducing the distinction between exploratory and confirmatory factor analysis, this example illustrates that a single regional index (cold-pool extent) can result in

SVC response

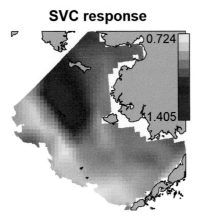

SE[SVC response]

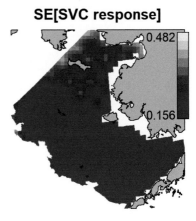

FIGURE 9.4: Illustrating confirmatory factor analysis by replacing a mode of variability in an exploratory factor analysis model with a time series covariate. In this case, we show the biomass-response associated with cold-pool extent for Alaska pollock.

SVC response

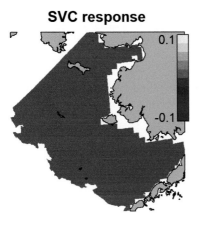

SE[SVC response]

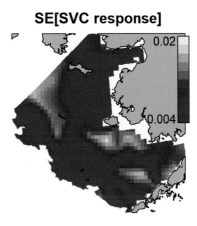

FIGURE 9.5: Illustrating confirmatory factor analysis when replacing the cold-pool index with a randomized time-series, where the response (left-hand-side) uses a colorbar from –0.1 to 0.1 rather than the miniscule range of the estimated response.

synchronous variation at multiple locations that are separated by large geographical distances. These types of ecological teleconnections are likely ubiquitous in ecological systems, but are not well represented by predicting local densities solely based on local environmental conditions.

9.4 Chapter Summary

In summary, we have showed that:

1. Teleconnections (i.e., synchrony in dynamics for geographically distant locations) occur in many physical systems. Teleconnections occurring in physical drivers as

well as ecological and evolutionary dynamics can given rise to ecological teleconnections, and these are dramatic and important in many ecosystems;

2. Teleconnections can be identified by fitting an empirical orthogonal function (EOF) model. EOF is conventionally estimated by fitting a principal components analysis to system measurements, but this definition has many drawbacks when applied to zero-inflated and right-skewed data from sampling designs that arise in ecology;

3. This conventional PCA analysis for EOF can be extended by fitting a generalized linear latent variable model (EOF-GLLVM). This involves estimating one or more time-series indices, where each index is associated with an estimated map of responses. Applying a "PCA rotation" to indices from an EOF-GLLVM then results in estimates that are directly comparable with PCA analysis;

4. In some cases, the EOF model can generate estimates of system dynamics that are similar to an unstructured spatio-temporal model. In these cases, the EOF model provides a "rank-reduced" and easy-to-interpret summary of system dynamics, while also estimating model components (e.g., average spatial patterns) that are typically excluded by the PCA analysis;

5. EOF can be interpreted as one type of *exploratory factor analysis*, and can be applied iteratively with *confirmatory factor analysis* to attribute modes of variability to hypothesized drivers. Specifically, confirmatory factor analysis can be applied by replacing an estimated EOF index with a known time-series that is then fitted with a spatially varying response. Iterating between exploratory and confirmatory factor models allows us to visualize dominant spatio-temporal patterns without any constraints (exploratory models), and then to see what portion of these patterns can be explained by hypothesized regional conditions (confirmatory models).

9.5   Exercise

The EOF-GLLVM (Eq. 9.2) and the spatio-temporal index model (Eq. 8.7) both estimate spatio-temporal variation, and differ in how the spatio-temporal term E is specified. Please fit the spatio-temporal index model to spatially balanced data (i.e., Arctic sea ice concentrations) and then apply the PCA algorithm for fitting EOF (Section 9.2) to the estimated densities. How does this two-stage approach differ from results obtained when fitting the EOF-GLLVM directly and doing both stages jointly? Now, please repeat the exercise while dropping data from a large portion of the spatial domain in every other year (i.e., using a spatially blocked cross-validation design [190]). Are the differences in estimates between these two approaches greater when analyzing spatially unbalanced data?

10

Population Movement and Habitat Selection

10.1 From Individual to Population Movement

In Chapter 4, we introduced the general importance of movement within ecological theory and applications. We then demonstrated how to fit a model for individual movement from a Lagrangian viewpoint, which tracks changes in the location $S_i(t)$ for one or more individuals i at each time t. We specifically solved functions describing dynamics using an Euler approximation that discretized continuous time into time-intervals with spacing Δ_t while tracking individual location $S_{i,t}$ in continuous space. We then decomposed movement over each interval into *taxis* (i.e., movement toward preferred habitats), *drift* (i.e., directional movement), and *diffusion* (i.e., otherwise unexplained movement) (see definitions in Section 4.1). We showed that we could model individual movement in this continuous-space and discrete-time framework by treating location as a random effect in a multivariate state-space model, and that we could model movement for multiple animals by expanding the set of state-space variables. We then showed that this had many practical advantages, e.g.:

1. The model can be used both to simulate dynamics and data resulting from individual movement, or efficiently fitted to those same data to estimate unknown movement parameters;

2. We could specify a preference function $h(s)$ and apply the *gradient operator* to this function $\nabla h(s)$, which results in a vector that points in the direction of increasing habitat preference. Individuals are then expected to move in the direction of this vector, i.e., toward preferred habitats (Eq. 4.7);

3. A fitted model can be used to predict animal location at times between sequential measurements. Predictive variance was then lowest when location was measured and was higher during times between measurements;

4. The model could be fitted simultaneously to data from multiple individuals, and movement that is correlated among individuals can be approximated by assembling the covariance among individuals using techniques derived from factor-analysis or structural equation models.

However, fitting a movement process as a state-space model from a Langrangian viewpoint becomes computationally expensive (or prohibitive) as the number of individuals increases.

DOI: 10.1201/9781003410294-10

To overcome computational limits when dealing with dynamics involving movement of many individuals, we now change from the Lagrangian (individual-based) to Eulerian (grid-based) viewpoint. We specifically track animal densities $d(s,t)$ across continuous space s and continuous time t, while still describing movement resulting from the same combination of taxis, drift, and diffusion. This can be written compactly as a partial differential equation (see Table B.1 for a summary of notation):

$$\frac{\partial}{\partial t}d(s,t) = \underbrace{D\nabla^2 d(s,t)}_{\text{diffusion}} - \underbrace{\mathbf{v}(s) \cdot \nabla d(s,t)}_{\text{drift}} - \underbrace{\nabla h(s) \cdot \nabla d(s,t)}_{\text{taxis}} \qquad (10.1)$$

where $\mathbf{v}(s)$ is a vector representing the direction of drift, $\nabla d(s,t)$ is the gradient of the density function, $\nabla h(s)$ is the gradient of the preference function (defined identically to Eq. 4.7), $\nabla^2 d(s,t)$ is the *Laplace operator* for the density function which is computed as the second derivative of density function $d(t)$ evaluated at s, and $D = \frac{\sigma^2}{2}$ is the diffusion coefficient where σ^2 is the variance in displacement that occurs per unit time from diffusion alone. We doubt that many ecologists will be familiar with this partial differential equation (PDE) notation, so we briefly summarize the components here:

- *Drift*: $\mathbf{v}(s) \cdot \nabla d(s,t)$ is the dot product of the vector $\mathbf{v}(s)$ that points in the direction of passive drift and the vector $\nabla d(s,t)$ that points toward higher densities. This dot product is positive if densities are increasing in the same direction as drift is pushing animals, and this will result in a local decrease in densities over time as the lower density of animals upstream is then advected to replace the animals currently at location s. Similarly, if the drift vector is pointing in one direction (e.g., north), and the vector of increasing density is orthogonal (e.g., east), then the dot product is zero. Passive drift will have no effect on local densities in this case because each packet of individuals that is advected northward is replaced by a packet of individuals with similar density;

- *Taxis*: $\nabla h(s) \cdot \nabla d(s,t)$ is the dot product of the gradient of the preference function and the gradient of the density function. This will be positive if animal densities are increasing in the same direction as the increase in preference, which will again cause local densities to decrease;

- *Diffusion*: $\nabla^2 d(s,t)$ is the Laplacian of the density function, calculated by taking the matrix of second derivatives of the density function at s and calculating the sum of diagonal entries (the trace of the Hessian matrix). This measures the curvature of the density function at location s; if $d(t)$ curves downward (i.e., $\nabla^2 d(s,t)$ is negative) then emigration exceeds immigration and diffusion causes local densities to decrease, while if it is positive then diffusion causes local densities to increase. To see this, imagine the case involving a single spatial dimension. If density function $d(t)$ has a local maximum at location s, then the second derivative is by definition negative. This will then result in a decrease in local densities as animals emigrate from that local maximum faster than immigration from nearby locations with lower densities. Similarly, if the second derivative is positive in this one-dimensional example, then densities will tend to increase from diffusion as immigration exceeds emigration to nearby locations.

Both taxis and drift represent different types of advective movement, so we can call Eq. 10.1 an advection-diffusion model. We previously referred to a random-walk process for location as diffusive movement, but note that the term diffusion refers specifically to a PDE (e.g., Eq. 10.1).

10.2   General Solution for Movement Processes

This PDE representing movement (Eq. 10.1) can be solved analytically for some specific density and preference functions[1]. In fact, we used these solutions already in Section 4.2 to calculate the expected density function in time $t + \Delta_t$ given a known location in time t, which we used to define the multivariate normal distribution used to simulate movement (e.g., Eq. 4.8).

For example, assuming that we know the location of an animal $\mathbf{s}_0$ at some initial time t_0 and that drift and taxis are absent, we can use Eq. 4.4 to compute a probability density for the animal being at any location $\mathbf{s}$ after interval Δ_t elapses:

$$d(\mathbf{s}, t_0 + \Delta_t) = \mathrm{dMVN}(\mathbf{s}_0, \Delta_t \sigma^2 \mathbf{I}) \qquad (10.2)$$

We can then use this *fundamental solution* to check how Eq. 10.1 works in the known case (Code 10.1) of diffusive movement. To do so, we select a starting location, diffusion covariance Sigma , and randomized location s0 and time t0 to evaluate. We then create a function Density that computes the known density function at that time and place (Eq. 10.2). Finally, we use package numDeriv [65] to compute finite-difference approximations to this density function, which we use to compare the time-gradient for this function (i.e., the left-hand-side of Eq. 10.1) and the Laplacian operator (i.e., the right-hand-side of Eq. 10.1). This confirms that Eq. 10.1 holds numerically in this known case.

CODE 10.1: R code for analytical solution to partial differential equation for diffusion.

```
1  # Define parameters and randomize locations
2  n <- 2   # dimensions (2 = spatial coordinates)
3  var <- runif(1) # diffusion rate for each dimension
4  Sigma <- var * diag(n)   # full diffusion matrix
5  s0 <- rnorm(n) # randomized location to evaluate
6  t0 <- runif(1) # randomized time to evaluate
7
8  ## Fundamental solution
9  Density <- function(s,t) mvtnorm::dmvnorm(s, mean=rep(0,n), sigma=Sigma*t)
10
11 ## diffusion -- time derivative
12 numDeriv::grad( function(t) Density(s0,t), t0 )
13 # output:   0.3069884
14
15 # diffusion -- trace of Hessian
16 # NOTE: only works for diagonal covariance
17 H = numDeriv::hessian( function(s) Density(s,t0), s0 )
18 var/2 * sum(diag(H)) # trace of matrix
19 # output:   0.3069884
```

However, we instead seek to solve the movement PDE generically for a wide range of ecological contexts. To do so, we discretize the model by switching to an Eulerian viewpoint (see Section 1.1 for discussion). This proceeds via the following steps:

1. *Discretize space*: partitioning a spatial domain $\mathcal{D}$ into a set of n_j non-overlapping spatial cells, such that we can approximate densities as being constant within each cell;

2. *Define adjacency*: defining an adjacency matrix $\mathbf{A}$ with dimension $n_j \times n_j$, having a value of one for any two cells that share an edge and zero otherwise;

---

[1]See https://github.com/james-thorson/Spatio-temporal-models-for-ecologists/Chap_10 for code associated with this chapter.

3. *Define abundance matrix*: replacing the population density function $d(s,t)$ from Eq. 10.1 with an abundance matrix $\mathbf{N}$ recording the number of individuals $n_{j,t}$ in each grid cell s in time t. For a single organism, we can define this abundance vector as an indicator where $n_{j,t} = 1$ for the cell j where it resides at time t and zero elsewhere;

4. *Define movement matrix*: representing movement probabilities over a time-interval from t to $t + \Delta_t$ via a movement matrix $\mathbf{M}_t$ with dimension $n_j \times n_j$;

5. *Project abundance via movement matrix*: calculating expected abundance $\hat{\mathbf{n}}_{t+\Delta_t}$ after interval Δ_t has elapsed as:

$$\hat{\mathbf{n}}^T_{t+\Delta_t} = \mathbf{n}^T_t \mathbf{M}_t \tag{10.3}$$

These steps calculate the expected abundance following the movement processes represented by $\mathbf{A}$, which we will construct to represent drift, taxis, and diffusion. However, these steps do not include the variance resulting from individual stochasticity (i.e., the directed random walk followed by each individual in Eq. 4.8). To see this, imagine that we use an indicator vector for density $\mathbf{n}_t$, i.e., a vector of zeros except for a single 1, representing the spatial cell where a single organism resides in time t. This indicator matrix is multiplied by $\mathbf{M}$, which results in a density $0 \leq d_j \leq 1$ that is spread among multiple cells in time $t + \Delta_t$. However, a single individual does not itself diffuse. Instead, it will move to a single new location, with expectation predicted by Eq. 10.3. To represent stochasticitiy, we can add another step to the algorithm:

6 *Simulate stochasticity*: simulating random variation in movement (i.e., stochasticity) by drawing a vector of realized abundance from a multinomial distribution with expected counts $\hat{\mathbf{n}}_{t+\Delta_t}$:

$$\mathbf{n}_{t+\Delta_t} \sim \text{Multinomial}(\hat{\mathbf{n}}_{t+\Delta_t}) \tag{10.4}$$

where the multinomial movement probabilities are calculated by dividing $\hat{\mathbf{n}}^T_{t+\Delta_t}$ by its sum. When sample sizes are large, this multinomial distribution will converge on its expectation $\mathbf{n}_{t+\Delta_t} =\approx \hat{\mathbf{n}}_{t+\Delta_t}$ such that stochasticity becomes unimportant for large groups of independent individuals.

For simplicity of presentation, we will describe individual movement from an Eulerian viewpoint in two dimensions using square grid cells with equal size and sides of length Δ_s. However, the same concepts can then be generalized with small changes to notation and code to more dimensions, other cell shapes (hexagons, triangles, etc.), or unequally sized cells.

The focus of interference then becomes: how do we specify a movement matrix $\mathbf{M}$ that provides a suitable discretization to Eq. 10.1 using an Eulerian viewpoint (i.e., Step-4 above)? Ideally, this matrix would have several properties:

A *Conservation of numbers*: we seek a movement-probability matrix $\mathbf{M}$ that represents only the effect of movement while conserving population size, such that a model can be combined with other components that represent size-transitions, survival, recruitment, and other demographics;

B *Habitat utilization as stationary distribution*: we seek a matrix such that, if it is repeatedly applied to a vector of initial abundance $\mathbf{n}_0$, then abundance will converge on a stationary distribution that represents their expected long-term or average habitat utilization;

C *Identical parameters using either Eulerian or Lagrangian viewpoint*: additionally, we seek to develop $\mathbf{M}$ in such a way that the same parameters can be used in a model using a Lagrangian viewpoint (e.g. Chap 4) to generate (nearly) identical dynamics;

D *Complicated dynamics using few parameters*: similarly, we seek a method that can include covariates, allows a wide range of model complexity (i.e., few or many estimated parameters), and can mimic complex behaviors with a small number of parameters;

E *Computational efficiency*: we seek a method for constructing $\mathbf{M}$ that remains computationally efficient even with a fine spatial resolution, and can be run quickly enough to allow Bayesian or maximum likelihood inference about parameters;

F *Scale-free discretization*: finally, we seek a model where the spatial scale and grid shape used during discretization has little effect on the estimated value of model parameters for a given data set;

G *Subdivision in time*: as a corollary of the preceding property, we seek a method where we can calculate the movement fractions arising for any subdivision of time, i.e., $n_{t+x\Delta_t}$ where $0 < x < 1$ is some fraction of the modeled time-step so that, e.g., we can project the likely path of movement between two modeled times t and $t + \Delta_t$ after the model is fitted.

In this chapter, we first introduce the theory of *continuous-time Markov Chains*, which can satisfy all of these desired criteria. We then discuss how to parameterize these to create a movement-probability matrix.

10.3 Assembling a Movement Probability Matrix

We specifically proceed by calculating movement matrix $\mathbf{M}$ by first assembling a matrix of instantaneous movement rates $\dot{\mathbf{M}}$. This movement-rate matrix represents the instantaneous rate at which individuals move from one cell g_1 to another g_2 per unit time. This movement-rate matrix therefore defines the instantaneous movement rate:

$$\frac{\partial}{\partial t}\mathbf{n}_t^T = \mathbf{n}_t^T \dot{\mathbf{M}} \tag{10.5}$$

We then seek to integrate the action of this movement-rate matrix over a time-interval Δ_t:

$$\mathbf{n}_{t+\Delta_t}^T = \int_t^{t+\Delta_t} \mathbf{n}_t^T \dot{\mathbf{M}} \delta t \tag{10.6}$$

This integral can be computed by an operation called the *matrix exponential*:

$$\mathbf{M} = e^{\Delta_t \dot{\mathbf{M}}}$$
$$\mathbf{n}_{t+\Delta_t}^T = \mathbf{n}_t^T \mathbf{M} \tag{10.7}$$

where this equation defines a *continuous-time Markov Chain* (CTMC)[79], representing first-order Markov transition rates among each grid cell. The computational methods for software implementing the matrix exponential are complicated. However, we believe that Eq. 10.7 will be conceptually familiar to ecologists, who are familiar with exponential population growth. In that well-known case, integrating population growth rate r over time Δ_t is solved by multiplying initial population size by $e^{\Delta_t r}$. The matrix exponential is also available in both R and TMB so we think that a conceptual understanding is sufficient for most users (see Appendix B.6 for more details regarding implementation). We note that there are many other avenues to integrate Eq. 10.5 or the original movement PDE (Eq. 10.1). In the following, we emphasize the matrix exponential precisely because we believe that many ecologists are already comfortable with text and code that applies the (scalar) exponential function to integrate time-series dynamics, such that the matrix exponential is a small expansion of this existing "minimal toolkit" for statistical ecology.

Ecological intuition immediately provides several rules for constructing movement-rate $\dot{\mathbf{M}}$:

- *No teleportation*: individuals can only move to an adjacent cell, i.e., where adjacency matrix $\mathbf{A}$ is nonzero. The movement-rate matrix $\dot{\mathbf{M}}$ then has the same sparsity pattern as $\mathbf{A}$;

- *Conservation of numbers*: individuals do not appear or disappear solely from the action of movement. Therefore, the rows of $\dot{\mathbf{M}}$ must sum to zero;

- *No negative abundance of animals*: the number of individuals in a given cell must be non-negative. To ensure that abundance after movement $\mathbf{n}_t^T \dot{\mathbf{M}}$ is non-negative for any possible initial abundance $\mathbf{n}_t^T$ that is also non-negative, it is sufficient to construct $\dot{\mathbf{M}}$ as a *Metzler matrix*, defined as a matrix with non-negative values in the off-diagonal.

Given conditions 2 (conservation of numbers) and 3 (no negative abundance), we see that the diagonal elements of $\dot{\mathbf{M}}$ must also be non-positive, with value equal to the negative-sum of values in that row.

We also note that the matrix exponential can also be approximated by dividing the time-interval Δ_t into N sub-intervals, and then approximating movement as a linear rate over each sub-interval. This *Euler method* (see Code 10.2) results in the following approximation to the matrix exponential:

$$\mathbf{M} \approx (\mathbf{I} + \frac{\Delta_t}{N}\dot{\mathbf{M}})^N \tag{10.8}$$

The resulting Euler approximation to $\mathbf{M}$ has sparsity structure of $\mathbf{A}^N$. Adjacency matrix $\mathbf{A}$ in turn is nonzero only for neighbors, so the Euler approximation is nonzero for any two cells only if you can get from one to the other by crossing at most N edges. This implies that the Euler approximation is essentially truncating movement beyond N cells per interval.

CODE 10.2: R code for Euler approximation to matrix exponential.

```
1  matexp <-
2  function( Mrate,
3            log2steps = 0, # Number of Euler-approximation steps
4            zap_small = FALSE ){
5
6    require(Matrix)
7    require(expm)
8    if( (log2steps <=0 ) || (log2steps > 100) ){
9      # Full version ... note that expm::expm is faster that Matrix::expm
10     out = expm(Mrate)
11     return( Matrix(out) )
12   }else{
13     # Euler approximation
14     Mrate = Diagonal(nrow(Mrate)) + Mrate / (2^log2steps)
15     for(stepI in seq(1,log2steps,length=log2steps)){
16       Mrate = Mrate %*% Mrate
17     }
```

```
18      if(zap_small) Mrate = zapsmall(Mrate)
19      return( Mrate )
20    }
21 }
```

Code 10.2 demonstrates a conceptually simple approach to solve this CMTC by exponentiating a sparse movement-rate matrix $\dot{\mathbf{M}}$ to calculate movement fractions $\mathbf{M}$, and we use the package expm [142] for efficient computation of the matrix exponential involving a sparse matrix. We therefore turn to discussing how to form the movement-rate matrix $\dot{\mathbf{M}}$. In particular, we decompose the movement-rate matrix into three additive components: diffusion $\dot{\mathbf{D}}$, taxis $\dot{\mathbf{Z}}$, and passive advection $\dot{\mathbf{V}}$ (recalling definitions in Section 4.1):

$$\dot{\mathbf{M}} = \dot{\mathbf{D}} + \dot{\mathbf{Z}} + \dot{\mathbf{V}} \tag{10.9}$$

where these components are all matrices with the same sparsity pattern as $\mathbf{A}$, such that $\dot{\mathbf{M}}$ also has this sparsity pattern. Each component then represents different mechanistic assumptions about dominant movement processes.

Similar to our movement models from a Lagrangian viewpoint, diffusion again represents a null assumption that movement is not otherwise explained by any directional process, whether passive (drift) or active (habitat selection). We present the simplified case of isotropic diffusion, where diffusion rates are identical in every direction.

$$\dot{d_{g_1,g_2}} = \begin{cases} D\frac{1}{\Delta_s^2}a_{g_1,g_2} & \text{if } g_1 \neq g_2 \\ -\sum_{g\neq g_2}\dot{d_{g_1,g}} & \text{if } g_1 = g_2 \end{cases} \tag{10.10}$$

where D is the same diffusion coefficient from Eq. 4.4 with units $distance^2/time$ in Eq. 4.2, Δ_s^2 is the area of square grid cells with sides of length Δ_s in the same distance units as D, and $\dot{a_{g_1,g_2}}$ is the corresponding element of the adjacency matrix $\mathbf{A}$ such that diffusion rate $\dot{d_{g_1,g_2}}$ is nonzero only for adjacent cells. The diffusion coefficient D will take the same value regardless of the spatial and temporal scale of discretization, but $D\frac{1}{\Delta_s^2}$ then calculates the diffusion-rate that corrects for the spatial scale of discretization being used. Diffusion rate $\dot{d_{g_1,g_2}}$ has units $1/time$, and $\dot{\mathbf{M}}$ is then multiplied by Δ_t in Eq. 10.7 which results in the scale-free value for diffusion.

Similarly, we again define taxis as advective movement toward preferred habitat, and define a habitat preference function h. We therefore calculate preference h_g for each grid cell g, and define instantaneous taxis as the difference in preference between two adjacent cells [97]:

$$\dot{z_{g_1,g_2}} = \begin{cases} (h_{g_2} - h_{g_1})\frac{1}{\Delta_s}a_{g_1,g_2} & \text{if } g_1 \neq g_2 \\ -\sum_{g\neq g_2}\dot{z_{g_1,g}} & \text{if } g_1 = g_2 \end{cases} \tag{10.11}$$

where differences in preference $h_{g_2} - h_{g_1}$ has units $distance/time$, but we include $(h_{g_2} - h_{g_1})\frac{1}{\Delta_s}$ such that taxis rates again correct for the spatial scale of discretization that is specified (and the time-scale again is corrected when multiplying $\dot{\mathbf{M}}$ by Δ_t).

In the following, we will specifically calculate habitat preference from covariates [97, 234]:

$$h_g = \gamma\mathbf{x}_g \tag{10.12}$$

where $\mathbf{x}_g$ is a vector of covariates for grid cell g and γ is covariate-response coefficients. Taxis depends only on the difference in preference $h_{g_2} - h_{g_1}$ between two grid cells, such that covariates $\mathbf{x}_g$ are specified to not include an intercept. Covariates $\mathbf{x}_g$ can themselves be constructed from a basis expansion (Section 5.2) to represent a dome-shaped or saturating relationship between preference and local habitat.

Finally, we define passive advection $\dot{\mathbf{V}}$ by pre-processing a known vector-field that is responsible for passive advection. For example, plankton near the ocean surface that is mixed by wind (the *mixed layer*) will drift at approximately the same rate as dominant ocean currents within that mixed layer, and these currents can often be reconstructed using a numerical ocean model. This numerical ocean model could therefore be used to simulate the fraction of particles in g_2 moving to g_1 in a sufficiently short time-interval, and this fraction be used to specify drift rate $\dot{\mathbf{V}}$ for plankton from passive drift. We note that drift can be used to specify a variety of interesting movement patterns that cannot be represented using taxis as we've defined it here (e.g., movement in cycles), but do not discuss drift further here.

To illustrate these concepts, we simulate diffusion-taxis movement for a hypothetical species that is introduced into the east coast of Madagascar (Code 10.3). We start by overlaying a square grid over the entire spatial domain (with cell size of 0.8 Lat $\times$ 0.8 Lon) using `st_make_grid` in package `sf` [175], extract the adjacency matrix representing rook adjacency using `st_relate`, assemble the sparse diffusion-rate matrix, and compute the Euler approximation or matrix exponential. This shows e.g., that the Euler approximation results in a movement matrix that is not non-negative when using only four sub-intervals given this specified diffusion rate (Fig. 10.1 top-right panel), and therefore this 4-step Euler approximation is not suitable. However, using eight sub-intervals is non-negative and appears to be a sparse approximation to the dense matrix exponential.

CODE 10.3: R code to illustrate Euler approximation and matrix exponential applied to a simple diffusion-rate matrix.

```
1  # make grid and exclude small boundary cells
2  cellsize = 0.8
3  sf_fullgrid = st_make_grid( sf_area, cellsize=cellsize )
4  sf_grid = st_make_valid(st_intersection( sf_fullgrid, sf_area ))
5  sf_grid = sf_grid[ st_area(sf_grid)>(0.2*max(st_area(sf_grid))) ]
6
7  # Download elevation and truncate negative values
8  grid = elevatr::get_elev_point( locations=st_point_on_surface(sf_grid), src =
       "aws" )
9  grid$elevation = ifelse( grid$elevation<1, 1, grid$elevation)
10
11 # Get adjacency
12 st_rook = function(m, ...) st_relate(m, m, pattern="F***1****", ... )
13 grid_A = st_rook( sf_grid, sparse=TRUE )
14 A = as(grid_A,"sparseMatrix")
15
16 # Diffusion rate
17 diffusion_coefficient = 1 / cellsize^2
18 D = diffusion_coefficient * A
19 diag(D) = -1 * colSums(D)
20
21 # Movement matrix at for different steps sizes
22 M = NULL
23 for( p in 1:3 ){
24   M[[p]] = matexp( D, log2steps=c(2,3,Inf)[p] )
25 }
```

We next extend this calculation by including movement toward preferred habitat (i.e., taxis), where we specify that habitat preference follows a lognormal distribution with optimal habitat at 100 meters above sea level. We specifically simulate movement for 1000 individuals that are introduced on the east coast of Madagascar, using this 0.8 Latitude

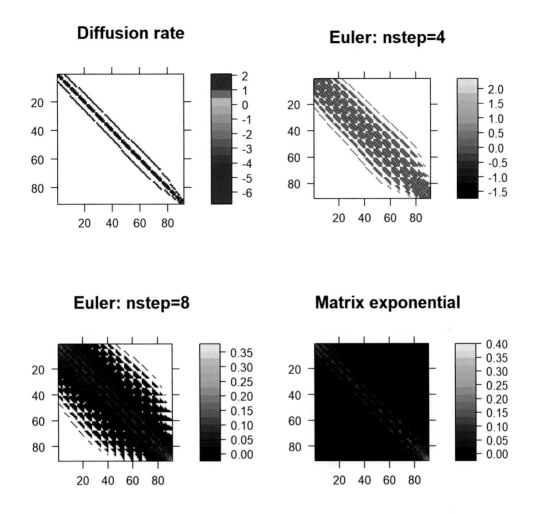

FIGURE 10.1: Euler approximation and matrix exponential for diffusion, showing the diffusion-rate matrix (top-left), using an Euler approximation with 4 or 8 sub-intervals, or the full matrix-exponential solution, generated using Code 10.3.

$\times$ 0.8 Longitude cell size initially. This shows that individuals are initially confined to the east coast due to the presence of mountains blocking westward movement. However, by time-interval 4 and 5, they have dispersed far enough south and north to then have viable paths to reach preferred habitat on the West Coast (Fig. 10.2).

For comparison, we also illustrate the same results using 16 times more cells (i.e., a 0.2 $\times$0.2 cell size). This high-resolution discretization results in a similar estimate of long-term habitat utilization. It also results in animals colonizing preferred habitats on the west coast by the 4th-5th time-interval (Fig. 10.3). We therefore conclude that the parameters have approximately similar interpretation regardless of the spatial scale used for discretization.

In summary, this process for constructing movement matrix **M** achieves all of our desired properties (Section 10.2):

A *Conservation of numbers*: as long as the rows of rate matrix $\dot{\mathbf{M}}$ sum to zero, then the rows of **M** will sum to one. This in turn implies that $\sum_{j=1}^{n_j} n_{j,t} = \sum_{j=1}^{n_j} n_{j,t+\Delta_t}$, i.e., that movement does not change abundance;

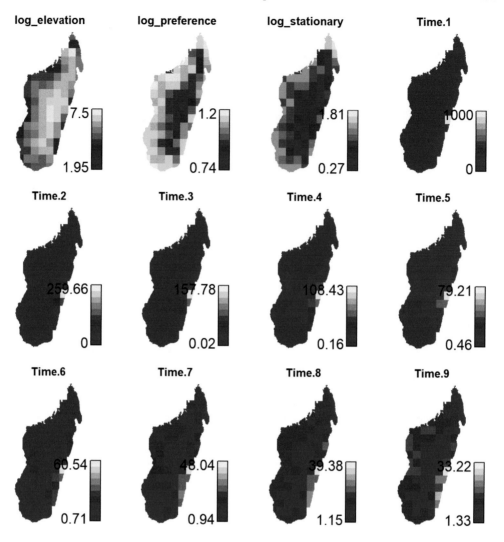

FIGURE 10.2: Projected movement for 1000 individuals introduced on the east coast of Madagascar, showing elevation, habitat preference, and the stationary habitat utilization as well as abundance for 9 time-intervals starting at introduction, using a coarse-scale discretization using 0.8 latitude by 0.8 longitude cells.

B *Habitat utilization as stationary distribution*: given that movement probabilities **M** conserve abundance, an eigendecomposition of **M** will have one or more eigenvalues equal to 1.0, and other eigenvalues will be smaller. The stationary distribution will then be a linear combination of the eigenvectors with an associated eigenvalue of 1.0. Furthermore, movement probabilities **M** and the movement rate matrix **Ṁ** have the same eigenvectors. Given that the rate matrix **Ṁ** is sparse, we can efficiently calculate its eigendecomposition using the igraph package [38] even for a large number of grid cells;

C *Identical parameters using either Eulerian or Lagrangian viewpoint*: by correcting for the temporal scale Δ_t and spatial scale Δ_s of discretization, we ensure that the Eq. 10.1 is represented using the same parameters using both viewpoints;

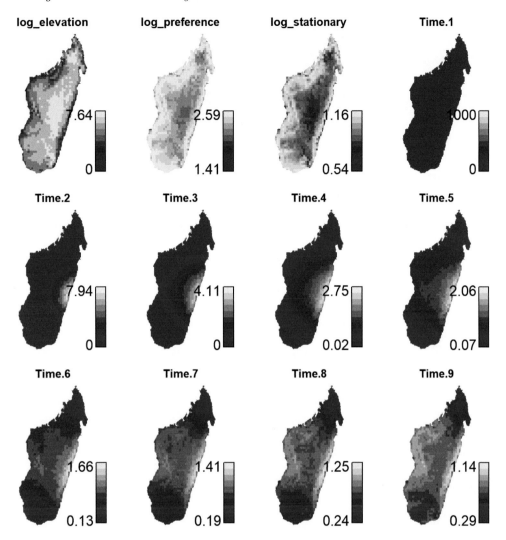

FIGURE 10.3: Projected movement for 1000 individuals introduced on the east coast of Madagascar, using a fine-scale discretization using 0.2 latitude by 0.2 longitude cells (see Fig. 10.2 for more details).

D   *Complicated dynamics using few parameters*: using a linear model to assemble the habitat-preference function from covariates ensures that we can specify taxis to describe complicated movement dynamics using few parameters. For example, we could use a polynomial or spline basis expansion to add degrees of freedom when calculating habitat preference (see Section 5.2);

E   *Computational efficiency*: finally, we can achieve computation efficiency by carefully choosing either the matrix exponential or Euler approximation, where the Euler approximation will likely be faster when a large number of grid cells are specified;

F   *Scale-free discretization*: as shown by comparing Figs. 10.2 and 10.3, the results are similar across a wide range of scales used for spatial discretization;

F *Subdivision in time*: given that we project between two times $\mathbf{n}_{t+\Delta_t}^T = \mathbf{n}_t^T e^{\Delta_t \dot{\mathbf{M}}}$, we can compute expected abundance at any intervening time as some fraction, e.g., $\mathbf{n}_{t+0.5\Delta_t}^T = \mathbf{n}_t^T e^{0.5\Delta_t \dot{\mathbf{M}}}$ for computing the expected density at the midpoint of interval Δ_t.

Given this general process for constructing movement matrix $\mathbf{M}$, we next show how we can use this process to estimate movement parameters from an Eulerian viewpoint.

10.4 Fitting Diffusion-taxis Movement to Individual Tracks

We first demonstrate the process of fitting this discretized movement model using an archival tag for a single tagged fish. We previously fitted similar data in continuous space from a Lagrangian viewpoint (Section 4.3) as a state-space model. However, this previous state-space model assumed that locations were measured following a normal distribution (Eq. 4.12), which then results in a close-to-normal distribution for random effects that can be approximately well using the Laplace approximation (recalling Section 2.2). By contrast, archival tags for fishes result in a distribution for measurement errors that is very different from a normal distribution (or any low-dimensional function). We therefore use an Eulerian viewpoint to represent the measurement process more precisely.

We specifically analyze data for a Pacific cod (*Gadus macrocephalus*) that was tagged with a pop-up satellite archival tag on Feb. 21, 2019, in the Aleutian Islands [17]. The satellite tag continuously recorded depth, temperature, and light-level data while attached to the fish. On May 23, 2019, the tag was programmed to release from the fish and subsequently floated to the water surface where it transmitted its archived data via the ARGOS satellite network. Locations of the tagged fish were known at the time of release (using a Global Positioning System measurement) and at the time the tag began transmitting data at the water surface (provided by ARGOS). The data transmitted by the tag was used to determine the likelihood that the tagged fish occupied any given location within a specified spatial domain (Fig. 10.4). The daily likelihood was calculated based on estimated longitude (derived from the time of local noon provided by light-level data) and matching the daily maximum depth of the tagged fish to bathymetric maps [163][2].

We therefore seek to use a known location on Feb. 21 and May 23, combined with this data likelihood for location at each day between these to infer movement parameters for winter-to-spring movement of Pacific cod in the Aleutian Islands. For this demonstration, we define a spatial domain composed of $90 \times 224 = 20160$ square grid cells, and define a daily probability $\pi_{g,t}$ that the animal is in cell g in day t. We know seafloor depth (*bathymetry*) for every grid cell, and use this as covariate to define the estimated habitat preference function.

To do so, we fit a *hidden Markov model* (HMM). This HMM involves defining the probability $\pi_{g,t}$ that the individual is in a given cell g in time t. This probability evolves over time based on a transition matrix $\mathbf{M} = e^{\dot{\mathbf{M}}\Delta_t}$, which again is calculated as the matrix exponential of a movement-rate matrix $\dot{\mathbf{M}}$ such that:

$$\pi_{t+1}^T = \pi_t^T \mathbf{M} \tag{10.13}$$

---

[2]Data were collected by the Groundfish Assessment Program, Alaska Fisheries Science Center, in an effort led by Susanne McDermott. Data likelihood and covariate data were formatted, obtained, and are reproduced with permission from Julie Nielsen on Dec. 5, 2022.

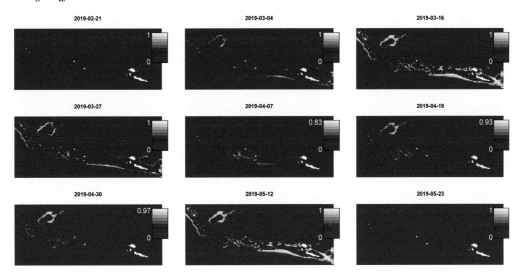

FIGURE 10.4: Data likelihood $p_{g,t}$ for evenly spaced intervals t between a Feb. 21 release and May 23 recovery for the tagged Pacific cod (panels, where white areas show land that has a zero likelihood a priori), obtained by comparing sub-daily vertical profile for the tagged individual with known bathymetry maps.

To fit this model, we have a *data likelihood* surface $p_{g,t}$ for each location g obtained from the archival tag, where this data likelihood $p_{g_{release},1} = 1$ for the known release location $g_{release}$ in time $t = 1$ and zero elsewhere, and again has a known recapture location $p_{g_{recapture},n_t} = 1$ at time of recapture n_t (see [176] for details). Importantly, $\dot{\mathbf{M}}$ has dimension 20160×20160, and it is impractical to directly compute all 400 million cells of $\mathbf{M}$. We therefore use a computational technique called *uniformization* [70, 206] to compute $\pi_t^T e^{\mathbf{M}}$ without actually computing $\mathbf{M}$ (see Section B.6).

To estimate parameters and predict state-variables, we apply a *forwards-backwards algorithm* that fits this hidden Markov model to the data likelihood $p_{g,t}$. We seek to define the likelihood $\mathcal{L}(\theta)$ of parameters θ so that we can calculate their maximum likelihood estimator $\hat{\theta}$, as well as the probability of a state X_t given data Y_t and estimated parameters $\hat{\theta}$. Recall that we outlined how to do this in Section 2.1.2, and it requires marginalizing across the value of random effects. When analyzing location in continuous space in Section 4.3, we marginalized across locations using the Laplace approximation. Here, however, we have discrete valued states, and the Laplace approximation cannot be used for discrete-valued states. We therefore use the alternative backwards-forwards algorithm to marginalize across latent states. To do so, we calculate:

- *Marginal likelihood*: we want to calculate the marginal likelihood $\mathcal{L}(\theta)$ of parameters θ, including those describing taxis, diffusion, and drift. To do so, we apply a forward algorithm, which calculates the probability of state X_t for any time t given all data prior to and including that time:

$$\mathbf{f}_t^T = \begin{cases} \mathbf{p}_t & \text{if } t = 1 \\ \mathbf{p}_t \odot \mathbf{f}_{t-1}^T \mathbf{M} & \text{if } t > 1 \end{cases} \tag{10.14}$$

where $\odot$ is the elementwise product of data likelihood $\mathbf{p}_t$ and projected states $\mathbf{f}_{t-1}^T \mathbf{M}$. The marginal likelihood is then calculated in the final time-interval by marginalizing across the location in that time:

$$\mathcal{L}(\theta) = \sum_{g=1}^{n_g} \mathbf{f}_{g,n_t} \tag{10.15}$$

The log-likelihood is then maximized with respect to parameters θ to identify preference and diffusion rates;

- *State probabilities*: we also want to calculate the probability for each state given all data. However, the quantity $\mathbf{f}_t$ calculated by the forward algorithm for time t does not include any information from the data likelihood $\mathbf{p}_{t^*}$ for later intervals $t^* > t$. To calculate the state-probability π_t that includes all information, we repeat the same algorithm but now proceeding backwards in time. In a lose sense, we are decomposing the data for each time into those before and including time t and those after time t, where the forward algorithm calculates the former and the backwards algorithm calculates the latter:

$$\Pr(X_t | \text{All data}) = \Pr(X_t, \text{Past or current data}) \Pr(\text{Future data} | X_t) \propto \mathbf{f}_t \mathbf{b}_t \tag{10.16}$$

This backwards algorithm involves projecting the state-probability $\mathbf{b}_t$ backwards from recovery $t = n_t$ to release time $t = 1$, while again calculating its product with the data-likelihood in each interval:

$$\mathbf{b}_t = \begin{cases} \mathbf{1} & \text{if } t = n_t \\ \mathbf{M}(\mathbf{p}_{t+1} \odot \mathbf{b}_{t+1}) & \text{if } t < n_t \end{cases} \tag{10.17}$$

We emphasize that we have defined the movement-rate matrix $\dot{\mathbf{M}}$ such that forward-projection involves *post-multiplying* the state-probability with the movement matrix $\mathbf{f}_t = \mathbf{f}_{t-1}^T \mathbf{M}$, while backwards-projection involves *pre-multiplying* the state probability with the movement matrix $\mathbf{b}_t = \mathbf{M}\mathbf{b}_{t+1}$. We then calculate the smoothed state-probability as the normalized product of forwards and backwards values:

$$\pi_{g,t} = \frac{f_{g,t} b_{g,t}}{\sum_{g'=1}^{n_g} f_{g',t} b_{g',t}} \tag{10.18}$$

where this state-probability then incorporates all data as well as the estimated parameters that are represented in movement-rate $\dot{\mathbf{M}}$.

However, model exploration also suggested that covariates often resulted in a movement rate $\dot{\mathbf{M}}$ that was not Metzler, i.e., where taxis was stronger than diffusion resulting in negative off-diagonal elements. To address this, we therefore reparameterize the instantaneous movement rate to ensure that it has positive offdiagonal elements for any set of diffusion and taxis parameters [79]:

$$\dot{m}_{g_1,g_2} = \begin{cases} a_{g_1,g_2} \dfrac{D}{\Delta_s^2} e^{\frac{h_{g_2} - h_{g_1}}{\Delta_s}} & \text{if } g_1 \neq g_2 \\ -\sum_{g \neq g_2} \dot{m}_{g_1,g} & \text{if } g_1 = g_2 \end{cases} \tag{10.19}$$

This essentially involves the same off-diagonal calculations, but done in log-space such that their exponentiated values are always positive. These two alternative forms for assembling movement-rates can both be implemented using sparse matrices in TMB (Code 10.4).

CODE 10.4: TMB code for assembling the instantaneous movement-rate matrix.

```
1  // expm_generator:  v^T %*% exp(M) ->  M( from, to ) AND rowSums(M) = 1
2  template<class Type>
3  Eigen::SparseMatrix<Type> make_M( int CTMC_version,
4                                    int n_g,
5                                    Type DeltaD,
6                                    matrix<int> At_zz,
7                                    Type ln_D,
8                                    vector<Type> h_g,
9                                    vector<Type> colsumA_g ){
10
11   int n_z = At_zz.rows();
12   Type D = exp( ln_D );
13   Eigen::SparseMatrix<Type> Mrate_gg( n_g, n_g );
14
15   // Standard approach
16   if( CTMC_version==0 ){
17     // Diffusion .. equal rate by cell, spread equally among neighbors
18     for(int z=0; z<n_z; z++){
19       Mrate_gg.coeffRef( At_zz(z,0), At_zz(z,1) ) += D / colsumA_g( At_zz(z
         ,0) ) / pow(DeltaD,2);
20       Mrate_gg.coeffRef( At_zz(z,0), At_zz(z,0) ) -= D / colsumA_g( At_zz(z
         ,0) ) / pow(DeltaD,2);
21     }
22     // Taxis
23     for(int z=0; z<n_z; z++){
24       Mrate_gg.coeffRef( At_zz(z,0), At_zz(z,1) ) += (h_g(At_zz(z,1)) - h_g(
         At_zz(z,0))) / DeltaD;
25       Mrate_gg.coeffRef( At_zz(z,0), At_zz(z,0) ) -= (h_g(At_zz(z,1)) - h_g(
         At_zz(z,0))) / DeltaD;
26     }
27   }
28
29   // Log-space to ensure Metzler matrix
30   if( CTMC_version!=0 ){
31     // Combined taxis and diffusion
32     for(int z=0; z<n_z; z++){
33       Mrate_gg.coeffRef( At_zz(z,0), At_zz(z,1) ) += D / pow(DeltaD,2) * exp(
         (h_g(At_zz(z,1)) - h_g(At_zz(z,0))) / DeltaD );
34       Mrate_gg.coeffRef( At_zz(z,0), At_zz(z,0) ) -= D / pow(DeltaD,2) * exp(
         (h_g(At_zz(z,1)) - h_g(At_zz(z,0))) / DeltaD );
35     }
36   }
37   return Mrate_gg;
38 }
```

We fit this model by maximizing the marginal likelihood in R. The latent states are integrated using the forwards-backwards algorithm and we do not specify any additional random effects, so the model is implemented using TMB without applying the Laplace approximation. We specify the habitat preference $h_g = \gamma_1 x_g + \gamma_2 x_g^2$ as a quadratic function of bathymetry x_g, and the estimated response function suggests a strong preference for shallow depths (Fig. 10.5). We can also visualize the estimated preference function on a map (Fig. 10.6). The map of bathymetry shows relatively shallow depths running along a band southeast to northwest, and Pacific cod clearly prefer this shallow band. Calculating habitat utilization as the stationary distribution (i.e., the eigendecomposition of $\dot{\mathbf{M}}$) indicates that this individual cod is expected to reside in larger discrete patches of preferred habitat, and therefore likely uses corridors of preferred habitats to rapidly cross from patch to patch.

Finally, the smoothed state-probabilities $\pi_{g,t}$ show that the animal released on Feb. 21 in the southeastern portion of the modeled domain likely moved west until early April, when it then made a rapid movement north toward the northern boundary (Fig. 10.7). It then likely resided in that northern area through the remainder of the tagged period.

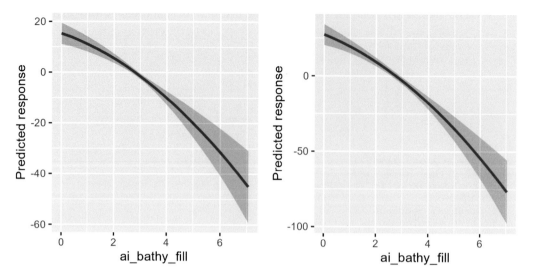

FIGURE 10.5: Habitat preference (y-axis) estimated as a quadratic function of seafloor depth (x-axis) based on an archival tag attached to a Pacific cod in the Aleutian Islands Feb. 21 and recovered May 23, 2019, showing estimates at the original 3km grid resolution (left panel) or a coarsened 6km grid resolution (right panel) visulized using the packages marginaleffects and ggplot2 (Section 1.7).

10.5   Fitting Diffusion-taxis Movement to Point-count Data

So far, we have demonstrated how to fit a movement model to location measurements for a single individual, using either a state-space model from a Lagrangian viewpoint (Section 4.3) or a Hidden Markov Model from an Eulerian viewpoint (Section 10.4). The Eulerian viewpoint seems particularly appropriate when the measurement likelihood is highly non-normal, but otherwise, the two can be constructed to fit the same underlying process (Eq. 10.1).

However, the Eulerian viewpoint has important advantages when fitting a movement process to point-count data such as those we analyzed from Chapter 5 onwards. We specifically explore a scenario where density is measured but tagging data are not available. We therefore use movement rates to inform changes in population density over time, similar to what is often done in species distribution/density models. Past studies have used a similar model to project population range expansion for invasive species [97], or to fit an integrated model that includes both density samples and tagging data [234].

We specifically revisit the bald eagle data set from Chapter 5, but now analyzing spatio-temporal data from 1968–2019 and from Alaska to California. We specifically seek to explore drivers for population range expansion as its population recovered following a ban on DDT [52]. Following the steps in Section 10.2, we first discretize the spatial domain into 93 square grid cells (each 250 km by 250 km), and calculate the rook-move adjacency matrix. We then obtain elevation, satellite measurements of green vegetation (the Normalized Difference Vegetation Index or *NDVI*), and distance to coastline as covariates, and model preference as a quadratic function of each covariate (i.e., 6 preference parameters total). We use these covariates to define the movement rate $\dot{\mathbf{M}}$ and integrate movement matrix $\mathbf{M}$.

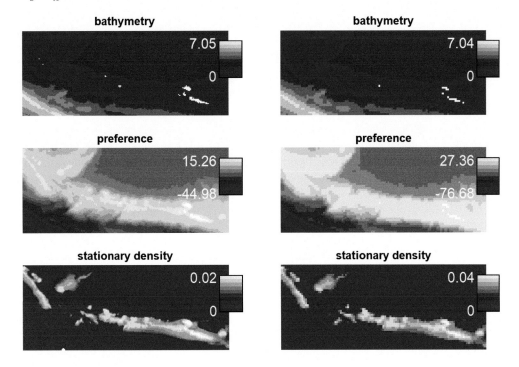

FIGURE 10.6: Bathymetry (top row), estimated habitat preference (middle row) and resulting stationary distribution (bottom row) based on fitting to an archival tag for Pacific cod at the original 3km grid resolution (left column) or a coarsened 6km grid resolution (right column). Note that areas with bathymetry less than 1 m are excluded (white areas in each map).

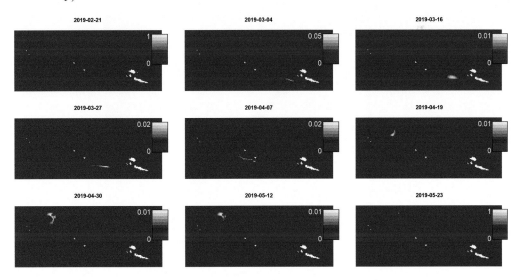

FIGURE 10.7: Smoothed state probabilities $\pi_{g,t}$ for evenly spaced intervals t between a Feb. 21 release and May 23 recovery for the tagged Pacific cod, estimated using 3 km × 3 km resolution but noting that 6km resolution results are similar.

We then treat $\log(d_{g,t})$ as a random effect while specifying:

$$\hat{\mathbf{d}}_t = \begin{cases} e^{\beta_1} \times \mathbf{1} & \text{if } t = 1 \\ e^{\beta_t} \times \mathbf{d}_{t-1}^T e^{\dot{\mathbf{M}}} & \text{if } t > 1 \end{cases}$$

$$\log(\mathbf{d}_t) \sim \text{MVN}\left(\log(\hat{\mathbf{d}}_t), \mathbf{Q}^{-1}\right)$$

(10.20)

where β_1 is the average log-abundance in the first modeled year, subsequent values β_t represent interannual changes in log-densities, and $\mathbf{Q}$ is the precision matrix derived from a conditional autoregressive process (see Eq. 5.12). Importantly, we specify the state variable $\log(d_{g,t})$ as random effect so that $\log(d_{g,t-1})$ is independent of $\log(d_{g,t+1})$ conditional upon the fixed value of $\log(d_{g,t})$. This then results in a sparse inner Hessian for log-density $\log(\mathbf{D})$ (recalling Section 2.4), as confirmed in Fig. 10.8. We specify a spatially correlated random effect in each modeled time as a series of conditional distributions (e.g., directly specifying Eq. 8.5). This results in a *non-separable model*, but it clearly serves a useful role by decomposing dynamics into separate components representing changes in log-density β_t versus spatial distribution shifts $\dot{\mathbf{M}}$.

We complete the model by specifying a Poisson distribution for counts. We are fitting point counts without any tagging data, and prior studies suggest that it is not feasible to separately estimate diffusion rates and the magnitude of habitat preferences without tagging data [234]. We therefore choose to fix diffusion D at a fixed value. To do so, we note that juvenile bald eagles typically disperse 80 km and the age of first reproduction is approximately 5 years, so mean-squared displacement (MSD) is $80^2 \ km^2$ when $\Delta_t = 5$ years. Remembering that the mean-squared displacement $MSD = 2nD\Delta_t$ (introduced in Eq. 4.5), we obtain $D = \frac{80^2}{2 \times 2 \times 5} \ km^2 \cdot year^{-1}$.

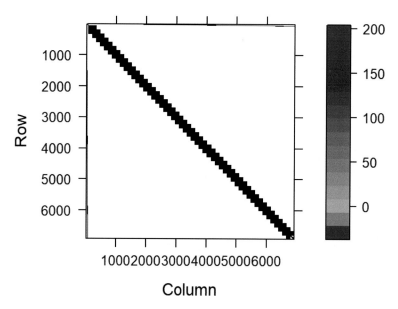

Dimensions: 6916 x 6916

FIGURE 10.8: Sparsity pattern for random effects in a model for eagle movement 1966–2019, where densities are block-tridiagonal because density $\log(d_{g,t-1})$ is conditionally independent of $\log(d_{g,t+1})$.

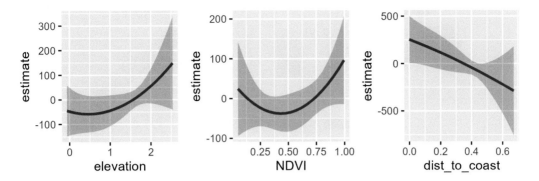

FIGURE 10.9: Habitat preference functions (y-axis) estimated as a quadratic function of elevation (left panel), Normalized Difference Vegetation Index (middle panel) or distance from coastline (right panel), and noting that the y-axis scale (and resulting magnitude of covariate effects) differs between panels, visualized using the `marginaleffects` package.

Inspecting results, we see that bald eagles prefer habitats that are close to the coastline and higher elevation all else equal (Fig. 10.9). We can again map these covariates and the resulting habitat preferences (Fig. 10.10) and confirm that the predicted preference is lower for areas that are relatively far from the coast. Finally, inspecting estimates of density show that bald eagles had a relatively constrained distribution in Southeast Alaska and British Columbia in 1968, but recolonized habitats in Washington and Oregon by 1978 and subsequently increased in California by 2008–2019. In particular, population density in later years is generally proportional to the estimated preference function, suggesting that the population has recolonized most of its preferred habitat in the western portion of North America.

10.6 Chapter Summary

In summary, we have showed that:

1. Individual movement can be represented using a partial differential equation that includes components for taxis, drift, and diffusion. This process can be represented from a Lagrangian (individual-based) viewpoint using a multivariate state-space model (Chap. 4) to estimate movement parameters and reconstruct tracks for a small set of individuals. It can also be discretized across space and modeled from an Eulerian (grid-based) viewpoint, by defining instantaneous movement rates and then solving for movement fractions by applying a matrix exponential operator. The movement-rate matrix can itself be constructed by adding together the effect of taxis, drift, and diffusion matrices. When tracking the probability that a single individual is in each spatial cell, parameters can be estimated using a Hidden Markov model (HMM) forwards algorithm, and location can be subsequently estimated using both the forwards filter and backwards smoother. Finally, the spatially discretized movement model can be applied to a vector representing population density for a population. This density or probability vector is then projected forward in time by multiplying it with a movement-fraction matrix.

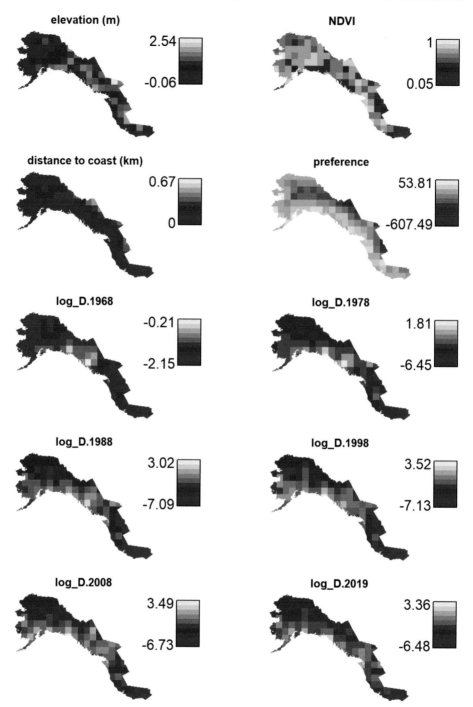

FIGURE 10.10: Mapped values of elevation, NDVI, and distance from coastline, the resulting prediction of habitat preference, and estimated bald eagle log-densities for evenly spaced years 1968-2019.

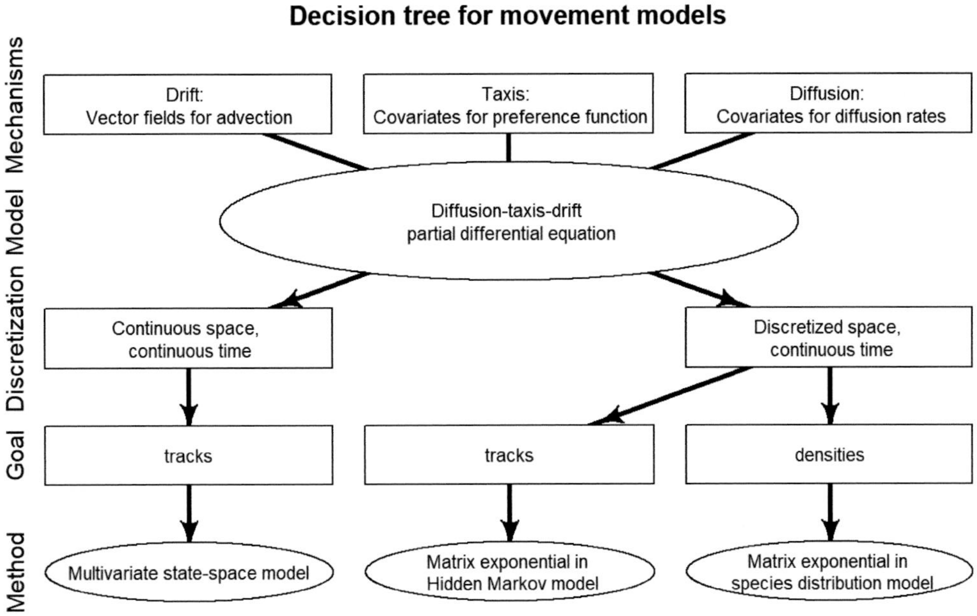

Decision tree for movement models

FIGURE 10.11: A simplified decision tree explaining how to use the partial differential equation that includes taxis, drift, and diffusion terms to explain movement (Eq. 10.1). Decisions are shown in boxes, and these include identifying vector-fields for drift, covariates for the habitat preference function, rates for diffusion, and decisions about the discretization that depend upon the intended usage.

These various options are summarized in Fig. 10.11;

2. This matrix exponential can be calculated efficiently either by using the Euler approximation (which pre-specifies the maximum number of grid-cells that can be moved in a single time-interval), or using uniformization (which calculates the product of the density vector and matrix exponential without directly computing the latter). Similarly, parameters can be specified without bounds if the movement-rate matrix is constructed with log-linked rates, such that directional movement from one to any other cell is guaranteed to be positive;

3. Using uniformization, it is feasible to estimate movement among thousands of spatial grid cells when fitting to individual animal tracks. This procedure results in parameter estimates that are largely independent of the spatial scale used for discretizing space;

4. Similarly, habitat preference parameters and resulting movement can be estimated directly from point-count data by specifying the diffusion rate a priori. This then allows spatio-temporal models to be fitted with covariates that inform habitat preferences rather than habitat-response functions in a conventional species distribution model. This results in a non-separable model, but the inner Hessian is still sparse such that the model can be fitted efficiently to data.

10.7 Exercises

1. In Eq. 10.1, we define the PDE for diffusion-taxis-drift movement. This includes the term $D\nabla^2 d(s,t)$ for isotropic diffusion, and Code 10.1 showed how to check this numerically using a diffusion-only process in R. However, the Laplacian operator can instead be generalized by writing $D\nabla \cdot \mathbf{\Sigma}\nabla d(s,t)$, where $\mathbf{\Sigma}$ is an *anisotropic diffusion* matrix, representing a different diffusion rate in different cardinal directions. How specifically would you code a transformation $\mathbf{\Sigma}$ to represent fastest diffusion in the northwest-southeast axis, and what ecological conditions might give rise to anisotropic diffusion?

2. In Section 10.5, we showed how to fit a diffusion-taxis movement model to point-count data. This is more conventionally done using regression-based species distribution models (SDM), such as the separable model with an autocorrelated interaction of space and time introduced (Eq. 8.9). Please refit the bald eagle data using that alternative spatio-temporal model, while also adding the quadratic effect of elevation, NDVI, and distance to the coastline as density covariates. How do the estimated covariate-response curves differ between the movement model and the regression-based SDM?

11

Multispecies Models for Community Diversity and Biogeography

11.1 Community Assembly

Naturalists have long recognized that particular groups of species are often found together at a given location, and that transitions between these *ecological communities* often involve simultaneous changes in the density of all species at major geographic boundaries [98]. This observation then contributed to the development of *community ecology* [35], which has subsequently grown to address a wide range of questions including:

1. *Relative dominance*: why are some species numerically abundant throughout a large range while others persist at low densities in a smaller number of habitat patches [94]?

2. *Meta-community dynamics*: what is the relative importance of different processes that allow ecological communities to persist over time, including fitness advantages in particular habitats, emigration from nearby habitats or regional species pools, or the dynamic balance of regional speciation and extirpation rates [128]?

3. *Community biogeography*: what environmental or ecological factors define the geographic boundaries of these distinct communities [259]?

4. *Functional traits*: what evolutionary innovations allow a given species to be well- or poorly-suited to a given ecological niche, and how well can we predict species fitness in a given habitat given measureable characteristics (called *functional traits*) for that species [150]?

These questions have special importance during the Anthropocene era, when changes wrought by humans are hugely increasing the rate at which ecological communities are changing worldwide. The relative importance of land development, climate change, incidental and intention mortality, and pollution in changing community structure and function varies among ecological communities. However, the general result is *community disassembly*, wherein co-occurring species are responding at different rates, thus resulting in novel species combinations and scrambling ecological interactions. This in turn results in winners (invasive species) and losers (species extirpation), and ecologists have the difficult challenge of informing limited conservation and policy efforts, including proposed efforts to protect 30% of land and ocean areas by 2030.

11.2 Biogeographic Analysis

Ecologists often obtain measurements of the density of multiple species from samples at different locations[1]. In the following, we use index $c \in \{1, 2, ..., n_c\}$ to identify species c from n_c total species, and we categorize different analytical goals for analyzing these data.

Most obviously, ecologists are often interested in predicting density $d_c(s)$ either at a set of habitat patches or sites, or for all locations $s \in \mathcal{D}$ within a prescribed spatial domain $\mathcal{D}$. This vector of density $\mathbf{d}(s)$ has been called an *essential biodiversity variable*, and it clearly has immediate use for national and international policies including the Convention on Biological Diversity and Sustainable Development Goals. For example, resulting maps are often visualized online as a simple explanation for the distribution of a given species [108]. As we saw in Section 7.4, we can infer density $d_c(s)$ from any combination of encounter, count, and biomass-sampling data, and the type of data available will likely differ based on the characteristics of each taxa.

As a simple extension, these densities can then be summarized in two main ways:

- *Cluster analysis*: the analyst could seek to identify two or more discrete communities, and represent the information in $\mathbf{d}(s)$ by associating each location s with one of those communities. This analysis generally involves defining a *distance metric* that can be used to compute the ecological distance between the vector of densities $\mathbf{d}(s)$ at every pair of sites. This ecological-distance matrix can then be analyzed using hierarchical clustering to identify the composition of clusters that best represents the original densities;

- *Ordination*: alternatively, the analyst might seek to describe density information $d_c(s)$ by defining a smaller set of synthetic variables $y_x(s)$ where the number of synthetic variables $n_x \leq n_c$, where these variables can then be back-transformed to predict the original data. The value $y_x(s)$ of synthetic variables is usually estimated such that they minimize some measure of lost information [148]. This is typically done using metric or nonmetric multidimensional scaling, principal components, or factor analysis, and this type of analysis is often called ordination because the original data $d_c(s)$ can be visualized along some reduced set of ordination axes y_x.

These two strategies differ in whether they result in a categorical (cluster analysis) or continuous (ordination) definition of ecological communities. However, they both provide a simplified description of the original density information.

Alternatively, ecologists often seek to attribute variation in species densities to different hypothesized mechanisms. There is a growing body of models (and associated software) to accomplish this, which are often called *joint species distribution models* (JSDMs). These JSDMs typically explain species densities using some combination of habitat variables, phylogenetic relatedness, species traits, and residual covariation [27, 169, 219].

In the following, we illustrate these concepts and analytical methods using site counts for twenty numerically abundant bird species obtained from the US Breeding Bird Survey [202], and restricting data to the western United States (westerward of Wyoming, Colorado, Montana, and New Mexico) in 2019. We specifically model counts for each site as a log-linked generalized linear mixed model that includes the log-linear effect of habitat, phylogeny, traits, and residual covariation, and define each of these in turn below.

---

[1] See https://github.com/james-thorson/Spatio-temporal-models-for-ecologists/Chap_11 for code associated with this chapter.

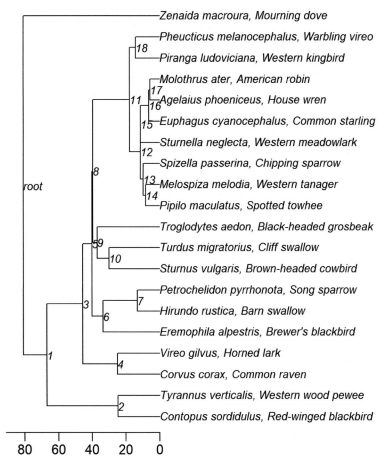

Time before present (million years)

FIGURE 11.1: Ultrametric phylogeny depicting time since speciation from nearest common ancestor (x-axis) for each of 20 bird species, obtained from AVONET [245] and visualized using R-package `ape` [171], and labeling all ancestral nodes as *root* (for the nearest common ancestor), or numbered from 1 to 18. The phylogeny has 38 edges, i.e., 20 *tips* ending at the 20 named species, and 18 ancestral edges ending in numbered ancestral nodes.

11.3 Phylogenetic Covariance

Before proceeding, however, we must introduce concepts for how evolutionary relatedness can be analyzed as a time-series model. We start by obtaining a simple representation of the evolutionary history for a given set of species, called a *phylogeny*. We specifically download the evolutionary history for birds from AVONET [78, 245]. AVONET specifically includes an *ultrametric phylogeny*, which represents the time prior to the present date when each bird species evolved from its nearest common ancestor. For the 20 selected bird species, evolutionary relatedness can be represented by 38 edges in a phylogeny (see Fig. 11.1 caption). This phylogeny shows e.g., that the nearest common ancestor of these 20 bird species was approximately 80 million years ago and, e.g., *Zenaida macroura* (mourning dove) speciated earliest of these selected species (Fig. 11.1).

Given a phylogeny with n_m edges, we will model the value of some characteristic δ_m at the endpoint of each edge. In the following, we assume that evolution follows a *Brownian motion* model. Given this assumption, we specify that characteristic δ_m follows some normally distributed deviation from its value δ_{p_m} for it's nearest ancestor p_m:

$$\delta_m \sim \text{Normal}(\delta_{p_m}, \Delta_t \sigma_\delta^2) \tag{11.1}$$

where Δ_t is the evolutionary time separating taxon g from its ancestor p_m and σ_δ^2 is the estimated evolutionary rate. Eq. 11.1 is identical to the diffusion model for individual movement (Eq. 4.1), and it arises from the same assumption that the characteristic δ_m follows a Wiener process over continuous time. This simple process is a typical *null model* against which alternative evolutionary models are often compared, and it can be extended to include first-order autocorrelations by using an Ornstein-Uhlenbeck process or the CIAR process (see Section 3.6 for the time-series versions that can extended to apply to a phylogenetic tree).

Given a phylogeny, an hypothesized evolutionary model, and parameters for that model, it is then typically feasible to directly construct the covariance in a given characteristic $\text{Cov}(\delta)$ among taxa. For example, using the Brownian motion model, we can calculate the evolutionary distance d_{c_1,c_2} from the nearest common ancestor of all taxa (called the *root*) to the nearest common ancestor for any pair of taxa c_1 and c_2, assemble this into distance matrix $\mathbf{D}$, and calculate the covariance as $\text{Cov}(\delta) = \sigma_\delta^2 \mathbf{D}$. Stated another way, two species have evolutionary covariance proportional to the length of shared evolutionary branches [83].

However, we adopt a different approach below and instead convert the phylogeny to a design matrix $\mathbf{C}$ containing one column for each of n_c species in the phylogeny and one row for each of n_m *edges* of the phylogeny (Code 11.1). Phylogenetic design matrix $C_{m,c}$ is nonzero only if the path of evolution from the root to taxa c passes through edge m. To construct this design matrix, we extract the path for each species to the root of the phylogeny using function nodepath in R-package ape [171]. However, ape defines the phylogenetic tree based on nodes (i.e., labels in Fig. 11.1) whereas we want to define a variable for each edge. We therefore remove the root of the phylogeny prior to constructing our design matrix. We then include nonzero values in the column of $\mathbf{C}$ corresponding to a given species only for edges that link connect that species to the root of the phylogeny (Code 11.1).

CODE 11.1: R code for reading a phylogenetic tree and constructing a phylogenetic design matrix.

```r
# Read tree
tree = read.tree( "Top20_Tree.tre" )
  tree$node.label = c( "root", seq_len(Nnode(tree)-1) )

# Extract path from species to root
root = Ntip(tree) + 1 # index for root
paths = sapply( match(taxa,tree$tip.label), FUN=nodepath, phy=tree, to=root )
names(paths) = taxa

# Convert to triplet form i=species; j=edge; x=edge_length
i = unlist(lapply(seq_along(paths), FUN=function(i){rep(i,length(paths[[i]]))
    }))
j = unlist(paths)
edge = match( j, tree$edge[,2] )
x = tree$edge.length[edge]
```

TABLE 11.1: Examples of publicly available, ultrametric phylogenies that can be used to replicate the phylogenetic component of the joint species distribution model; see `https://vertlife.org/data/` for additional sources.

Taxa		Reference	URL
Mammals (Mammalia)		[61]	`https://doi.org/10.1111/j.1461-0248.2009.01307.x`
Ray-finned fishes (Actinopterygii)		[185]	`https://datadryad.org/stash/dataset/doi:10.5061/dryad.fc71cp4`
Seed plants (Spermatophyta)		[210]	`https://github.com/FePhyFoFum/big_seed_plant_trees`
Cartilaginous fishes (Chondrichthyes)		[214]	`https://vertlife.org/sharktree/downloads/`
Birds (aves)		[245]	`https://figshare.com/s/b990722d72a26b5bfead`
Amphibians		[107]	`https://datadryad.org/stash/dataset/doi:10.5061/dryad.cc3n6j5`
Crabs		[265]	`https://datadryad.org/stash/dataset/doi:10.5061/dryad.tmpg4f52z`

```
15
16 # remove root and renumber
17 ijx = na.omit(cbind(i,j,x))
18 ijx[,1:2] = ifelse( ijx[,1:2]>Ntip(tree), ijx[,1:2]-1, ijx[,1:2] )
19
20 # Convert to dense design matrix
21 PhyloDesign_gc = matrix(0, nrow=Nedge(tree), ncol=length(taxa) )
22 PhyloDesign_gc[ijx[,2:1]] = ijx[,3]
```

Later we will treat δ_m as a random effect, estimate evolutionary variance σ_δ^2, and predict δ_m conditional upon this estimated variance. We will then calculate the net effect of phylogeny for each species c by summing δ_m for all edges connecting that species to the root, i.e., as $\mathbf{C}\delta$. This phylogenetic effect $\mathbf{C}\delta$ can then be included in the linear predictor for a generalized linear model, and used to predict variation among species in any modeled variable. This approach requires estimating δ_m for each edge m as a random effect, and hence becomes computationally expensive as the phylogeny becomes large. However, it also allows us to extract the predicted phylogenetic effect δ_m for each edge (i.e., all labeled values in Fig. 11.1), and therefore provides us with extra ecological insight about evolutionary patterns. For example, we might re-plot Fig. 11.1 but color coding each edge based on the value δ_m to visually check where in the evolutionary history a given characteristic changed rapidly. Importantly, it is feasible to include a phylogenetic design matrix in a generalized linear model for many other taxa, given the growing number of high-quality ultrametric phylogenies that are now publicly available (Table 11.1).

11.4   Trait-based Ecology

To analyze meta-community dynamics, we must also introduce the concept of *functional traits*. There has been a growing interest in *trait-based ecology* for at least 20 years [150], as ecologists have used this approach to investigate why particular species are associated with particular habitats, what changes are associated with speciation, and how variation among individuals is maintained across generations (among many other questions).

We here define *ecological traits* as those characteristics of an individual organism that can be measured at a particular moment. Clearly, individuals have many traits, including body coloration, age, body mass and shape, reproductive status, feeding behaviors, energy reserves, chemical composition, etc. The specific list of ecological traits will vary among taxa, e.g., where stomach contents are a functional measurement of recent energy acquisition for animals, while other traits will measure the same process in fungi and plants. Some ecological traits will be highly relevant to describing an individual's ecological status and function (e.g., size) while others may be less functionally relevant (e.g., eye color). We therefore use the term *functional trait* to describe ecological traits that are highly correlated to some ecological rate or output, and note that we often select functional traits (e.g., tree height) that are a proxy for some larger set of correlated ecological traits (tree mass, total photosynthetic rates, etc.).

By definition, these functional traits are all measurable for individual organisms, and can be treated as marks in an individual-based model (thinking back to Section 1.1). Defining ecological traits for individuals then allows us to explore how traits vary for the same individual across their life, and to compare the variance in traits among individuals within a given species vs. the variance among species [7]. Beyond this definition of functional traits as measurable characteristics of each individual at a specific time, it may also be useful to define *species traits* as the average across individuals for a given species.

Obtaining species traits then allows an analyst to investigate which traits are associated with specific habitats, and thereby generate hypotheses about how functional traits contribute to species densities given local environmental conditions. This type of analysis has been called the *fourth corner problem*, because ecologists typically measure the density of species in different habitat patches (the 1st corner), the environmental conditions at those patches (the 2nd corner), and species traits for each taxon (the 3rd corner), and must use these three measurements to infer the underlying ecological relationship between traits and environmental conditions [127].

Trait-based ecological analysis is increasingly feasible for a wide range of taxa given the ubiquity of high-quality databases compiling measurements of species traits (Table 11.2). However, these databases are often incomplete, both in terms of missing measurements of one or more species traits for a given taxon, or entirely excluding a taxon and therefore having no trait measurements for it. Given that species trait databases are typically incomplete, there is also a large and growing literature on *phylogenetic trait imputation* methods, which typically predict these missing values based on correlations among traits as well as taxonomic or phylogenetic information [178]. Although we do not discuss trait imputation in detail here, it is worth knowing that a complete list of traits may require some type of informal (expert judgement) or formal (statistical) trait imputation.

In the following, we extract from AVONET three continuous and/or categorical species traits for our 20 bird species:

- *Body size*, i.e., a continuous variable measuring average adult body mass;

TABLE 11.2: Examples of publicly available trait databases that can be used to replicate the trait-based component of the joint species distribution model; see the Open Traits Network `https://opentraits.org/datasets.html` for additional sources.

Taxa	Database name	Reference	URL
Mammals	PanTHERIA	[110]	`https://doi.org/10.6084/m9.figshare.c.3301274.v1`
Fishes	FishBase	[62]	`https://fishbase.org/`
Birds	AvoNET	[245]	`https://doi.org/10.1111/ele.13898`
Plants	TRY	[114]	`https://www.try-db.org/TryWeb/Home.php`

- *Primary lifestyle,* i.e., a categorical variable that describes their predominant locomotory niche, with levels for "Aerial", "Generalist", "Insessorial" (i.e., perching), or "Terrestrial";

- *Hand Wing Index,* i.e., a continuous variable that measures flight efficiency (and therefore dispersal ability) based on a morphological measurement of wing shape [204].

We specifically convert "Primary lifestyle" to a design-matrix, with one row per species and four columns, containing a 1 for the measured level for a given species and 0s otherwise. Similarly, we log-transform "Body size" and "Hand Wing Index". We compile these in trait matrix $\mathbf{T}$, with one row per $n_c = 20$ species and $n_h = 6$ columns, four of which are indicator (0 or 1) variables and the other two log-transformed and continuous (Fig. 11.2). The effect of these traits could be included in the linear predictor for a generalized linear model by estimating a vector of trait response coefficients γ, where the trait response is $\mathbf{T}\gamma$.

11.5   Joint Species Distribution Model

So far, we have claimed that phylogenetic and trait information can be included in a joint species distribution model that is specified as a generalized linear mixed model. We then constructed a matrix $\mathbf{C}$ representing phylogeny and a separate matrix $\mathbf{T}$ representing traits. Here, we specifically aim to understand how these affect species densities $d_{s,c}$ for each location s and species c. Trait and phylogeny matrices are specified as constant across space, so using these to predict spatial variation in densities $d_{s,c}$ requires that we estimate a spatially varying response to traits and phylogeny. In the following, we refer to this estimated response as spatially varying coefficients (see Section 8.3.2) [239], but they could be estimated in other models (e.g., generalized additive models) by including a tensor spline interaction of traits (or the phylogenetic design matrix) and spatial coordinates (Section 5.3).

We follow past chapters in defining a distribution for available species counts. We here use a negative binomial distribution as a null model for the observation process. The negative binomial arises as a compound gamma-Poisson process (Eq. 2.5), and we estimate the

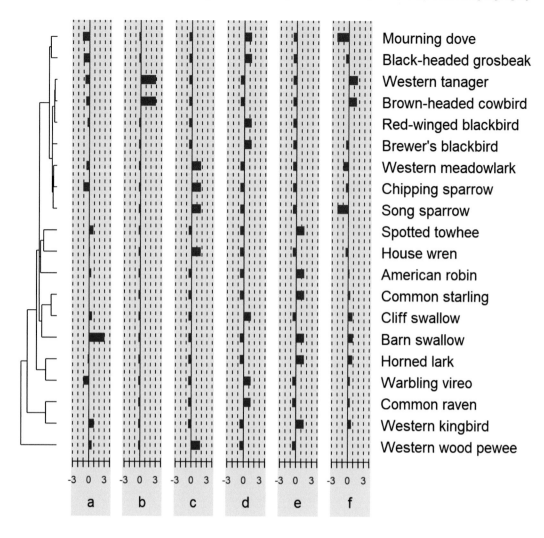

FIGURE 11.2: Six species traits representing different bioenergetic and behavioral characteristics of 20 analyzed bird species, including log-body mass (grey box labeled a), indicator variables for Primary Lifestyle (b=Aerial, c=Generalist, d=Insessorial, e=Terrestrial), and log-Hand Wing Index (f), and also showing the ultrametric phylogeny. Variables are then centered and scaled to have a mean of zero and a standard deviation of 1. The plot is generated using R package phylosignal [115].

additional gamma-distributed variance to represent local clustering that causes the variance for nearby samples to exceed the expected variance of a simple Poisson process:

$$y_i \sim \text{NegBin}\left(d_{s_i,c_i}, d_{s_i,c_i}(1 + \sigma_M^2)\right) \qquad (11.2)$$

where we fit to count y_i at each location s_i for each taxon c_i. We use the TMB function dnbinom2 to compute this negative binomial likelihood, which is parameterized such that d_{s_i,c_i} is the mean and $d_{s_i,c_i}(1 + \sigma_M^2)$ is the variance for replicated samples occurring at a given location [133]. We then constraint $\sigma_M^2 > 0$ such that this parameter is the estimated magnitude of overdispersion.

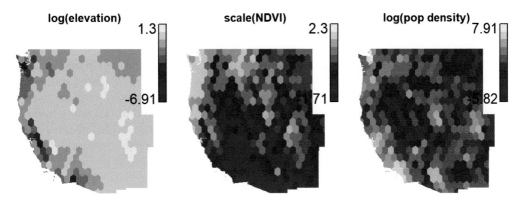

FIGURE 11.3: Three covariates used to explain habitat utilization for 20 bird species in the western US states, including log-scaled elevation, normalized difference vegetation index, and log-scaled human population density, where we apply a quadratic basis expansion (i.e., include each variable and its squared value) to construct matrix $x_{s,k}$ containing six columns.

We specifically define a linear predictor for log-density $\log(d_{s,c})$ that combines multiple components:

- *Local habitat characteristics*: we first might seek to attribute variation in community composition to local habitat conditions. We include environmental variables $x_{s,k}$ as covariates with estimated response $\beta_{k,c}$ for each basis-expanded covariate k and species c

$$\beta_{s,c}^* = \sum_{k=1}^{n_k} x_{s,k} \beta_{k,c} \tag{11.3}$$

where we use $\beta_{s,c}^*$ to indicate the estimated response for each location and species (and similar notation for other components). This specification assumes that each species has its own distinct response to the specified set of environmental drivers. For demonstration, we include a quadratic response (recalling Fig. 5.2) to three covariates: elevation from Amazon Web Service Terrain Tiles downloaded using R-package elevatr [93]; normalized difference vegetation index (NDVI) measured by the Copernicus Global Land Services program and extracted from R-package rasterdiv [191]; and human population density using the UN WPP-Adjusted Population Density, v4.11 [24][2] (Fig. 11.3). Specifying a quadratic response to three covariates involves estimating six fixed effects for each of 20 species, or 120 total parameters in $\beta_{k,c}$;

- *Spatially correlated trait responses*: after accounting for measured habitat conditions, there might remain some spatial variation resulting from community responses to latent (unmeasured) ecological conditions. These spatial residuals might be explained by species traits where, e.g., some latent variables affect all terrestrial birds while others primarily affect migratory birds. We include species traits $T_{h,c}$ with a spatially varying coefficient $\gamma_{s,h}$ for each trait h at each location s:

$$\gamma_{s,c}^* = \sum_{h=1}^{n_h} \gamma_{s,h} T_{h,c} \tag{11.4}$$

---

[2]Accessed from the NASA Socioeconomic Data and Applications Center on July 22, 2022.

This specification assumes that trait responses are constant across species, and this allows us to separate the trait response from other model terms;

- *Spatially correlated phylogenetic responses*: similarly, after accounting for measured habitat, we might find spatial patterns that arise predictably for closely related species. Over continental scales, a phylogeny-by-space interaction could arise due to barriers to dispersal for some lineages and not others [194]. On smaller spatial scales, this effect could arise from responses to latent environmental variables that affect species that share some trait that is evolutionarily conserved but not itself modeled. We include phylogeny $C_{m,c}$ with a spatially varying coefficient $\delta_{s,m}$ for each edge in the phylogeny m and each species c;

$$\delta^*_{s,c} = \sum_{m=1}^{n_m} \delta_{s,m} C_{m,c} \tag{11.5}$$

The phylogenetic effect for a given species is then the sum of spatial responses for edges connecting that species to the root of the phylogeny;

- *Residual spatial variation*: finally, we might estimate spatial residuals that are not correlated with either traits or phylogenetic information. We include this term to account for spatially autocorrelated residuals, which allows subsequent predictions of species density to be conditioned upon these residual patterns;

$$\omega^*_{s,c} = \sum_{j=1}^{n_j} \omega_{s,j} \lambda_{j,c} \tag{11.6}$$

where $\mathbf{\Lambda}$ is an estimated loadings matrix (see Section 4.4.1), where element $\lambda_{j,c}$ associates each of n_j spatial factors j with each species c. In the following, we specify that $\mathbf{\Lambda}$ is a diagonal and unequal matrix, where the absolute value of each diagonal element $|\lambda_{c,c}|$ is the standard deviation of spatial residuals for species c. However, estimating loadings $\mathbf{\Lambda}$ as a lower-triangle matrix then estimates ω^*_c for each species c as a linear combination of estimated spatial factors ω_j, and this has been termed *spatial factor analysis* [219]. This spatial factor model is then one example of a larger class of *coregionalization* models for multivariate spatial data, which has a long history in geostatistics [69].

We combine these four components to define the linear predictor that defines the joint species distribution model, while also including a species-specific intercept α_c that controls for differences in average log-density among species:

$$\log(d_{s,c}) = \alpha_c + \beta^*_{s,c} + \gamma^*_{s,c} + \delta^*_{s,c} + \omega^*_{s,c} \tag{11.7}$$

The model is then completed by specifying a distribution for spatial variables:

$$\gamma_h \sim \mathrm{MVN}(\mathbf{0}, \sigma^2_{\gamma_h} \mathbf{Q}^{-1})$$
$$\delta_m \sim \mathrm{MVN}(\mathbf{0}, \sigma^2_\delta \mathbf{Q}^{-1}) \tag{11.8}$$
$$\omega_j \sim \mathrm{MVN}(\mathbf{0}, \mathbf{Q}^{-1})$$

where we use the SPDE approach to construct precision $\mathbf{Q}$ and projection matrices $\mathbf{A}$ to implement bilinear interpolation for spatial random effects (see Section 5.5.2). For this demonstration, we assume that the Matérn correlation function has the same decorrelation rate κ for all spatial variables, such that $\mathbf{Q}$ is identical across terms. Spatial variables are then treated as random effects, and we marginalize across them when calculating the log-likelihood that is then optimized. We also observe that fixed effects (e.g., covariate-response parameters $\beta_{k,c}$ and intercepts α_c) are highly correlated with spatial variables that are treated as random effects, and this causes the inner and outer optimization steps to converge slowly (see discussion around Eq. 2.32). We therefore treat non-variance fixed effects as if they are random using a flat distribution (and hence integrate across them using the Laplace approximation). This technique is sometimes called *restricted maximum likelihood* (REML), and it ensures that these correlated variables are all identified simultaneously during the inner optimization step. REML also has a side-benefit of mitigating a small bias in the maximum-likelihood estimate of variance parameters, which subsequently improves our inference when partitioning model variance into different components [85]. Finally, we visualize results by projecting spatial variables to integration points (Section 6.2.2) that are defined by overlaying a hexagonal grid over the spatial domain and extracting the centroid of each hexagon.

This model has a deceptively simple structure, but generates a multitude of interpretable output. For example, the quadratic response to NDVI (Fig. 11.4) shows that the Western wood pewee has higher densities in areas with dense vegetation (i.e., a high value for scaled NDVI), while the Western meadowlark is associated with intermediate levels of vegetation.

The partial effect of habitat is then computed by summing across the quadratic effect of each covariate. For example, the Western wood pewee has a positive partial effect of covariates in coastal Oregon and Washington as well as Idaho (Fig. 11.5), resulting from the positive estimated response to NDVI and the high NDVI in those states.

We can also visualize the effect of evolutionary relatedness by mapping the spatial term $\delta_{s,m}$ associated with each edge $C_{m,c}$ of the phylogeny. For example, we inspect the shared ancestors for two closely related species (barn and cliff swallows) and their nearest relative (horned lark). Edge-7 is shared by barn and cliff swallows but not horned lark (Fig. 11.1), and the spatial effect for Edge-7 shows decreased densities in the southwest (Fig. 11.6 bottom-right panel). This then results in a small difference in the phylogenetic effect for these three species, where the partial effect of phylogeny for barn and cliff swallows is more negative in this southwest area than horned lark (Fig. 11.7).

Similarly, we can visualize the predicted response γ_h for each modeled trait h, as well as the net trait effect γ_c^* that is calculated by summing across all traits for each species. In this application, modeled traits appear to have a small spatial effect relative to covariates (Fig. 11.8). In fact, three of the six traits have an estimated spatial response that has a standard deviation $\sigma_{\gamma_h}^2$ approaching zero, such that these traits are estimated to have no spatial effect (similar to the outcome when randomizing an environmental index in Section 9.3). However, terrestrial birds appear to have higher densities in the northeastern habitats than otherwise expected from covariates alone.

The cumulative effect of covariates, traits, phylogeny, and residual variation then predicts density for each species (Fig. 11.9). Several species (including song, cliff, and barn sparrows) have lower density in the southern portion of the spatial domain. Meanwhile, black-headed grosbeak and spotted towhee have a predominantly coastal distribution.

11.6   Community Variance Partitioning

Importantly, we can calculate the covariance among species at a fixed location $\text{Var}(\log(\mathbf{d}_s))$ associated with each term in Eq. 11.7. Specifically, the covariance due to similar or different responses to covariates depends on the covariance among the covariates themselves $\text{Var}(\mathbf{X})$, as well as the matrix $\mathbf{B}$ containing the covariate response $\beta_{k,c}$ for each species c to each covariate k:

$$\text{Var}_{\log(\mathbf{d}_s)|\beta} = \mathbf{B}\text{Var}(\mathbf{X})\mathbf{B}^t \tag{11.9}$$

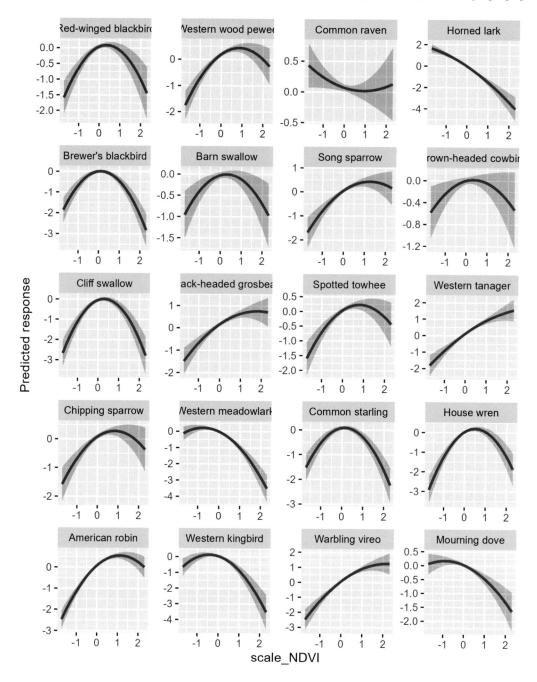

FIGURE 11.4: Partial effect (blue line) and estimated 95% confidence intervals (blue shared area) for the normalized difference vegetation index (NDVI) for each species, which is specified using a quadratic basis expansion resulting in an estimated hump-shaped response. Plots are generated using marginaleffects and ggplot2 packages. The NDVI was scaled and centered prior to analysis, such that -2 corresponds to little vegetation, and +2 corresponds to dense vegetation.

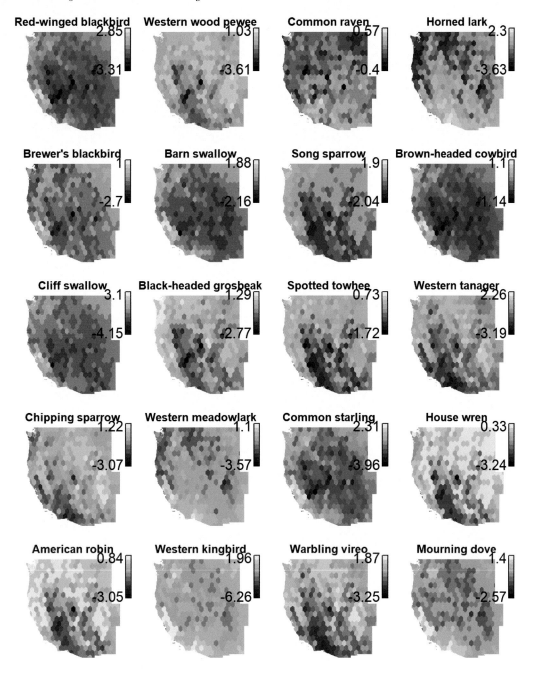

FIGURE 11.5: Spatial effect $\beta^*_{s,c}$ from Eq. 11.3 resulting from the estimated quadratic response to three habitat covariates for 20 bird species.

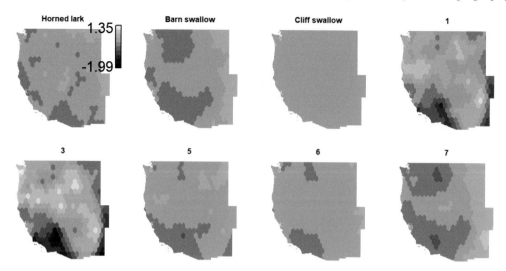

FIGURE 11.6: Spatial effect $\delta_{s,m}$ associated with edges m of the ultrametric phylogeny that are ancestors of horned lark, barn swallow, or cliff swallow, where edges are labeled based on descendent node label and see Fig. 11.1 for definitions.

Similarly, the covariance due to traits depends on the variance of spatially varying responses $\sigma^2_{\gamma_h}$ for each trait h, which we assemble into a diagonal matrix $\text{diag}(\sigma^2_\gamma)$, as well as the covariance among traits $\mathbf{T}$:

$$\text{Var}_{\log(\mathbf{d}_s)|\gamma} = \mathbf{T}\text{diag}(\sigma^2_\gamma)\mathbf{T}^t \tag{11.10}$$

Next, the covariance due to phylogeny results from the phylogenetic design matrix $\mathbf{C}$ as well as the estimated magnitude of the phylogenetic response σ^2_δ:

$$\text{Var}_{\log(\mathbf{d}_s)|\delta} = \sigma^2_\delta \mathbf{C}\mathbf{C}^t \tag{11.11}$$

Finally, the variance due to residual covariance is representing using a factor decomposition (see Section 4.4.1). The resulting covariance is entirely described by the factor loadings matrix $\mathbf{\Lambda}$:

$$\text{Var}_{\log(\mathbf{d}_s)|\omega} = \mathbf{\Lambda}\mathbf{\Lambda}^t \tag{11.12}$$

These variances can then be summed across components to calculate the total variance in log-density:

$$\text{Var}(\log(d_{s,c})) = \text{Var}_{\log(\mathbf{d}_s)|\beta} + \text{Var}_{\log(\mathbf{d}_s)|\gamma} + \text{Var}_{\log(\mathbf{d}_s)|\delta} + \text{Var}_{\log(\mathbf{d}_s)|\omega} \tag{11.13}$$

The diagonal elements of this covariance represent the variance for a given species. We can then visualize the proportion of variance associated with habitat covariates, traits, phylogeny, or residual spatial variation for each species (Fig. 11.10) [180, 169]. As expected from Fig. 11.8, we see that traits explain a smaller portion of variance than covariates. We also see that, for many species, the residual spatial term represents the majority of model variance. However, other species (e.g., warbling vireo, American robin, and Western tanager) have a relatively large fraction of spatial variance that is explained by the quadratic response

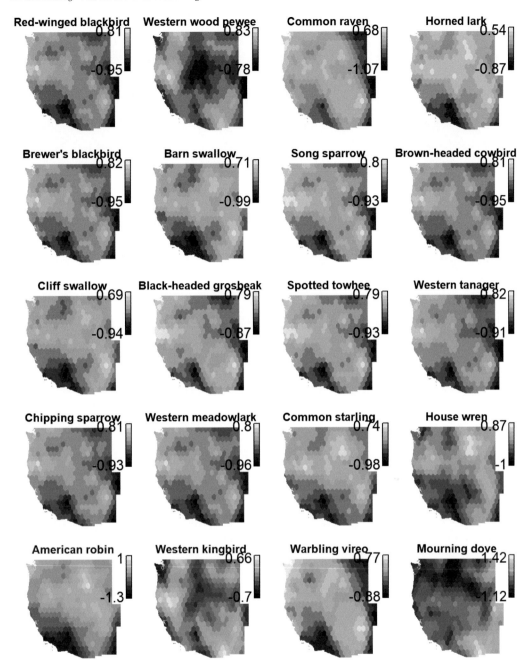

FIGURE 11.7: Spatial effect $\delta^*_{s,c}$ from Eq. 11.5 resulting from all phylogenetic variables for each bird species c.

to the three habitat covariates. Variance partitioning also provides a basis to compare among species. For example, we see that the common raven has a relatively low variance in log-density across the model area, while cliff swallow has a high variance in habitat utilization relative to other species.

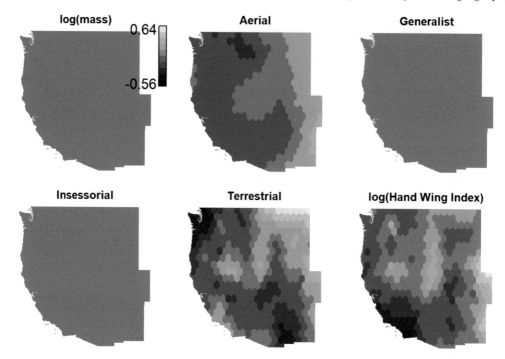

FIGURE 11.8: Spatial effect γ_h estimated for each modeled trait $T_{h,c}$, using the same colorbar (top-right panel) for all traits to highlight differences in estimated scale among traits.

11.7   Community Description

Given these estimates of population density and their association with habitat, phylogeny, traits, and residual spatial predictors, we next explore how results can be used to describe ecological communities.

11.7.1   Biogeographic Clusters

As a simple first step, we explore how we can identify k discrete communities based on estimates of log-density $\log(d_{s,c})$. This involves three steps:

1. *Distance metric*: we first compute the ecological distance D_{s_1,s_2} between every pair of locations s_1 and s_2. In the following we use *Ward distance* [254], calculated as the squared-difference between log-density vectors:

$$D_{s_1,s_2} = \sum_{c=1}^{n_c} \left(\log(d_{s_1,c}) - \log(d_{s_2,c})\right)^2 \tag{11.14}$$

where this is sometimes called method="WardD2" in R. We note that ecologists have developed a wide range of alternative distance metrics, including the *Bray-Curtis dissimilarity* [14] or *Mahalanobis distance* [143] (as two widespread

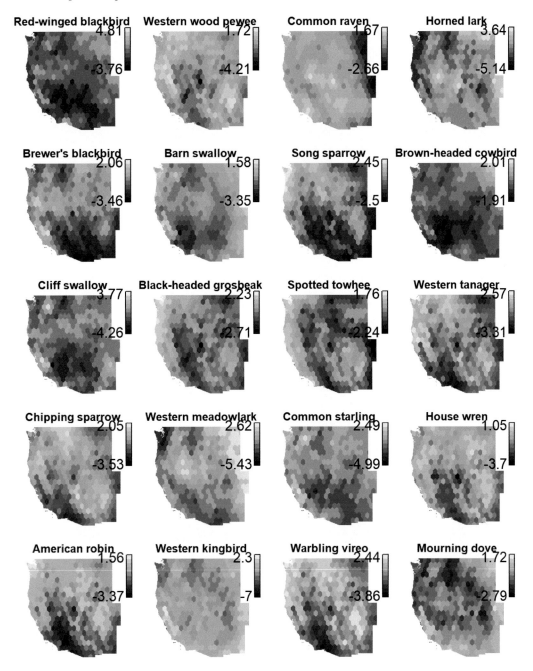

FIGURE 11.9: Estimated log-density $\log(d_{s,c})$ from Eq. 11.7 for each bird species, resulting from partial effect of covariates, phylogeny, traits, and residual spatial variation.

examples). The former is specifically useful for raw sampling data (which includes many zeros), while the latter accounts for correlations in log-density among species. In the following we use Ward distance to instead weight the contribution of all species based on their spatial variance, knowing that our estimated log-density does not include any estimates of exactly zero [218];

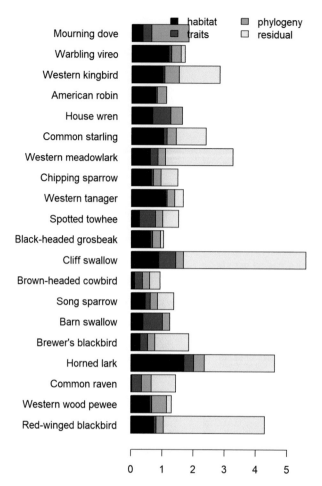

FIGURE 11.10: Variance in log-density (calculated as the diagonal of Eq. 11.13) associated with covariates, traits, phylogeny, or residual spatial terms (x-axis) for each of 20 bird species (y-axis).

2. *Hierarchical clustering*: we then apply a hierarchical clustering algorithm to the ecological distance matrix (Eq. 11.14). This algorithm proceeds by merging the two sites that have the minimum distance, recalculating the distance matrix based on their average distance from other sites, and then proceeding iteratively until the algorithm has merged all sites. We here apply hierarchical clustering to a large number of sites and species, and therefore use R-package fastcluster [158] for computational efficiency;

3. *Selecting the number of clusters*: after hierarchical clustering is complete, the user can display the clusters that were identified at any step in the hierarchical clustering algorithm. They can then select how many clusters are parsimonious or convenient to display, and extract the set of clusters that arise from that chosen number;

4. *Calculating species averages in each cluster*: finally, after selecting the number of clusters and associating each location with a cluster, the user can calculate the average log-density for each species and cluster. These averages can then be

used to interpret clusters, e.g., by identifying *indicator species* that are strongly associated with a given cluster [49].

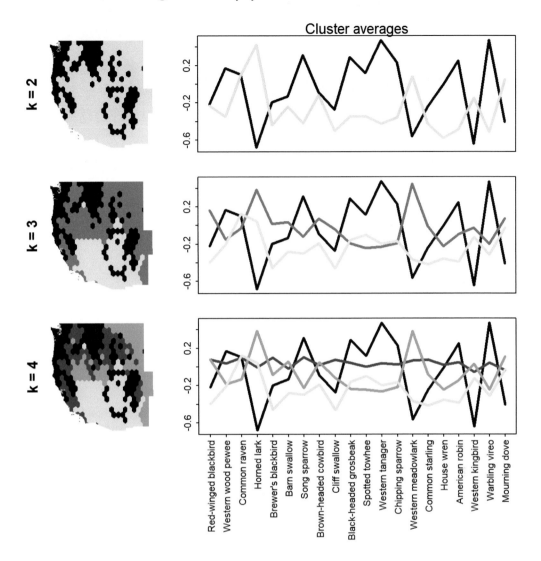

FIGURE 11.11: Hierarchical cluster analysis applied to Ward distance calculated using log-densities for all twenty species (top-left), showing results using two, three, or four clusters (rows from top to bottom) and displaying the resulting cluster maps (left column) and average species log-densities in each cluster (right column), where the cluster colors are the same between both columns for each row.

Applying Ward distance and hierarchical clustering to our estimated log-densities (Fig. 11.11) shows that two major clusters are strongly discriminated, corresponding generally to the (1) Pacific northwest through western Colorado, vs (2) southwest and eastern Montana. This biogeographic division somewhat resembles the map of NDVI (see Fig. 11.3), and is apparent in estimated densities (Fig. 11.9) for the perching (Incessorial) birds (Fig. 11.2) which have lower density in the southern cluster. Adding a third cluster then distinguishes the southwestern habitats from Wyoming and eastern Montana, and adding a fourth cluster further distinguishes within this cluster.

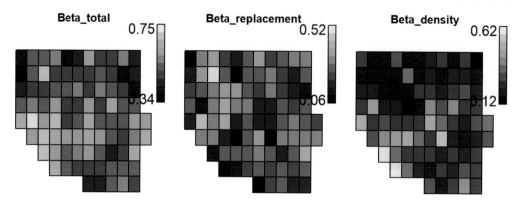

FIGURE 11.12: Beta diversity (measured as Jaccard dissimilarity) applied to predicted densities, including total diversity, the portion due to species replacement, or the portion due to overall differences in bird density.

11.7.2   Species Turn-over

Alternatively, we can also visualize spatial patterns of species (or functional trait) diversity. For illustration, we calculate *beta-diversity* from predicted densities $d_{s,c}$, specifically using Jaccard dissimilarity [105] to measure the difference between two sites s_1 and s_2:

$$J(s_1, s_2) = \frac{\sum_{c=1}^{n_c} X_c + Y_c}{\sum_{c=1}^{n_c} X_c + Y_c + Z_c} \tag{11.15}$$

where $Z_c = \min(d_{s_1,c}, d_{s_2,c})$ is the density for species c that is shared in common between two locations, $X_c = d_{s_1,c} - Z_c$ is the increase in density for s_1 relative to this shared density, and $Y_c = d_{s_2,c} - Z_c$ is the increase in density for s_2. When calculating beta-diversity directly from species counts, it is common to apply *rarefaction* (i.e., resample from the available counts to achieve a fixed sample size at all locations) to control for differences in sample sizes across sites [201]. However, this is not necessary here, given that we have already controlled for spatial differences in sampling density by estimating population density across the spatial domain. We calculate these using the function **BAT::raster.beta** [23] to provide an easy-to-replicate demonstration, but the calculation could be easily replicated using other spatial packages and dissimilarity metrics.

Using **BAT::raster.beta**, we first re-project from our hexagonal grid of integration points to a square grid, to match the expected input of that function. For a given focal site, we then identify the set of adjacent sites, and we here use queen adjacency (i.e., the eight square cells that are adjacent or diagonal to the focal cell, where fewer cells are used near boundaries). We then calculate Jaccard similarity (Eq. 11.15) between the focal site s_1 and each adjacent site s_2, and then average $J(s_1, s_2)$ across those adjacent sites. In plain language, this calculates the change in species densities for a site relative to its neighbors. We then further partition Jaccard dissimilarity into the portion that occurs because s_1 has proportionally greater or less density for all species (called *beta density*), or the portion that controls for these differences and only measures changes in the species proportions in s_1 (called *beta replacement*).

Results show that bird beta-diversity is highest in the transition from California to Nevada (Fig. 11.12). This is largely due to high beta-density in this area, where densities of many species (Barn swallow, Cliff swallow, Western meadowlark, etc.) are much lower

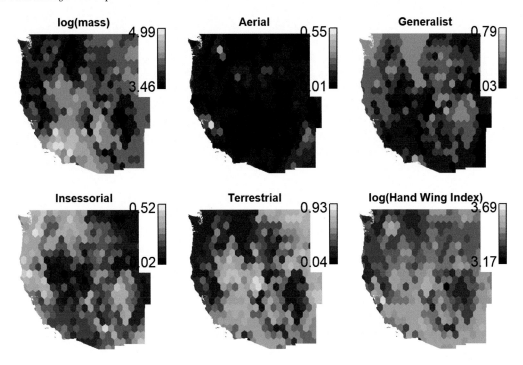

FIGURE 11.13: Average trait-values $\bar{T}_s$ from Eq. 11.16 calculated from bird densities across the Western US states and the trait-matrix **T** shown in Fig. 11.2.

in this transition area than southwestward towards the coast. However, changes in species composition after controlling for differences in bird density (labeled Beta_replacement in Fig. 11.13) are highest along the southern border of the modeled area. We note that it is also possible to calculate beta-diversity as the turnover in species traits (rather than the turnover of species densities as we show here), and this emphasizes spatial changes in community trait composition. However, we do not explore the topic further here.

11.7.3 Community Traits and Functional Diversity

Finally, we can visualize how species traits vary across the landscape. We have estimated the density of different species, and each species is associated with a species trait. We can therefore calculate a frequency distribution of species traits at each location, and can summarize this by calculating the mean (i.e., first moment of the frequency distribution) and variance (i.e., second moment) at each location. For simplicity of presentation, we use the function FD::dbFD , which provides a high-level interface to calculate common metrics of functional diversity [124, 125].

To calculate the trait average, we specifically calculate the *community-weighted trait* $\bar{T}_s$ at each location s as the weighted average of species trait T_c weighted by local density $d_{s,c}$ for each species:

$$\bar{T}_s = \sum_{c=1}^{n_c} T_c w_{s,c}$$

$$w_{s,c} = \frac{d_{s,c}}{\sum_{c'=1}^{n_c} d_{s,c'}}$$

(11.16)

where $w_{s,c}$ is the proportion of local density for species c. We here include traits T_c extracted from the columns from the trait-matrix that was used during parameter estimation (Fig. 11.2). The map of $\bar{T}_s$ then shows how traits are distributed across a landscape. For categorical or binary traits, the local trait average then represents the proportion of the local community that belongs to a given trait category. For this bird community (Fig. 11.13), these maps suggest that large-bodied birds with high dispersal ability (i.e., high Hand-Wing index) are relatively abundant in the arid southwestern area where NDVI is low (see Fig. 11.3). Similarly, Terrestrial birds represent the majority (approaching 90%) of the bird community in southeast Oregon and northern Nevada, while perching (i.e., Insessorial) birds are relatively abundance in the Pacific northwest and coastal areas that have dense vegetation (high NDVI).

We also calculate the variance of species traits to identify locations where species occupy a narrow or wide range of ecological traits (termed *functional diversity*). Here, we estimate spatial hotspots for functional diversity, and calculate several related measures of functional diversity:

- *Functional evenness*: whether individuals are evenly distributed across the range of values for a given trait;

- *Functional divergence*: whether individuals at a given location tend to cluster at high or low trait values, or are clustered near the spatially averaged trait value;

- *Functional dispersion*: whether any pair of individuals tends to have similar or different trait values.

When applied to our estimates of bird densities, functional divergence and dispersion are highly correlated and areas where Terrestrial birds are dominant also have relatively low functional diversity (Fig. 11.14). There is a growing interest in whether functional diversity is decreasing over time, called *biotic homogenization* [29], and a dynamic version of this same model could be used to calculate how the landscape of functional diversity has changed over time.

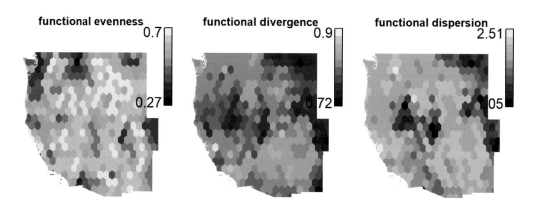

FIGURE 11.14: Measures of functional diversity including evenness, divergence, and dispersion.

11.8 Chapter Summary

In summary, we have showed that:

1. Evolutionary relatedness can be easily measured using an ultrametric phylogeny, which is available for many major taxa, and can be converted to a design matrix when assuming a Brownian motion model for evolution. Similarly, species traits are publicly available for many taxa, and can be converted to a similar design matrix;

2. Point-count data for multiple species can be fitted using a joint species distribution model (JSDM) that includes additive effects of phylogeny, traits, nonlinear responses to environmental covariates, as well as spatial residuals that are either independent or correlated among species. Using the JSDM, we can visualize the nonlinear effect of each covariate, the partial effect of each edge of the phylogenetic tree, and the spatial association for each species trait. This additive structure then allows a simple calculation of the variance in species density associated with each model component. Similarly, the spatial effect of each component can be mapped to visualize the contribution of evolution, traits, and environmental drivers to community composition;

3. Resulting densities can be summarized to answer different ecological questions. For example, hierarchical clustering can be used to identify ecological communities, and visualize what species are associated with each community. Alternatively, we can calculate beta diversity to identify areas where community density or composition rapidly changes. Finally, we can calculate community-weighted traits to identify how species traits vary across the landscape, or calculate community diversity to visualize which habitats support a wide or narrow set of traits.

11.9 Exercises

1. We introduce a phylogenetic tree for birds in Section 11.3, and we obtain trait measurements for 20 species that are present in that tree in Section 11.4. We also introduce *phylogenetic trait imputation* as a method to predict missing trait values, and here elaborate how this can be done in practice. Using the tree from Code 11.1, we see that `tree$edge` lists the edges, where each row indexes the nearest ancestory $p(g)$ for each node g, and `tree$edge.length` lists the length for each edge. Assuming that traits follow a Brownian motion model (Eq. 11.1), we can specify the conditional distribution for the trait value for each node given its ancestor. Please write TMB code (perhaps modifying Code 3.2) for the evolution of a continuous trait, and fit it to the hand-wing index (HWI) data for these 20 species. Then conduct a *jacknife experiment* (i.e., fitting to 19 of the 20 measurements, predicting the value for the excluded measurement, and repeating this 20 times to get the predicted HWI for each species), and compare performance with an alternative model (either prediction using the mean HWI, or fitting a linear model to other available traits). Does phylogenetic trait imputation improve performance for predicting HWI?

2. In Section 11.6, we compute the proportion of spatial variance in log-density that is attributed to phylogeny, traits, covariate responses, or residuals. However, our analysis ignored the degree to which phylogeny might itself explain traits or covariate responses. To explore this further, please modify the code used in that section to specify a linear response to NDVI and no other covariates. Then, using the conditional distribution defined in Exercise 1, please modify the code to specify that the covariate response β_c evolves along the bird phylogeny following a Brownian motion model, such that the covariate response is itself explained by phylogeny. How do estimates of the covariate responses β_c change when specifying a phylogenetic structure relative to the estimates without this? How might you decide whether phylogeny is a useful predictor of habitat responses based on these results?

12

A Decadal Forecast for Spatio-Temporal Models

12.1   Additional Tools for a Minimal Toolbox

The biological realism, computational efficiency, and breadth of application for spatio-temporal analysis is rapidly developing. Rapid advancement is exciting for researchers, but has obvious disadvantages when trying to categorize, summarize, and simplify methods. Most obviously, any textbook on spatio-temporal models risks being out-of-date by the time it is published. Additionally, current spatio-temporal research proceeds in parallel in multiple disciplines, and it is difficult to decide what to include in a *minimal toolbox* during early stages of development. To address these concerns, we conclude by noting a few topics that deserved inclusion but which we ultimately did not transform into chapters.

12.1.1   Non-separable Models

Throughout the book, we introduced a variety of simple ways to construct a precision or covariance matrix, which then represents correlations among ecological variables arising from different mechanisms (Table 12.1). These elementary building-blocks can be swapped in and out easily when building ecological models. They can also be used to represent alternative ecological assumptions, and model performance can then compared for a given data set. For example, an analyst can compare structural equation and factor-model construction of the covariance in movement for multiple animals (Section 4.5). Ecologists can then combine these building-blocks (using a Kronecker product or the SEPARABLE function in the density namespace) to construct the joint covariance across space, time, phylogeny, individuals, and species. For example, an analyst can build a spatio-temporal model by combining a spatial covariance with either temporal autocorrelation (the index model in Section 8.4) or a temporal factor model (the generalization of empirical orthogonal function models in Section 9.2). These techniques allow a model-building workflow that is:

- *Modular*, allowing researchers to quickly swap in new model elements, explore the consequence, and continue with model development. Modularity then allows the iterative model-building process to be guided by research goals rather than software constraints. We suspect that modularity contributes to the growing popularity of integrated population models [116], and facilitates the research-driven model development that ecologists expect when building statistical software;

- *Computationally efficient*, allowing researchers to use a single software platform (using automatic differentiation and the Laplace approximation) for statistical inference across huge range of model complexities. This then avoids any need to rewrite the model in new software as study goals change.

DOI: 10.1201/9781003410294-12

TABLE 12.1: Alternative ways to construct a sparse precision matrix, or project variables in a way that results in a specified covariance, where these can then be combined in a modular fashion to construct separable multivariate spatio-temporal models.

Name	Equation	Ecological use
1D autoregressive	3.15	Time-series or one-dimensional space
Conditional autoregressive	5.12	Areal (e.g., gridded multidimensional) space
Simultaneous autoregressive	B.13	Areal (e.g., gridded multidimensional) space
SPDE	5.14	Continuous two-dimensional space
Multiplying (projecting) by basis functions	5.3	Continuous variation among multiple dimensions
Factor model	4.15	Covariance among categories (individuals, species, etc.) resulting from ordination (a.k.a. rank reduction)
Structural equation model	4.23	Covariance among categories resulting from specified linear effects

We suspect that combining different precision or covariance constructors to create separable models will be sufficient for most researchers and practitioners.

However, we have avoided any comprehensive discussion of non-separable spatio-temporal processes, which arise from interactions among variables that are more complicated than a separable process can represent. Instead, we have emphasized a few ecological processes that might cause non-separable dynamics (e.g., movement in Chapter 10), and have noted how these can be specified as a Markov process by defining the expected value for spatial variables in time t based on a transformation of their value in time $t-1$ (e.g., Section 10.5). However, there is a large body of research regarding specific classes of non-separable models where the precision matrix can be constructed efficiently and then approximate a wide range of real-world mechanisms [181]. Many of these non-separable models are both mathematically elegant and require few modifications to the sparse precision matrices that we have emphasized throughout our presentation [137]. We agree that accurate and mechanistic forecasting for ecological dynamics likely requires developing a toolbox for simple non-separable models, and encourage ongoing research using a wide variety of statistical and computational approaches. However, we do not yet know which methods to include in a *minimal toolbox* for non-separable spatio-temporal models for ecological systems. A decade from now, we anticipate that any minimal toolbox for ecologists will include several efficient specifications for non-separable processes.

12.1.2   Size, Age, and Stage-structured Models

Demographic rates (e.g., metabolic rate, growth, and consumption) scale in predictable ways based on individual size, both across taxa and among individuals within a given taxon

[16]. For this reason, ecologists have included information regarding individual age, size, or stage in ecological models for over one-hundred years [138]. Such models are widely used by applied ecologists where, e.g., the specified age or size dynamics provides a *demographic motherboard* for an integrated population model, and all observations are then hardwired to this motherboard by defining how their distribution is linked to size and age-structured variables [116].

Information regarding size, age, stage, and sex is particularly interesting within a spatio-temporal model for several reasons. For example, many organisms show striking differences in habitat utilization between juveniles and adults. As an extreme example, Chinook salmon (*Oncorhynchus tshawytscha*) start their lives in freshwater habitats, migrate to ocean habitats for adult feeding and growth, and complete their lifecycle by spawning and then dying in freshwater. This results in an ecological teleconnection that affects ecosystem function in both marine and freshwater habitats [3, 211]. Similarly, migratory monarch butterflies complete several generations per year but still perform an annual migration that links winter habitats in Mexico and summer habitats in the eastern United States. For these and other species, analyzing spatial distribution jointly for multiple stages is essential both for evaluating the likely outcome of potential conservation policies, as well as for understanding the ecological mechanisms that underlie observed patterns.

Given this ecological importance, there is a growing literature regarding multivariate spatial and spatio-temporal models that treat different sizes or ages as separate spatial variables (i.e., combining methods from Chapters 8 and 11). The simplest way to proceed is by defining a separable multivariate spatio-temporal model, e.g., using a factor model (Section 4.4.1) to represent covariance among sizes or ages [111]. However, this approach does not use demographic information about size or age transitions. Some studies have added size- and age-dependent growth and survival within spatio-temporal models, and these studies confirm that many classic size and age-structured models can be efficiently embedded within a spatio-temporal analysis. The simplest way to do so is to replace one or more scalar variables (e.g., abundance-at-age) with vector variables (e.g., density-at-age for a discretized set of areas), and then define a spatial precision matrix that represents spatial covariance among those areas. In some cases, the specified model for time-series age- or size-structured dynamics results in a simple expression for covariance among ages and over time, for example, when assuming individual size increases linearly with age [121]. In these cases, adding a spatial covariance results in a separable multivariate spatio-temporal covariance, which then allows efficient computation. In other cases, the specified time-series dynamics for ages/sizes does not have a closed-form covariance [219]. These cases result in a non-separable spatio-temporal process, but they can still be specified and fitted by specifying dynamics as a series of conditional distributions (Eq. 8.5). This specification still results in a sparse precision matrix (and reasonably efficient computation) when dynamics are Markovian.

Despite progress regarding age- and size-structured spatio-temporal models, there has been less research on combining structured dynamics with other spatial processes such as individual movement when estimating spatially correlated variables. This lack of research is perhaps unsurprising, given that efficient approaches to analyze animal movement are also an ongoing topic for research (e.g., Chapter 10). We therefore recommend ongoing research to add mechanistic detail including movement to size and age-structured spatio-temporal models. In some cases, there may be useful special cases or analytical techniques that result in efficient computation [137].

12.1.3 Empirical Dynamic Models

Throughout this textbook, we have typically assumed that analysts can specify a function that closely approximates system dynamics (e.g., Eq. 3.1 or 8.5), which can then be extended by estimating spatio-temporal correlations in ecological variables. For simplicity of presentation, we have typically approximated this dynamical function as a linear model, such that we can fit a Gompertz or other autoregressive process (e.g., Chapter 3). Specifying a parametric function (whether linear or nonlinear) then allows us to efficiently fit the hypothesized dynamics to data, and subsequently predict densities across space or time.

However, fitting a parametric function representing system dynamics will perform poorly when ecological processes are either:

1. *Poorly known*: in many cases, ecologists do not know a simplified equation that provides an accurate approximation to ecological dynamics. This occurs both because important system variables have not been identified or measured, and because variables may interact in ways that have not been studied for that system. In both cases, this then results in *ecological surprise* [44];

2. *Chaotic*: in other cases, ecological systems involve rapid and strong linkages among system variables (e.g., fast population growth rates in short-lived taxa). In these cases, ecological dynamics may become chaotic [192], i.e., where small perturbations are expected to grow in importance over time such that forecasts cannot be reliably computed past a certain time-horizon. Chaotic dynamics complicate any simple process for fitting or interpreting the output of linear state-space models [179].

We suspect that one or both cases will apply to many important problems in ecology, earth sciences, and public health.

Fortunately, both unknown and chaotic dynamics can be addressed by reconstructing system dynamics from lagged measurements of previous dynamics. This involves estimating a nonlinear function $g(.)$ representing system changes:

$$\mathbf{n}_{t+1} - \mathbf{n}_t = g(\mathbf{n}_t, \mathbf{n}_{t-1}, \mathbf{n}_{t-2}, ...) \tag{12.1}$$

where the function $g(.)$ representing system dynamics treats past system state as a covariate while adding flexibility by applying a lag operator for basis expansion (see Section 5.2). Eq. 12.1 then uses this lag basis-expansion to predict the next system change, and theory shows that this lag-basis expansion can capture important properties of the original system [217]. Function g can be estimated as a Gaussian process or using nonlinear estimation methods, where the resulting estimator is currently called *empirical dynamic modelling* (EDM) [159].

There are many productive avenues to combine EDM and spatio-temporal analysis. The most obvious involves discretizing space and defining system variables in each area as a separate variable in the EDM. Such an approach has revealed linkages in blue-crab dynamics along the eastern United States [193]. Alternatively, it is conceptually straightforward to introduce a small number of spatial basis functions to define spatial projection matrix $\mathbf{A}$ (e.g., analogous to the interpolation matrix used in the SPDE method, see Section 5.5.2). Analysis would then specify local measurements using the spatial projection of EDM variables, $\mathbf{An}$. Parameters defining this projection matrix could then be estimated as the same time as parameters defining the EDM function g. Whether using EDM or other methods, we recommend ongoing research regarding *semi-parametric* models that can estimate system dynamics flexibly at the same time as spatio-temporal variables [229].

12.2   Addressing Barriers to Adoption

Throughout this book, we argue that spatio-temporal models are ready for real-world applications to inform public policy, and will offer useful insights or improvements over conventional approaches. However, in policy-relevant fields such as econometrics, public health, and fisheries science, we anticipate that adoption of spatio-temporal methods will be rate-limited by three inter-connected steps:

1. *Evaluation*: in many governments, scientific information is discussed by elected or appointed representatives who decide on real-world regulatory changes. These decisions are typically made using incomplete information, where the importance of scientific evidence is weighed based on its perceived credibility. In these venues, it is important to allow a broad range of researchers to independently evaluate and debate the relative merits of different methodological approaches. Criteria for evaluation vary among disciplines, and methods must often be adapted for each field of public policy. Ongoing research is badly needed to evaluate spatio-temporal model performance within the context of specific policy goals for use in real-world applications;

2. *Education*: to both apply and progressively improve spatio-temporal models, we anticipate needing ideas and perspectives from a broader range of researchers than currently conduct spatio-temporal research. We therefore recommend that more graduate programs (not restricted to statistics departments) offer some coursework regarding the use of spatio-temporal methods in their discipline;

3. *Simplification*: to support these tasks (evaluating and teaching spatio-temporal methods), we see a continuing need to identify a minimal toolbox for spatio-temporal modelling. This toolbox should perform well on average for new real-world problems, be sufficiently general to cover most use-cases, and minimize barriers-to-entry for scientists seeking to understand, review, or adapt methods. This textbook represents one perspective on defining such a minimal toolbox, but we encourage other researchers to propose alternative approaches to simplify the presentation of spatio-temporal vocabulary and techniques (e.g., [262]). Presumably, educators could then compare these alternatives based on student outcomes and simplicity of presentation.

These three rate-limiting steps are being addressed in myriad different ways by researchers worldwide. We're excited to see the next decade of research in spatio-temporal models, and hope you are too.

A

Acknowledgments

This work would not have happened without contributions from many colleagues over the past decade. We list some of these contributions, and apologize to those colleagues who we forget to mention. Specifically, we thank Hans Skaug for discussions regarding the Laplace approximation, how to include the SPDE method in TMB, and the implementation and performance of restricted maximum likelihood. We also thank Steve Munch for discussing phylogenetic comparative methods and empirical dynamical models. Thanks to Eli Holmes, Eric Ward, and Mark Scheuerell for past discussions regarding Gompertz density dependence, Jay Ver Hoef for discussing simultaneous autoregressive processes, and Yumi Arimitsu and Bill Sydeman for discussing climate velocity. Finally, we thank Sergey Feldman for discussions regarding causal analysis and do-calculus, and Julie Nielsen, Kevin Siwicke, and Devin Johnson for discussing the hidden Markov model analysis of the archival tagging data.

We also feel lucky to include data that were collected and compiled by many different teams worldwide. We thank Carey Kuhn and Jeremy Stirling for providing data for the northern fur seal case study (as well as discussion of results), and Susanne McDermott, Julie Nielsen, and many team members who collected and provided the Pacific cod archival tag data set. We thank Richard Condit, Stephen Hubbell, and many others collecting, distributing, and permission to use Barro Colorado data, and the huge number of contributors to the Breeding Bird survey. We thank the Groundfish Assessment Program at the Alaska Fisheries Science Center for conducting the Bering Sea shelf bottom trawl survey annually since 1982, and the SEAMAP Groundfish Trawl Survey, NMFS Pelagic Acoustic Trawl Survey, and NMFS Red Snapper/Shark Bottom Longline Surveys for collecting the red snapper case-study data (as well as Arnaud Grüss for compiling these data). Finally, thanks to Kevin Siwicke for guidance in accessing the sea ice concentration data, and Connie Okasaki for guidance in accessing the ozone concentration data.

Finally, we're grateful for ideas and editorial contributions from Baptiste Alglave, Cheryl Barnes, Jie Cao, Matt Cheng, Curry Cunningham, Tim Essington, Devin Johnson, Sami Kivelä, Cole Monnahan, Julie Nielsen, and Sam Urmy. Remaining errors are obviously our own. We also thank Anders Nielsen, Casper Berg, Tim Essington, Devin Johnson, and many others for encouragement to write this book, and Nobel Hendrix for co-teaching the initial version of a class that inspired this book. Finally, we thank countless co-authors, reviewers, journal editors, R-package developers, and colleagues for guiding our research regarding spatio-temporal models over the past decade.

DOI: 10.1201/9781003410294-A

B

Appendices

B.1 Mathematical notation

We strive to use mathematical notation that is consistent with previous publications and across chapters, while also adhering to a few notational conventions. We generally use:

1. Vector-matrix notation, i.e., bold-nonitalic-lowercase for vector (e.g., $\mathbf{y}$), bold-nonitalic-uppercase for matrices (e.g., $\mathbf{Y}$), and plainface-italic-lowercase for scalars (e.g., y);

2. Greek symbols (e.g., β) for estimated parameters and Roman symbols (e.g., b) for data;

3. Parentheses for indicating the value of a continuous function at a given space and/or time (e.g., $h(s)$), and subscripts for indicating the value when the function is being evaluated at a fixed set of modeled locations (e.g., h_s).

Given these constraints, we typically use the following notation for the following common expressions (Table B.1), variables (Table B.2) and indices (Table B.3). Individual chapters then augment these conventions, based upon specifics of each topic.

We note in particular that we use, e.g., $Y \sim \text{Poisson}(\lambda)$ to indicate that a variable follows the Poisson distribution (or whatever distribution is named on the right-hand-side of an expression), and also use $\Pr[Y = y] = \text{dPoisson}(y|\lambda)$ for evaluating the probability mass or density function itself. We will use both notations depending on context, and therefore use both when referring to distributions.

B.2 Common Link Functions and Distributions

We believe that ecologists are broadly familiar with Generalized Linear Models (GLMs) (see Section 1.4), and therefore typically discuss spatio-temporal models using a vocabulary associated with GLMs. In this vocabulary, measurement y_i for each sample i is assumed to follow a distribution with mean μ_i, and this mean in turn is calculated by applying an inverse-link function g^{-1} to a linear predictor p_i that is calculated by adding together some linear transformation of estimated fixed and random effects.

$$y_i \sim f(\mu_i, \sigma^2)$$
$$g(\mu_i) = p_i \tag{B.1}$$

We therefore provide a brief summary of common link functions (Table B.4) and distributions (Table B.5). We acknowledge that there are many more distributions and link functions that are useful and convenient, but restrict attention to those that arise in the main text.

DOI: 10.1201/9781003410294-B

TABLE B.1: A brief summary of common mathematical expressions used throughout the presentation.

Symbol	Definition
$\Pr[X\|\theta]$	Probability of event X given parameters θ, or (using loose notation) the probability density for such an event
$\mathcal{L}(\theta; Y)$	Likelihood for parameters θ given data Y
$X \sim \text{Normal}(\mu, \sigma^2)$	Random variable X follows a normal distribution with mean μ and variance σ^2
$\omega \sim \text{MVN}(\mathbf{0}, \boldsymbol{\Sigma})$	Random vector ω follows a multivariate normal distribution with mean $\mathbf{0}$ and covariance $\boldsymbol{\Sigma}$
$C \sim \text{Poisson}(\lambda)$	Count C follows a Poisson distribution with intensity λ
$\Pr[C = c] = \text{dPoisson}(c\|\lambda)$	The probability that variable C takes value c is calculated using a Poisson probability mass function with parameter λ
$\nabla h(\mathbf{s})$	The gradient of a function at location $\mathbf{s}$, which results in a vector $\mathbf{v}(s)$ with length equal to the number spatial dimensions $\mathbf{s}$. This gradient reduces to a standard derivative calculation (e.g., $\frac{d}{ds}h(s)$) when space is defined in one dimension, and is calculated sequentially for each dimension otherwise
$\nabla^2 h(s)$	The Laplace operator of a function h evaluated at location $\mathbf{s}$, representing curvature in h around that location. This Laplace operator reduces to the standard second-derivative calculation (e.g., $\frac{d^2}{ds^2}h(s)$) when space is defined in one dimension. Otherwise, it is calculated as the sum of second partial derivatives for each individual dimension (*unmixed partial derivatives*): $$\nabla^2 h(\mathbf{s}) = \sum_{i=1}^{n} \frac{\partial^2}{\partial s_i^2} h(\mathbf{s}) \qquad (B.2)$$ where n is the number of dimensions involved.
$\mathbf{u} \cdot \mathbf{v}$	The dot-product $\mathbf{u}\cdot\mathbf{v}$ of two vectors $\mathbf{u}$ and $\mathbf{v}$ is calculated by taking the sum of the product of each dimension individually, $\mathbf{u} \cdot \mathbf{v} = \sum_{s=1}^{n_s} u_s v_s$ where n_s is the length of each vector. This arises, e.g., when calculating the dot-product $\mathbf{v}(s) \cdot \nabla d(s,t)$ of a gradient field $\nabla d(s,t)$ and an estimated response in each dimension $\mathbf{v}(s)$

TABLE B.2: A partial list of symbols used for common variables, quantities, or functions.

Symbol	Definition
$\boldsymbol{\Sigma}$	Spatial covariance matrix
$\mathbf{V}$	Covariance for non-spatial variables, e.g., among individuals or species
$\mathbf{Q}$	Precision (inverse covariance) matrix
$d_{s,t}$	Population density at location s in time t
$\pi(s)$	Sampling or inclusion probability density, i.e., the relative probability that location s will have a sample available
β_t	Variable indexed over time (i.e., temporal main effect in GLMM)
ω_s	Variable correlated across space (i.e., spatial main effect in GLMM)
$\epsilon_{s,t}$	Variable correlated across space and time (i.e., spatio-temporal interaction in GLMM)
$S_j(t)$	Random variable representing location for individual j at time t
$\mathcal{D}$	The spatial domain for which a model is defined, $s \in \mathcal{D}$
δ	Random variable drawn from a normal distribution, i.e., representing variation in location over time
Δ_t	A specified time-interval used when discretizing time
Δ_s	A specified distance used when discretizing space
$\mathbf{A}$	Adjacency matrix, representing whether two spatial cells share an edge, where instantaneous movement is nonzero only for adjacent cells
$h(S)$ and h_s	Habitat preference function either defined continuously across space $h(S)$ or discretized on a grid h_s
$\boldsymbol{\Lambda}$	Estimated loadings matrix, representing some subset of columns of the Cholesky matrix associated with covariance $\boldsymbol{\Sigma} = \boldsymbol{\Lambda}\boldsymbol{\Lambda}^t$
$g(\mu) = X\gamma$	Link function g used to transform the continuous and unbounded domain of a linear predictor $X\gamma$ to the domain for the central tendency μ of a random variable

TABLE B.3: A list of symbols used when indexing elements from a matrix or array.

Symbol	Definition
s	Location of sample or event
g	Location used for new prediction, e.g., approximating an integral using integration points
t	Time
c	Category in a multivariate model (e.g., species)
i	Sample
j	Individual
k	Covariate
l	Factor
m	Ancestor

TABLE B.4: A brief summary of common link functions used when specifying a generalized linear model, where μ is the central tendency parameter used subsequently in a probability distribution and p_i is the linear predictor computed from fixed and random effects.

Link function	Syntax	Inverse-link	Interpretation
log	$\log(\mu) = p$	$\mu = e^p$	Constrains μ to be positive, and is therefore appropriate when modelling population densities
logit	$\text{logit}(\mu) = p$	$\mu = \frac{e^p}{1+e^p}$	Constrains μ to be between 0 and 1 (i.e., the range for modelling the probability of an event) in a way that is symmetric
Complementary log-log	$\text{cloglog}(\mu) = p$	$\mu = 1 - e^{-e^p}$	Constrains μ to be between 0 and 1 in a way that is nonsymmetric and specifically mimics the probability that a count greater than zero arises from a Poisson distribution: $$\begin{aligned} C &\sim \text{Poisson}(\lambda) \\ \log(\lambda) &= p \end{aligned} \quad \text{(B.3)}$$ where $\Pr(C > 0) = 1 - e^{-\lambda}$ and hence $\text{cloglog}(\Pr(C > 0)) = p$

TABLE B.5: A brief summary of common distributions for use in generalized linear models, where Y is the resulting random variable, μ is the central tendency parameter used subsequently in a probability distribution, and other parameters are defined.

Distribution	Syntax	Range	Interpretation
Normal	$Y \sim \text{Normal}(\mu, \sigma^2)$	$-\infty < Y < \infty$	Response arises as the sum of many different processes that cumulatively have mean μ and variance σ^2
Bernoulli	$Y \sim \text{Bernoulli}(\mu)$	$Y \in \{0, 1\}$	$\Pr(Y = 1) = \mu$ and $\Pr(Y = 0) = 1 - \mu$
Binomial	$Y \sim \text{Binomial}(N, \mu)$	$Y \in \{0, 1, ..., N\}$	Response arises as the sum from N independent Bernoulli distributions each having probability μ
Poisson	$Y \sim \text{Poisson}(\mu)$	$Y \in \{0, 1, 2, ...\}$	Response arises from a Binomial distribution with a very large size N and low probability for each, where $\mu = \pi N$
Gamma	$Y \sim \text{Gamma}(\sigma^{-2}, \mu\sigma^2)$	$Y > 0$	Response is continuous and positive (e.g., animal size) with mean μ and coefficient of variation $\sigma = \frac{\sqrt{\text{Var}(Y)}}{\text{Mean}(Y)}$, using shape σ^{-2} and scale $\mu\sigma^2$
Lognormal	$\log(Y) \sim \text{Normal}(\mu, \sigma^2)$	$Y > 0$	Response is continuous and positive, with mean $\mathbb{E}(Y) = e^{\mu + 0.5\sigma^2}$ and coefficient of variation $\frac{\sqrt{\text{Var}(Y)}}{\text{Mean}(Y)} = \sqrt{e^{\sigma^2} - 1}$, using meanlog μ and sdlog σ, where the lognormal distribution has more density in the tails than the gamma distribution
Tweedie	$Y \sim \text{Tweedie}(\mu, \phi, \psi)$	$Y \geq 0$	Response is continuous and positive but also has some density at $Y = 0$, where this latter is not within the range of the lognormal or gamma distributions. The variance $\text{Var}(Y) = \phi\mu^\psi$, where $1 < \psi < 2$, and the Tweedie can be interpreted as arising from a compound Poisson-gamma distribution

B.3 Rejection Sampling

We use *rejection sampling* to sample values from low-dimensional functions. It typically becomes less efficient for a higher number of dimensions, and therefore doesn't have as much practical use for higher dimensions (e.g., integrating across random effects). Rejection sampling seeks to generate samples from a function $f(x)$, called the "target distribution". To do so:

1. Calculate the maximum $M = \text{argmax}_x(f(x))$ of the target distribution;

2. Draw a sample x^* from a function $g(x)$, which is called the "proposal distribution";

3. Evaluate the target and proposal distribution at that sample, $F = f(x^*)$ and $G = g(x^*)$;

4. Calculate the "acceptance probability", $P = F/(MG)$;

5. Draw a sample p^* from a uniform distribution from 0 to 1;

6. If the uniform sample is less than the acceptance probability, $p^* < P$, accept and record the sample x^* or otherwise reject it;

7. Repeat steps 2 through 6 until the desired number of samples have been accepted, where a larger number of samples will be a more precise approximation to the target distribution.

In the main text, we typically use a uniform distribution for the proposal distribution, $g(s) = 1$. Rejection sampling is more efficient (i.e., a smaller fraction of samples are rejected) when the target and proposal distributions are similar, and will be perfectly efficient (i.e., all samples are accepted) when $g(x) = f(x)$.

B.4 Matrix Decomposition

In the main text, we introduce models that include covariance across locations s, times t, and species c. These involve specifying a covariance Σ or its matrix inverse $\mathbf{Q} = \Sigma^{-1}$, termed the *precision matrix*. Both Σ and $\mathbf{Q}$ are square and symmetric with dimension $n_i \times n_i$, and we here review some techniques that are useful to interpret or efficiently compute these important matrices.

B.4.1 Eigendecomposition

Arguably the most important technique for decomposing a covariance or precision matrix is the *eigendecomposition*. This function is accessible in R using function `eigen`, and it returns a matrix of eigenvectors $\mathbf{V}$ and a vector of eigenvalues λ defined such that:

1. *Decomposition*: we can reconstitute the original matrix $\Sigma = \mathbf{V}\Lambda\mathbf{V}^{-1}$, where $\Lambda = \text{diag}(\lambda)$ is a diagonal matrix of eigenvalues;

2. *Orthogonal and unit-length eigenvectors*: each column of $\mathbf{V}$ is an eigenvector $\mathbf{v}_j$, and all eigenvectors are orthogonal and have "length" of one. This means that $\sum_{i=1}^{n_i} v_{i,j_1} v_{i,j_2} = 0$ for any pair of eigenvectors $\mathbf{v}_{j_1}$ and $\mathbf{v}_{j_2}$, and $\sum_{i=1}^{n_i} v_{i,j}^2 = 1$;

3. *Ordered eigenvalues*: the eigenvalues are ordered from largest to smallest, i.e., $\lambda_i > \lambda_{i+1}$ for all n_i.

The function eigen becomes very slow as matrix size increases. For sparse matrices, however, we can instead apply the eigendecomposition using the igraph package [38], and this remains computationally feasible.

In the main text, we use the eigendecomposition in several ways:

- *Evaluate convergence*: fitting a maximum likelihood model using the Laplace approximation involves both inner and outer optimizers (Section 2.3). Both optimizers must converge for the model to be fitted, and convergence is assessed by identifying the vector of values that minimizes the joint negative log-likelihood (for the inner optimizer) or the log-marginal likelihood (for the outer optimizer), and that the matrix of second derivatives (the Hessian matrix) evaluated at that vector of values is positive definite. We can check that the Hessian matrix is positive definite by applying an eigendecomposition and confirming that all eigenvalues are positive. If any of these eigenvalues are sufficiently close to zero, then the Hessian matrix cannot be inverted and either the Laplace approximation cannot be computed (for the inner optimizer) or the estimation covariance is not defined (for the outer optimizer). Alternatively, if any of these eigenvalues are negative, then the optimizer did not identify the minimum of the function and was not converged. Furthermore, we can identify the eigenvalues that are problematic (either zero or negative), and then extract the eigenvectors associated with those problematic eigenvalues. Those eigenvectors then have a nonzero value for those fixed and/or random effects that are associated with convergence issues, and we can focus our attention on identifying problems for those parameters. We use these properties to diagnose convergence issues in Section 2.5;

- *Visualize basis functions*: if we have a multivariate normal distribution with a given covariance:

$$\omega \sim \text{MVN}(\mathbf{0}, \boldsymbol{\Sigma}) \tag{B.4}$$

then can instead express it as a series of normal distributions:

$$\omega = \sum_{i=1}^{n_i} \mathbf{v}_i \sqrt{\lambda_i} \delta_i \tag{B.5}$$

where each $\delta_i \sim \text{Normal}(0, 1)$ is drawn from a standard normal distribution. Each eigenvector $\mathbf{v}_i$ projects a normally distributed error with standard deviation of 1 to the domain of the specified covariance, while the square-root of each eigenvalue $\sqrt{\lambda_i}$ defines the standard deviation for that projected vector. We use this property extensively in Section 5.5 (e.g., Fig 5.9) to visualize the eigenvectors with spatial basis functions that are contained within a given spatial covariance $\boldsymbol{\Sigma}$;

- *Diagonal form*: the eigendecomposition provides a diagonal form for any matrix. Given the eigendecomposition of the covariance $\boldsymbol{\Sigma} = \mathbf{V}\boldsymbol{\Lambda}\mathbf{V}^{-1}$, we can then calculate the eigendecomposition of the precision as $\mathbf{Q} = \mathbf{V}\boldsymbol{\Lambda}^{-1}\mathbf{V}^{-1}$, where $\boldsymbol{\Lambda}^{-1} = \text{diag}(\lambda^{-1})$. This shows that the basis functions $\mathbf{V}$ of a given process are the same when expressed as a covariance or precision matrix, although the eigenvalues are inverted and also ordered from smallest to largest. In Section 5.5, we actually calculate the eigendecomposition of the precision matrix $\mathbf{Q}$ (which is sparse and hence it is cheaper to calculate the eigendecomposition), and then invert the smallest eigenvalues to extract the basis vectors for a given covariance;

Similarly, in Section 10.3 we define a continuous-time Markov chain for movement rates with rate matrix $\dot{\mathbf{M}}$, such that integrated movement $\mathbf{M}$ over interval Δ_t is calculated as $\mathbf{M} = e^{\dot{\mathbf{M}}\Delta_t}$. Rate matrix $\dot{\mathbf{M}}$ is typically sparse (i.e., non-zero only for adjacent areas), whereas integrated movement $\mathbf{M}$ is dense (i.e., animals can move anywhere as long as all areas are connected). Using this property of diagonal matrices, we can calculate the eigen-decomposition of the sparse rate matrix $\dot{\mathbf{M}} = \mathbf{V}\dot{\mathbf{\Lambda}}\mathbf{V}^{-1}$ relatively cheaply, and this has the same eigenvectors as the integrated movement matrix $e^{\dot{\mathbf{M}}\Delta_t} = \mathbf{V}\mathbf{\Lambda}\mathbf{V}^{-1}$ where $\lambda_i = e^{\dot{\lambda}_i \Delta_t}$. The stationary distribution (representing expected long-term habitat utilization) is the basis formed from any eigenvectors with an eigenvalue of 1. So we can calculate this stationary distribution even for an enormous number of locations by calculating the dominant eigenvectors of the sparse movement rate matrix, e.g., using R function `igraph::arpack`.

B.4.2 Cholesky Decomposition

We introduce *factor models* in Section 4.4.1 and use them repeatedly in the book. Here, we provide a brief background on the Cholesky decomposition, which is one interpretation of these factor models.

The Cholesky decomposition is available for dense matrices in R using `chol`, or using the `Matrix` package to access various options for sparse matrices. Applied to a covariance $\mathbf{\Sigma}$, it returns a matrix $\mathbf{L}$ with zeros above the diagonal (i.e., a *lower-triangle matrix*) such that $\mathbf{\Sigma} = \mathbf{L}\mathbf{L}^T$. If we have a multivariate normal distribution with a given covariance:

$$\omega \sim \text{MVN}(\mathbf{0}, \mathbf{\Sigma}) \tag{B.6}$$

then can instead express it as a series of normal distributions:

$$\omega = \mathbf{L}\delta \tag{B.7}$$

where each $\delta_i \sim \text{Normal}(0, 1)$ is drawn from a standard normal distribution. In TMB we can easily specify an estimated loadings matrix $\mathbf{L}$ to be lower-triangle, and estimating the product of this with a series of independent and normally distributed errors then yields a random effect that has covariance $\mathbf{\Sigma} = \mathbf{L}\mathbf{L}^T$.

B.5 Simultaneous Autoregressive Process

A *simultaneous autoregressive (SAR) process* involves re-arranging a spatial variable or model to express it as a simultaneous equation [251]:

$$\mathbf{x} = \rho\mathbf{B}\mathbf{x} + \epsilon$$
$$\epsilon = \text{MVN}(\mathbf{0}, \mathbf{\Sigma}) \tag{B.8}$$

where matrix $\mathbf{B}$ represents spatial dependencies in an areal or point-process model, typically calculated from an adjacency matrix. Subtracting $\rho\mathbf{B}\mathbf{x}$ from both sides yields:

$$(\mathbf{I} - \rho\mathbf{B})\mathbf{x} = \epsilon$$
$$\epsilon = \text{MVN}(\mathbf{0}, \mathbf{\Sigma}) \tag{B.9}$$

Then multiplying both sides by $(\mathbf{I} - \rho\mathbf{B})^{-1}$ yields:

$$\mathbf{x} = (\mathbf{I} - \rho\mathbf{B})^{-1}\epsilon$$
$$\epsilon = \text{MVN}(\mathbf{0}, \mathbf{\Sigma}) \tag{B.10}$$

We can then multiply $(\mathbf{I} - \rho\mathbf{B})^{-1}$ into the covariance for ϵ to yield:

$$\mathbf{x} \sim \text{MVN}\left(\mathbf{0}, (\mathbf{I} - \rho\mathbf{B})^{-1}\mathbf{\Sigma}((\mathbf{I} - \rho\mathbf{B})^{-1})^{T})\right) \tag{B.11}$$

This expression therefore defines the covariance for x:

$$\mathbf{V} = \text{Var}(\mathbf{x}) = (\mathbf{I} - \rho\mathbf{B})^{-1}\mathbf{\Sigma}((\mathbf{I} - \rho\mathbf{B})^{-1})^{T} \tag{B.12}$$

Alternatively, the inverse covariance ("precision") can then be computed directly as:

$$\mathbf{V}^{-1} = (\mathbf{I} - \rho\mathbf{B})\mathbf{\Sigma}^{-1}(\mathbf{I} - \rho\mathbf{B})^{T} \tag{B.13}$$

and constructing the precision matrix does not require matrix inversion if Σ is diagonal.

For example, recall that the state-space Gompertz model defines a 1st-order autoregressive process for $x_t = \log(b_t)$:

$$x_t = \begin{cases} \frac{\alpha}{1-\rho^2} + \epsilon_t & \text{if } t = 1 \\ \alpha + \rho x_{t-1} + \epsilon_t & \text{if } t > 1 \end{cases} \tag{B.14}$$
$$\epsilon_t \sim \text{Normal}(0, \sigma^2)$$

This can then be rewritten as a simultaneous equation:

$$\mathbf{x} = \frac{\alpha}{1-\rho^2} + \mathbf{z}$$
$$\mathbf{z} = \rho\mathbf{B}\mathbf{z} + \epsilon \tag{B.15}$$
$$\epsilon = \text{MVN}(\mathbf{0}, \mathbf{\Sigma})$$

where:

$$\mathbf{B} = \begin{bmatrix} 0 & 0 & 0 & 0 \\ -1 & 0 & 0 & 0 \\ 0 & -1 & 0 & 0 \\ 0 & 0 & -1 & 0 \end{bmatrix} \tag{B.16}$$

we can then use the SAR form to construct the precision matrix $\mathbf{V}^{-1}$ directly from $\mathbf{B}$ and ρ using Eq. B.13, and then plug that into the expression [240]:

$$\mathbf{x} = \mathbf{z} + \frac{\alpha}{1-\rho^2}$$
$$\mathbf{z} = \text{MVN}(\mathbf{0}, \mathbf{V}^{-1}) \tag{B.17}$$

Furthermore, we can gain some intuition about the sparsity of $\mathbf{V}^{-1}$ by seeing that:

$$\mathbf{I} - \rho\mathbf{B} = \begin{bmatrix} 1 & 0 & 0 & 0 \\ -\rho & 1 & 0 & 0 \\ 0 & -\rho & 1 & 0 \\ 0 & 0 & -\rho & 1 \end{bmatrix} \tag{B.18}$$

and

$$(\mathbf{I} - \rho\mathbf{B})^T = \begin{bmatrix} 1 & -\rho & 0 & 0 \\ 0 & 1 & -\rho & 0 \\ 0 & 0 & 1 & -\rho \\ 0 & 0 & 0 & 1 \end{bmatrix} \tag{B.19}$$

Assuming that $\Sigma = \sigma^2\mathbf{I}$, this then yields:

$$(\mathbf{I} - \rho\mathbf{B})\Sigma^{-1}((\mathbf{I} - \rho\mathbf{B})^T) = \sigma^{-2} \begin{bmatrix} 1 & -\rho & 0 & 0 \\ -\rho & 1+\rho^2 & -\rho & 0 \\ 0 & -\rho & 1+\rho^2 & -\rho \\ 0 & 0 & -\rho & 1+\rho^2 \end{bmatrix} \tag{B.20}$$

where this closely resembles the joint precision presented without derivation in Eq. 3.15. We note however, that the two differ somewhat in the variance of the first term x_1 corresponding to the 1st row of $\mathbf{V}^{-1}$, due to differences in boundary effects. Importantly, this construction yields the tri-diagonal structure that was automatically detected by TMB using the conditional form of the Gompertz model (Fig. 3.5).

B.6 Matrix Exponential Computation

In Chapter 10, we repeatedly use the matrix exponential operator to solve a discretized movement process. The matrix exponential arises naturally when solving a differential equation:

$$\frac{\partial}{\partial t}\mathbf{n}_t^T = \mathbf{n}_t^T \dot{\mathbf{M}} \tag{B.21}$$

where this represents a continuous-time Markov Chain with instantaneous transition-rate matrix $\dot{\mathbf{M}}$ [79, 234]. The solution is then:

$$\mathbf{n}_{t+\Delta_t}^T = \mathbf{n}_t^T e^{\Delta_t \dot{\mathbf{M}}} \tag{B.22}$$

where this arises from integrating the instantaneous rate from time t to $t + \Delta_t$.

We highlight three convenient methods for computing this solution in practice:

1. *Existing software*: good statistical software will generally have a matrix exponential function. We recommend expm::expm in R, or expm in TMB. These are generally sufficient when state vector $\mathbf{n}$ is low-dimensional;

2. *Euler approximation*: we introduce the Euler approximation in Eq. 10.8. There, we claim that it approximates an exponential function with a series of N linear calculations. To see this further, we take Eq. 10.8, replace $\dot{\mathbf{M}}$ with a scalar (intrinsic growth rate r), and replace vector $\mathbf{n}$ with a scalar (initial population size b_0):

$$b(t) = b_0 e^{rt} \approx b_0 \prod_{n=1}^{N} \left(1 + \frac{rt}{N}\right) \tag{B.23}$$

We can easily calculate the Euler approximation and visualize how it converges on the true function $b(t) = b_0 e^{rt}$ as the number of sub-intervals N increases. In

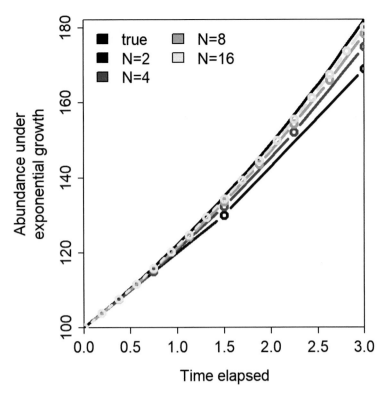

FIGURE B.1: A demonstration of the Euler approximation, applied to the exponential growth function $\frac{d}{dt}b(t) = rb(t)$ and showing either the known solution $b(t) = b_0 e^{rt}$ as black line, or the Euler approximation (Eq. B.23) with an increasing number of sub-intervals N as colored lines and dots at the breakpoints between linear segments.

Fig. B.1 we can see that even $N = 2$ can roughly approximate the exponential growth function, but higher values of N approximate the true function more closely. This same intuition applies to a vector exponential operator, although results with two or more variables are somewhat harder to visualize than the simple one-dimensional example in Fig. B.1.

3. *Uniformization*: we also introduce *uniformization* in Section 10.4 without providing any details. This approach involves calculating the series approximation to the exponential, with the number of series terms set to achieve a desired accuracy. Recall that the scalar exponential can be approximated as a series:

$$e^x \approx \sum_{n=0}^{N} \frac{x^n}{n!} \tag{B.24}$$

where N is set to some sufficiently high number. We can approximate the matrix exponential using this same series:

$$e^{\mathbf{A}} \approx \sum_{n=0}^{N} \frac{\mathbf{A}^n}{n!} \tag{B.25}$$

If matrix $\mathbf{A}$ is sparse we can then approximate this summation using a series of sparse matrix calculations, without ever computing $e^{\mathbf{A}}$ itself, using a recursive formula:

$$\mathbf{b}^T e^{\mathbf{A}} \approx \sum_{n=0}^{N} \tilde{\mathbf{b}}_n^T \tag{B.26}$$

where:

$$\tilde{\mathbf{b}}_n^T = \begin{cases} \mathbf{b}^T & \text{if } n = 0 \\ \frac{1}{n}\tilde{\mathbf{b}}_{n-1}^T \mathbf{A} & \text{if } n > 0 \end{cases} \tag{B.27}$$

This approach is used in TMB when constructing the object `expm_series` using the `sparse_matrix_exponential` library, and it remains computationally efficient even when the number of states is very large.

Bibliography

[1]   H Akaike. "New look at statistical-model identification". In: *IEEE Transactions on Automatic Control* AC19.6 (1974), pp. 716–723. ISSN: 0018-9286.

[2]   Vincent Arel-Bundock. "marginaleffects: Marginal effects, marginal means, predictions, and contrasts". In: *R package version 0.3* 1 (2022).

[3]   Jonathan B. Armstrong et al. "Resource waves: phenological diversity enhances foraging opportunities for mobile consumers". en. In: *Ecology* 97.5 (May 2016), pp. 1099–1112. ISSN: 1939-9170.

[4]   Kevin M. Bailey. *Billion-dollar fish: the untold story of Alaska pollock*. English. Chicago: University of Chicago Press, May 2013. ISBN: 978-0-226-02234-5.

[5]   Haakon Bakka et al. "Non-stationary Gaussian models with physical barriers". en. In: *Spatial Statistics* 29 (Mar. 2019), pp. 268–288. ISSN: 2211-6753.

[6]   Haakon Bakka et al. "Spatial modeling with R-INLA: A review". en. In: *WIREs Computational Statistics* 10.6 (2018), e1443. ISSN: 1939-0068.

[7]   Lewis A. K. Barnett et al. "Realizing the potential of trait-based approaches to advance fisheries science". en. In: *Fish and Fisheries* 20.5 (2019), pp. 1034–1050. ISSN: 1467-2979.

[8]   Douglas Bates, Martin Maechler, and Mikael Jagan. *Matrix: sparse and dense matrix classes and methods*. 2023.

[9]   Douglas Bates et al. "Fitting linear mixed-effects models using lme4". In: *Journal of Statistical Software* 67.1 (2015), pp. 1–48.

[10]   Atilim Gunes Baydin et al. "Automatic differentiation in machine learning: a survey". In: *Journal of Marchine Learning Research* 18 (2018). Publisher: Microtome Publishing, pp. 1–43.

[11]   Raymond J. H. Beverton and Sidney J. Holt. *On the dynamics of exploited fish populations*. London: Chapman & Hall, 1957. ISBN: 1-930665-94-6.

[12]   Carl E. Bock and Terry L. Root. "The Christmas bird count and avian ecology". In: *Studies in Avian Biology* 6 (1981), pp. 17–23.

[13]   Peter Brandt et al. "Annual and semiannual cycle of equatorial Atlantic circulation associated with basin-mode resonance". EN. In: *Journal of Physical Oceanography* 46.10 (Oct. 2016), pp. 3011–3029. ISSN: 0022-3670, 1520-0485.

[14]   J. Roger Bray and J. T. Curtis. "An ordination of the upland forest communities of southern Wisconsin". en. In: *Ecological Monographs* 27.4 (1957), pp. 325–349. ISSN: 1557-7015.

[15]   David R. Brillinger. "Learning a potential function from a trajectory". en. In: *Selected Works of David Brillinger*. Ed. by Peter Guttorp and David Brillinger. Selected Works in Probability and Statistics. Springer, New York, NY, 2012, pp. 361–364. ISBN: 978-1-4614-1343-1.

[16] James H. Brown et al. "Toward a metabolic theory of ecology". In: *Ecology* 85.7 (2004), pp. 1771–1789.

[17] David R. Bryan et al. "Seasonal migratory patterns of Pacific cod (*Gadus macrocephalus*) in the Aleutian Islands". In: *Animal Biotelemetry* 9.1 (July 2021), p. 24. ISSN: 2050-3385.

[18] Kenneth P. Burnham and David Anderson. *Model selection and multi-model inference*. 2nd. New York: Springer, July 2002. ISBN: 0-387-95364-7.

[19] Kenneth P. Burnham and David R. Anderson. "Multimodel inference: understanding AIC and BIC in model selection". en. In: *Sociological Methods & Research* 33.2 (Nov. 2004), pp. 261–304. ISSN: 0049-1241.

[20] Michael T. Burrows et al. "Ocean community warming responses explained by thermal affinities and temperature gradients". en. In: *Nature Climate Change* 9.12 (Dec. 2019), pp. 959–963. ISSN: 1758-6798.

[21] Michael T. Burrows et al. "The pace of shifting climate in marine and terrestrial ecosystems". en. In: *Science* 334.6056 (Nov. 2011), pp. 652–655. ISSN: 0036-8075, 1095-9203.

[22] Gregory S. Butcher et al. "An evaluation of the Christmas bird count for monitoring population trends of selected species". In: *Wildlife Society Bulletin (1973-2006)* 18.2 (1990), pp. 129–134. ISSN: 0091-7648.

[23] Pedro Cardoso, François Rigal, and José C. Carvalho. "BAT – Biodiversity Assessment Tools, an R package for the measurement and estimation of alpha and beta taxon, phylogenetic and functional diversity". en. In: *Methods in Ecology and Evolution* 6.2 (2015), pp. 232–236. ISSN: 2041-210X.

[24] Center for International Earth Science Information Network—CIESIN. *Gridded population of the World, Version 4 (GPWv4): Population density*. Tech. rep. Palisades. NY: NASA Socioeconomic Data and Applications Center (SEDAC), 2016.

[25] Peter Chesson. "Mechanisms of maintenance of species diversity". In: *Annual review of Ecology and Systematics* 31.1 (2000), pp. 343–366.

[26] Megan A. Cimino et al. "Long-term patterns in ecosystem phenology near Palmer Station, Antarctica, from the perspective of the Adélie penguin". en. In: *Ecosphere* 14.2 (2023), e4417. ISSN: 2150-8925.

[27] James S. Clark. "Why species tell more about traits than traits about species: predictive analysis". en. In: *Ecology* 97.8 (Aug. 2016), pp. 1979–1993. ISSN: 1939-9170.

[28] Lucia Clarotto et al. *The SPDE approach for spatio-temporal datasets with advection and diffusion*. arXiv:2208.14015 [math, stat]. Mar. 2023.

[29] Joanne Clavel, Romain Julliard, and Vincent Devictor. "Worldwide decline of specialist species: toward a global functional homogenization?" en. In: *Frontiers in Ecology and the Environment* 9.4 (2011), pp. 222–228. ISSN: 1540-9309.

[30] William G. Cochran. *Sampling Techniques, 3rd Edition*. 3rd. John Wiley & Sons, Jan. 1977. ISBN: 0-471-16240-X.

[31] J. M. Colebrook. "Continuous plankton records-zooplankton and environment, northeast Atlantic and North-Sea, 1948-1975". In: *Oceanologica acta* 1.1 (1978), pp. 9–23.

[32] Josefino C. Comiso and Fumihiko Nishio. "Trends in the sea ice cover using enhanced and compatible AMSR-E, SSM/I, and SMMR data". In: *Journal of Geophysical Research: Oceans* 113.C2 (2008). Publisher: Wiley Online Library.

[33] Richard Condit et al. *[dataset:] Barro Colorado forest census plot data (version 2012)*. Tech. rep. Smithsonian Libraries and Archives, 2012.

[34] Paul B. Conn, James T. Thorson, and Devin S. Johnson. "Confronting preferential sampling when analysing population distributions: diagnosis and model-based triage". en. In: *Methods in Ecology and Evolution* 8.11 (Nov. 2017), pp. 1535–1546. ISSN: 2041-210X.

[35] Henry Chandler Cowles. "The ecological relations of the vegetation on the sand dunes of Lake Michigan. part I.-geographical relations of the dune floras." In: *Botanical gazette* 27.2 (1899), pp. 95–117.

[36] Noel Cressie. *Statistics for Spatial Data*. English. Revised edition. New York: Wiley-Interscience, Sept. 1993. ISBN: 978-0-471-00255-0.

[37] Noel Cressie and Christopher K. Wikle. *Statistics for spatio-temporal data*. Hoboken, New Jersey: John Wiley & Sons, 2011.

[38] Gabor Csardi and Tamas Nepusz. "The igraph software package for complex network research". In: *InterJournal, complex systems* 1695.5 (2006), pp. 1–9.

[39] Andrew Curry. "Wildlife energy: Survival of the fittest". en. In: *Nature* 513.7517 (Sept. 2014), pp. 157–159. ISSN: 1476-4687.

[40] William Deacy et al. "Kodiak brown bears surf the salmon red wave: direct evidence from GPS collared individuals". en. In: *Ecology* 97.5 (2016). _eprint: https://onlinelibrary.wiley.com/doi/pdf/10.1890/15-1060.1, pp. 1091–1098. ISSN: 1939-9170.

[41] Brian Dennis and José Miguel Ponciano. "Density-dependent state-space model for population-abundance data with unequal time intervals". en. In: *Ecology* 95.8 (2014), pp. 2069–2076. ISSN: 1939-9170.

[42] Brian Dennis et al. "Estimating density dependence, process noise, and observation error". In: *Ecological Monographs* 76.3 (2006), pp. 323–341.

[43] Peter Diggle and Paulo Justiniano Ribeiro. *Model-based geostatistics*. en. New York: Springer, May 2007. ISBN: 978-0-387-48536-2.

[44] Daniel F. Doak et al. "Understanding and predicting ecological dynamics: are major surprises inevitable". en. In: *Ecology* 89.4 (Apr. 2008), pp. 952–961. ISSN: 1939-9170.

[45] Theodosius Dobzhansky. *Genetics and the Origin of Species*. en. Columbia University Press, 1982. ISBN: 978-0-231-05475-1.

[46] Theodosius Dobzhansky. "Nothing in biology makes sense except in the light of evolution". en. In: *The American Biology Teacher* 35.3 (Mar. 1973), pp. 125–129. ISSN: 0002-7685.

[47] Carsten F. Dormann et al. "Collinearity: a review of methods to deal with it and a simulation study evaluating their performance". en. In: *Ecography* 36.1 (2013), pp. 27–46. ISSN: 1600-0587.

[48] Carsten F. Dormann et al. "Correlation and process in species distribution models: bridging a dichotomy". en. In: *Journal of Biogeography* 39.12 (2012), pp. 2119–2131. ISSN: 1365-2699.

[49] Marc Dufrêne and Pierre Legendre. "Species assemblages and indicator species: the need for a flexible asymmetrical approach". en. In: *Ecological Monographs* 67.3 (1997), pp. 345–366. ISSN: 1557-7015.

[50] Peter K. Dunn and Gordon K. Smyth. "Randomized quantile residuals". In: *Journal of Computational and Graphical Statistics* 5.3 (1996), pp. 236–244.

[51] Robert R. Dunn et al. "The internal, external and extended microbiomes of hominins". In: *Frontiers in Ecology and Evolution* 8 (2020). ISSN: 2296-701X.

[52] Wade L. Eakle et al. "Wintering bald eagle count trends in the conterminous Cnited States, 1986–2010". In: *Journal of Raptor Research* 49.3 (Sept. 2015), pp. 259–268. ISSN: 0892-1016.

[53] Bradley Efron and Carl N. Morris. "Stein's paradox in statistics". In: *Scientific American* 236.5 (1977), pp. 119–127.

[54] Stephen P. Ellner et al. "An expanded modern coexistence theory for empirical applications". en. In: *Ecology Letters* 22.1 (2019), pp. 3–18. ISSN: 1461-0248.

[55] Felipe Elorrieta, Susana Eyheramendy, and Wilfredo Palma. "Discrete-time autoregressive model for unequally spaced time-series observations". en. In: *Astronomy & Astrophysics* 627 (July 2019), A120. ISSN: 0004-6361, 1432-0746.

[56] Andrew O. Finley. "Comparing spatially-varying coefficients models for analysis of ecological data with non-stationary and anisotropic residual dependence". en. In: *Methods in Ecology and Evolution* 2.2 (2011), pp. 143–154. ISSN: 2041-210X.

[57] R. A. Fisher. "On the mathematical foundations of theoretical statistics". In: *Philosophical Transactions of the Royal Society of London. Series A, Containing Papers of a Mathematical or Physical Character* (1922).

[58] Scott D. Foster and Mark V. Bravington. "A Poisson–Gamma model for analysis of ecological non-negative continuous data". en. In: *Environmental and Ecological Statistics* 20.4 (2013), pp. 533–552. ISSN: 1352-8505, 1573-3009.

[59] John Fox, Zhenghua Nie, and Jarrett Byrnes. *sem: Structural equation models. R package version 3.1-11*. 2020.

[60] Alexa Fredston et al. "Range edges of North American marine species are tracking temperature over decades". en. In: *Global Change Biology* n/a.n/a (2021). ISSN: 1365-2486.

[61] Susanne A. Fritz, Olaf R. P. Bininda-Emonds, and Andy Purvis. "Geographical variation in predictors of mammalian extinction risk: big is bad, but only in the tropics". en. In: *Ecology Letters* 12.6 (2009), pp. 538–549. ISSN: 1461-0248.

[62] R. Froese, D. Pauly, et al. *FishBase 2000: concepts, design and data sources*. Vol. 1594. 2000.

[63] Alan E Gelfand et al. "Spatial modeling with spatially varying coefficient processes". In: *Journal of the American Statistical Association* 98.462 (June 2003), pp. 387–396. ISSN: 0162-1459.

[64] Thomas Giesecke et al. "From early pollen trapping experiments to the Pollen Monitoring Programme". en. In: *Vegetation History and Archaeobotany* 19.4 (Aug. 2010), pp. 247–258. ISSN: 1617-6278.

[65] Paul Gilbert and Ravi Varadhan. *numDeriv: accurate numerical derivatives*. 2019.

[66] Dale D. Goble et al. "Local and national protection of endangered species: an assessment". en. In: *Environmental Science & Policy* 2.1 (Feb. 1999), pp. 43–59. ISSN: 1462-9011.

[67] Benjamin R. Goldstein and Perry de Valpine. "Comparing N-mixture models and GLMMs for relative abundance estimation in a citizen science dataset". en. In: *Scientific Reports* 12.1 (July 2022), p. 12276. ISSN: 2045-2322.

[68] Benjamin Gompertz. "XXIV. On the nature of the function expressive of the law of human mortality, and on a new mode of determining the value of life contingencies. In a letter to Francis Baily, Esq. F. R. S. &c". In: *Philosophical Transactions of the Royal Society of London* 115 (1825). Publisher: Royal Society, pp. 513–583.

[69] M. Goulard and M. Voltz. "Linear coregionalization model: Tools for estimation and choice of cross-variogram matrix". en. In: *Mathematical Geology* 24.3 (Apr. 1992), pp. 269–286. ISSN: 1573-8868.

[70] W. K. Grassmann. "Transient solutions in markovian queueing systems". en. In: *Computers & Operations Research* 4.1 (Jan. 1977), pp. 47–53. ISSN: 0305-0548.

[71] Andreas Griewank and Andrea Walther. *Evaluating derivatives: principles and techniques of algorithmic differentiation.* SIAM, 2008.

[72] Volker Grimm and Steven F. Railsback. *Individual-based modeling and ecology:* Princeton, New Jersey: Princeton University Press, July 2005. ISBN: 0-691-09666-X.

[73] M. Grimmer. "The space-filtering of monthly surface temperature anomaly data in terms of pattern, using empirical orthogonal functions". en. In: *Quarterly Journal of the Royal Meteorological Society* 89.381 (July 1963), pp. 395–408. ISSN: 1477-870X.

[74] Arnaud Grüss and James T. Thorson. "Developing spatio-temporal models using multiple data types for evaluating population trends and habitat usage". en. In: *ICES Journal of Marine Science* 76.6 (Dec. 2019), pp. 1748–1761. ISSN: 1054-3139.

[75] Arnaud Grüss et al. "Synthesis of interannual variability in spatial demographic processes supports the strong influence of cold-pool extent on eastern Bering Sea walleye pollock (Gadus chalcogrammus)". en. In: *Progress in Oceanography* 194 (June 2021), p. 102569. ISSN: 0079-6611.

[76] Nicolás L. Gutiérrez et al. "Eco-label conveys reliable information on fish stock health to seafood consumers". In: *PLoS ONE* 7.8 (Aug. 2012), e43765.

[77] Peter Guttorp and Tilmann Gneiting. "Studies in the history of probability and statistics XLIX On the Matérn correlation family". In: *Biometrika* 93.4 (Dec. 2006), pp. 989–995. ISSN: 0006-3444.

[78] Shannon J. Hackett et al. "A phylogenomic study of birds reveals their evolutionary history". eng. In: *Science (New York, N.Y.)* 320.5884 (June 2008), pp. 1763–1768. ISSN: 1095-9203.

[79] Ephraim M. Hanks, Mevin B. Hooten, and Mat W. Alldredge. "Continuous-time discrete-space models for animal movement". In: *The Annals of Applied Statistics* 9.1 (Mar. 2015), pp. 145–165. ISSN: 1932-6157, 1941-7330.

[80] I. Hanski et al. "Checkerspots as a model system in population biology". In: *On the wings of checkerspots: a model system for population biology. Oxford University Press, Oxford* (2004), pp. 245–263.

[81] Ilkka Hanski et al. "Ecological and genetic basis of metapopulation persistence of the Glanville fritillary butterfly in fragmented landscapes". en. In: *Nature Communications* 8.1 (Feb. 2017), p. 14504. ISSN: 2041-1723.

[82] Achaz von Hardenberg and Alejandro Gonzalez-Voyer. "Disentangling evolutionary cause-effect relationships with phylogenetic confirmatory path analysis". eng. In: *Evolution; International Journal of Organic Evolution* 67.2 (Feb. 2013), pp. 378–387. ISSN: 1558-5646.

[83] Luke J. Harmon. *Phylogenetic comparative methods: learning from trees.* English. 1.0 edition. CreateSpace Independent Publishing Platform, May 2018. ISBN: 978-1-71958-446-3.

[84] Florian Hartig. "DHARMa: residual diagnostics for hierarchical (multi-level/mixed) regression models". In: *R package version 0.1* 5 (2017).

[85] David A. Harville. "Bayesian inference for variance components using only error contrasts". In: *Biometrika* 61.2 (1974), pp. 383–385.

[86] T. J Hastie and R. J Tibshirani. *Generalized additive models.* Chapman & Hall/CRC, 1990.

[87] Trevor Hastie and Robert Tibshirani. "Varying-coefficient models". en. In: *Journal of the Royal Statistical Society: Series B (Methodological)* 55.4 (1993), pp. 757–779. ISSN: 2517-6161.

[88] L. R. Haury, J. A. McGowan, and P. H. Wiebe. "Patterns and processes in the time-space scales of plankton distributions". en. In: *Spatial Pattern in Plankton Communities.* Ed. by John H. Steele. NATO Conference Series. Boston, MA: Springer US, 1978, pp. 277–327. ISBN: 978-1-4899-2195-6.

[89] Elliott L. Hazen et al. "Scales and mechanisms of marine hotspot formation". en. In: *Marine Ecology Progress Series* 487 (July 2013), pp. 177–183. ISSN: 0171-8630, 1616-1599.

[90] Robert J. Hijmans et al. "Package 'terra'". In: *Maintainer: Vienna, Austria* (2022).

[91] Matthew D. Hoffman and Andrew Gelman. "The No-U-Turn sampler: adaptively setting path lengths in Hamiltonian Monte Carlo." In: *J. Mach. Learn. Res.* 15.1 (2014), pp. 1593–1623.

[92] Ary A. Hoffmann and Loren H. Rieseberg. "Revisiting the impact of inversions in evolution: from population genetic markers to drivers of adaptive shifts and speciation?" In: *Annual Review of Ecology, Evolution, and Systematics* 39.1 (2008), pp. 21–42.

[93] Jeffrey Hollister et al. *elevatr: access elevation data from various apis.* 2022.

[94] Alison R. Holt, Kevin J. Gaston, and Fangliang He. "Occupancy-abundance relationships and spatial distribution: A review". en. In: *Basic and Applied Ecology* 3.1 (Jan. 2002), pp. 1–13. ISSN: 1439-1791.

[95] R. D. Holt. "Community modules". In: *Multitrophic interactions in terrestrial ecosystems, 36th Symposium of the British Ecological Society.* Blackwell Science Oxford, 1997, pp. 333–349.

[96] M. B. Hooten and N. T. Hobbs. "A guide to Bayesian model selection for ecologists". en. In: *Ecological Monographs* 85.1 (2015), pp. 3–28. ISSN: 1557-7015.

[97] Mevin B. Hooten and Christopher K. Wikle. "Statistical agent-based models for discrete spatio-temporal systems". In: *Journal of the American Statistical Association* 105.489 (Mar. 2010), pp. 236–248. ISSN: 0162-1459.

[98] Alejandro de Humboldt. *Ansichten der Natur.* 1808.

[99] Christine M. Hunter and Hal Caswell. "Rank and redundancy of multistate mark-recapture models for seabird populations with unobservable states". en. In: *Modeling Demographic Processes In Marked Populations.* Environmental and Ecological Statistics. Springer, Boston, MA, 2009, pp. 797–825. ISBN: 978-0-387-78150-1.

[100] Stuart H. Hurlbert. "Pseudoreplication and the design of ecological field experiments". en. In: *Ecological Monographs* 54.2 (Feb. 1984), pp. 187–211. ISSN: 1557-7015.

[101] Rolf A. Ims, Nigel G. Yoccoz, and Siw T. Killengreen. "Determinants of lemming outbreaks". In: *Proceedings of the National Academy of Sciences* 108.5 (Feb. 2011), pp. 1970–1974.

[102] IUCN Standards and Petitions Subcommittee. *Guidelines for using the IUCN red list categories and criteria, version 15.1*. Tech. rep. Prepared by the Standards and Petitions Committee, 2022.

[103] A. R. Ives et al. "Estimating community stability and ecological interactions from time-series data". In: *Ecological monographs* 73.2 (2003), pp. 301–330.

[104] Anthony R. Ives. "Random errors are neither: On the interpretation of correlated data". en. In: *Methods in Ecology and Evolution* 13.10 (2022), pp. 2092–2105. ISSN: 2041-210X.

[105] Paul Jaccard. "The distribution of the flora in the alpine zone.1". en. In: *New Phytologist* 11.2 (1912), pp. 37–50. ISSN: 1469-8137.

[106] John A. Jacquez and Peter Greif. "Numerical parameter identifiability and estimability: Integrating identifiability, estimability, and optimal sampling design". In: *Mathematical Biosciences* 77.1-2 (1985). Publisher: Elsevier, pp. 201–227.

[107] Walter Jetz and R. Alexander Pyron. "The interplay of past diversification and evolutionary isolation with present imperilment across the amphibian tree of life". en. In: *Nature Ecology & Evolution* 2.5 (May 2018), pp. 850–858. ISSN: 2397-334X.

[108] Walter Jetz et al. "Essential biodiversity variables for mapping and monitoring species populations". en. In: *Nature Ecology & Evolution* 3.4 (Apr. 2019), pp. 539–551. ISSN: 2397-334X.

[109] G. M Jolly. "Explicit estimates from capture-recapture data with both death and immigration-stochastic model". In: *Biometrika* 52.1-2 (1965), p. 225. ISSN: 0006-3444.

[110] Kate E. Jones et al. "PanTHERIA: a species-level database of life history, ecology, and geography of extant and recently extinct mammals". en. In: *Ecology* 90.9 (2009), pp. 2648–2648. ISSN: 1939-9170.

[111] Mikihiko Kai et al. "Predicting the spatio-temporal distributions of pelagic sharks in the western and central North Pacific". en. In: *Fisheries Oceanography* 26.5 (2017), pp. 569–582. ISSN: 1365-2419.

[112] D. Kaplan. "Structural equation modeling". en. In: *International Encyclopedia of the Social & Behavioral Sciences*. Ed. by Neil J. Smelser and Paul B. Baltes. Oxford: Pergamon, Jan. 2001, pp. 15215–15222. ISBN: 978-0-08-043076-8.

[113] Robert E. Kass and Duane Steffey. "Approximate Bayesian inference in conditionally independent hierarchical models (parametric empirical bayes models)". In: *Journal of the American Statistical Association* 84.407 (1989), pp. 717–726. ISSN: 0162-1459.

[114] J. Kattge et al. "TRY: a global database of plant traits". en. In: *Global Change Biology* 17.9 (2011), pp. 2905–2935. ISSN: 1365-2486.

[115] François Keck et al. "phylosignal: an R package to measure, test, and explore the phylogenetic signal". en. In: *Ecology and Evolution* 6.9 (2016), pp. 2774–2780. ISSN: 2045-7758.

[116]   Marc Kéry and Michael Schaub. *Integrated population models: theory and ecological applications with R and JAGS*. English. 1st edition. S.l.: Academic Press, July 2021. ISBN: 978-0-12-820564-8.

[117]   John W. Kidson. "Eigenvector analysis of monthly mean surface data". In: *Monthly Weather Review* 103.3 (Mar. 1975), pp. 177–186. ISSN: 0027-0644.

[118]   Ruben Klein. "A representation theorem on stationary gaussian processes and some local properties". In: *The Annals of Probability* 4.5 (1976), pp. 844–849. ISSN: 0091-1798.

[119]   Jonas Knape and Perry de Valpine. "Are patterns of density dependence in the Global Population Dynamics Database driven by uncertainty about population abundance?" en. In: *Ecology Letters* 15.1 (Jan. 2012), pp. 17–23. ISSN: 1461-0248.

[120]   John R. Krebs. "Optimal foraging: decision rules for predators". In: *Behavioural Ecology: an evolutionaly approach* (1978), pp. 23–63.

[121]   Kasper Kristensen et al. "Estimating spatio-temporal dynamics of size-structured populations". In: *Canadian Journal of Fisheries and Aquatic Sciences* 71.2 (2014), pp. 326–336. ISSN: 0706-652X.

[122]   Kasper Kristensen et al. "TMB: Automatic differentiation and Laplace approximation". In: *Journal of Statistical Software* 70.5 (2016), pp. 1–21.

[123]   Carey E. Kuhn et al. "Test of unmanned surface vehicles to conduct remote focal follow studies of a marine predator". en. In: *Marine Ecology Progress Series* 635 (Feb. 2020), pp. 1–7. ISSN: 0171-8630, 1616-1599.

[124]   Etienne Laliberté and Pierre Legendre. "A distance-based framework for measuring functional diversity from multiple traits". In: *Ecology* 91.1 (2010). Publisher: Wiley Online Library, pp. 299–305.

[125]   Etienne Laliberté et al. "Measuring functional diversity from multiple traits, and other tools for functional ecology". In: *R Package FD* (2014).

[126]   Robert Russell Lauth and J. Conner. *Results of the 2013 eastern Bering Sea continental shelf bottom trawl survey of groundfish and invertebrate resources*. NOAA Technical Memorandum NMFS-AFSC-331. Seattle, WA: Alaska Fisheries Science Center, 2016.

[127]   Pierre Legendre, René Galzin, and Mireille L. Harmelin-Vivien. "Relating behavior to habitat: solutions to the fourth-corner problem". In: *Ecology* 78.2 (Mar. 1997), pp. 547–562. ISSN: 0012-9658.

[128]   M. A. Leibold et al. "The metacommunity concept: a framework for multi-scale community ecology". en. In: *Ecology Letters* 7.7 (2004), pp. 601–613. ISSN: 1461-0248.

[129]   S. A. Levin. "The problem of pattern and scale in ecology: the Robert H. MacArthur award lecture". In: *Ecology* 73.6 (1992), pp. 1943–1967.

[130]   Simon A. Levin and Stephen W. Pacala. "Theories of simplification and scaling of spatially distributed processes". In: *Spatial ecology: the role of space in population dynamics and interspecific interactions*. Ed. by D Tilman and P Kereiva. Princeton, New Jersey: Princeton University Press, 1997, pp. 271–296.

[131]   Qianxiao Li et al. "Spatial feedbacks and the dynamics of savanna and forest". en. In: *Theoretical Ecology* 12.2 (June 2019), pp. 237–262. ISSN: 1874-1746.

[132]   Che-Hung Lin et al. "Moonrise timing is key for synchronized spawning in coral Dipsastraea speciosa". In: *Proceedings of the National Academy of Sciences* 118.34 (Aug. 2021), e2101985118.

[133]   Andreas Lindén and Samu Mäntyniemi. "Using the negative binomial distribution to model overdispersion in ecological count data". en. In: *Ecology* 92.7 (2011), pp. 1414–1421. ISSN: 1939-9170.

[134]   Lindgren. "Continuous domain spatial models in R-INLA". In: *The ISBA Bulletin* 19.4 (2012), pp. 14–20.

[135]   Finn Lindgren. *fmesher: Triangle Meshes and Related Geometry Tools*. 2023.

[136]   Finn Lindgren, Håvard Rue, and Johan Lindström. "An explicit link between Gaussian fields and Gaussian Markov random fields: the stochastic partial differential equation approach". en. In: *Journal of the Royal Statistical Society: Series B (Statistical Methodology)* 73.4 (Sept. 2011), pp. 423–498. ISSN: 1467-9868.

[137]   Finn Lindgren et al. *A diffusion-based spatio-temporal extension of Gaussian Matérn fields*. arXiv:2006.04917 [stat]. Apr. 2023.

[138]   Alfred J. Lotka. "Relation between birth rates and death rates". In: *Science* 26.653 (July 1907), pp. 21–22.

[139]   Olli J. Loukola et al. "Observed fitness may affect niche overlap in competing species via selective social information use." In: *The American Naturalist* 182.4 (Oct. 2013), pp. 474–483. ISSN: 0003-0147.

[140]   Robert MacArthur and Edward O. Wilson. *The theory of island biogeography*. REV - Revised. Princeton University Press, 1967. ISBN: 978-0-691-08836-5.

[141]   Darryl I. MacKenzie et al. *Occupancy estimation and modeling: inferring patterns and dynamics of species occurrence*. Burlington, MA: Academic Press, Dec. 2005. ISBN: 0120887665.

[142]   Martin Maechler, Christophe Dutang, and Vincent Goulet. *expm: matrix exponential, log, 'etc'*. 2023.

[143]   Prasanta Chandra Mahalanobis. "On the generalised distance in statistics". In: *Proceedings of the National Institute of Sciences of India* 2 (1936). Issue: 1, pp. 49–55.

[144]   Nathan J. Mantua et al. "A pacific interdecadal climate oscillation with impacts on salmon production*". In: *Bulletin of the American Meteorological Society* 78.6 (June 1997), pp. 1069–1080. ISSN: 0003-0007.

[145]   Ryan A. Martin et al. "In a nutshell, a reciprocal transplant experiment reveals local adaptation and fitness trade-offs in response to urban evolution in an acorn-dwelling ant". en. In: *Evolution* 75.4 (2021), pp. 876–887. ISSN: 1558-5646.

[146]   Aurore A. Maureaud et al. "Are we ready to track climate-driven shifts in marine species across international boundaries? - A global survey of scientific bottom trawl data". en. In: *Global Change Biology* 27.2 (2021), pp. 220–236. ISSN: 1365-2486.

[147]   Brett T. McClintock et al. "An integrated path for spatial capture–recapture and animal movement modeling". en. In: *Ecology* 103.10 (2022), e03473. ISSN: 1939-9170.

[148]   Bruce McCune, James B. Grace, and Dean L. Urban. *Analysis of ecological communities*. Vol. 28. MjM software design Gleneden Beach, OR, 2002.

[149]   Bruce McCune and Heather T. Root. "Origin of the dust bunny distribution in ecological community data". en. In: *Plant Ecology* 216.5 (May 2015), pp. 645–656. ISSN: 1573-5052.

[150] Brian J. McGill et al. "Rebuilding community ecology from functional traits". en. In: *Trends in Ecology & Evolution* 21.4 (Apr. 2006), pp. 178–185. ISSN: 0169-5347.

[151] Michael C. Melnychuk et al. "Fisheries management impacts on target species status". In: *Proceedings of the National Academy of Sciences* 114.1 (Jan. 2017), pp. 178–183.

[152] Cory Merow et al. "On using integral projection models to generate demographically driven predictions of species' distributions: development and validation using sparse data". en. In: *Ecography* 37.12 (Dec. 2014), pp. 1167–1183. ISSN: 1600-0587.

[153] Cóilín Minto et al. "Productivity dynamics of Atlantic cod". In: *Canadian Journal of Fisheries and Aquatic Sciences* 71.2 (2014), pp. 203–216.

[154] Cole C. Monnahan and Rebecca Haehn. *Assessment of the Flathead Sole-Bering flounder Stock in the Bering Sea and Aleutian Islands*. Tech. rep. Seattle WA: Alaska Fisheries Science Center, 2020.

[155] Cole C. Monnahan and Kasper Kristensen. "No-U-turn sampling for fast Bayesian inference in ADMB and TMB: Introducing the adnuts and tmbstan R packages". en. In: *PLOS ONE* 13.5 (May 2018), e0197954. ISSN: 1932-6203.

[156] Cole C. Monnahan, James T. Thorson, and Trevor A. Branch. "Faster estimation of Bayesian models in ecology using Hamiltonian Monte Carlo". en. In: *Methods in Ecology and Evolution* 8.3 (Mar. 2017), pp. 339–348. ISSN: 2041-210X.

[157] Irini Moustaki and Martin Knott. "Generalized latent trait models". en. In: *Psychometrika* 65.3 (Sept. 2000), pp. 391–411. ISSN: 1860-0980.

[158] Daniel Müllner. "Fastcluster: fast hierarchical, agglomerative clustering routines for R and Python". In: *Journal of Statistical Software* 53.9 (2013), pp. 1–18.

[159] Stephan B. Munch, Tanya L. Rogers, and George Sugihara. "Recent developments in empirical dynamic modelling". en. In: *Methods in Ecology and Evolution* 14.3 (2023). _eprint: https://onlinelibrary.wiley.com/doi/pdf/10.1111/2041-210X.13983, pp. 732–745. ISSN: 2041-210X.

[160] Ran Nathan et al. "A movement ecology paradigm for unifying organismal movement research". en. In: *Proceedings of the National Academy of Sciences* 105.49 (Dec. 2008), pp. 19052–19059. ISSN: 0027-8424, 1091-6490.

[161] Philippe Naveau, Alexis Hannart, and Aurélien Ribes. "Statistical methods for extreme event attribution in climate science". In: *Annual Review of Statistics and Its Application* 7.1 (2020). _eprint: https://doi.org/10.1146/annurev-statistics-031219-041314, pp. 89–110.

[162] John Ashworth Nelder and Robert WM Wedderburn. "Generalized linear models". In: *Journal of the Royal Statistical Society Series A: Statistics in Society* 135.3 (1972), pp. 370–384.

[163] J.K. Nielsen et al. "Geolocation of a demersal fish (Pacific cod) in a high-latitude island chain (Aleutian Islands, Alaska)". In: *Animal Biotelemetry* (2023).

[164] NOAA National Centers for Environmental Information. *ETOPO 2022 15 Arc-Second Global Relief Model*. 2022.

[165] Cecilia A O'Leary et al. "Understanding transboundary stocks' availability by combining multiple fisheries-independent surveys and oceanographic conditions in spatiotemporal models". In: *ICES Journal of Marine Science* 79.4 (May 2022), pp. 1063–1074. ISSN: 1054-3139.

[166] Cecilia A. O'Leary et al. "Adapting to climate-driven distribution shifts using model-based indices and age composition from multiple surveys in the walleye pollock (Gadus chalcogrammus) stock assessment". en. In: *Fisheries Oceanography* 29.6 (2020), pp. 541–557. ISSN: 1365-2419.

[167] Akira Okubo et al. "Some examples of animal diffusion". In: *Diffusion and Ecological Problems: Modern Perspectives* (2001). Publisher: Springer, pp. 170–196.

[168] Aaron Osgood-Zimmerman and Jon Wakefield. "A statistical review of Template Model Builder: a flexible tool for spatial modelling". en. In: *International Statistical Review* 91.2 (2023), pp. 318–342. ISSN: 1751-5823.

[169] Otso Ovaskainen et al. "How to make more out of community data? A conceptual framework and its implementation as models and software". en. In: *Ecology Letters* 20.5 (May 2017), pp. 561–576. ISSN: 1461-0248.

[170] Eric Pante, Benoit Simon-Bouhet, and and Jean-Olivier Irisson. *Marmap: import, plot and analyze bathymetric and topographic data.* May 2020.

[171] Emmanuel Paradis and Klaus Schliep. "ape 5.0: an environment for modern phylogenetics and evolutionary analyses in R". In: *Bioinformatics* 35.3 (Feb. 2019), pp. 526–528. ISSN: 1367-4803.

[172] K. L. Pardieck et al. "North American breeding bird survey dataset 1966-2018, version 2018.0". In: *US Geological Survey, Patuxent Wildlife Research Center, Beltsville, Maryland, USA* (2019).

[173] Judea Pearl. "Causal inference in statistics: An overview". EN. In: *Statistics Surveys* 3 (2009), pp. 96–146. ISSN: 1935-7516.

[174] Judea Pearl. *Causality.* Cambridge, UK: Cambridge University Press, 2009.

[175] Edzer Pebesma. "Simple features for R: standardized support for spatial vector data". en. In: *The R Journal* 10.1 (2018), pp. 439–446. ISSN: 2073-4859.

[176] M. W. Pedersen et al. "Geolocation of North Sea cod (Gadus morhua) using hidden Markov models and behavioural switching". In: *Canadian Journal of Fisheries and Aquatic Sciences* 65.11 (Nov. 2008), pp. 2367–2377. ISSN: 0706-652X.

[177] Johanne Pelletier, Davy Martin, and Catherine Potvin. "REDD+ emissions estimation and reporting: dealing with uncertainty". en. In: *Environmental Research Letters* 8.3 (July 2013), p. 034009. ISSN: 1748-9326.

[178] Caterina Penone et al. "Imputation of missing data in life-history trait datasets: which approach performs the best?" en. In: *Methods in Ecology and Evolution* 5.9 (2014), pp. 961–970. ISSN: 2041-210X.

[179] Charles T. Perretti, Stephan B. Munch, and George Sugihara. "Model-free forecasting outperforms the correct mechanistic model for simulated and experimental data". In: *Proceedings of the National Academy of Sciences* 110.13 (Mar. 2013), pp. 5253–5257.

[180] Laura J. Pollock et al. "Understanding co-occurrence by modelling species simultaneously with a Joint Species Distribution Model (JSDM)". en. In: *Methods in Ecology and Evolution* 5.5 (2014), pp. 397–406. ISSN: 2041-210X.

[181] Marcos O. Prates et al. "Non-separable spatio-temporal models via transformed multivariate gaussian markov random fields". In: *Journal of the Royal Statistical Society Series C: Applied Statistics* 71.5 (Nov. 2022), pp. 1116–1136. ISSN: 0035-9254.

[182]   Haiganoush K. Preisler et al. "Modeling animal movements using stochastic differential equations". en. In: *Environmetrics* 15.7 (2004), pp. 643–657. ISSN: 1099-095X.

[183]   Nicholas D. Pyenson. "The ecological rise of whales chronicled by the fossil record". en. In: *Current Biology* 27.11 (June 2017), R558–R564. ISSN: 0960-9822.

[184]   R Core Team. *R: a language and environment for statistical computing*. Vienna, Austria: R Foundation for Statistical Computing, 2021.

[185]   Daniel L. Rabosky et al. "An inverse latitudinal gradient in speciation rate for marine fishes". en. In: *Nature* 559.7714 (July 2018), pp. 392–395. ISSN: 1476-4687.

[186]   Vishaal Ram and Laura P. Schaposnik. "A modified age-structured SIR model for COVID-19 type viruses". en. In: *Scientific Reports* 11.1 (July 2021), p. 15194. ISSN: 2045-2322.

[187]   C. E. Rasmussen and C. K. I. Williams. *Gaussian processes for machine learning*. 1st ed. Cambridge, MA: MIT press, 2006.

[188]   Paulo Justiniano Ribeiro et al. *geoR: analysis of geostatistical data*. 2022.

[189]   Robert E. Ricklefs. *The economy of nature*. Macmillan, 2008.

[190]   David R. Roberts et al. "Cross-validation strategies for data with temporal, spatial, hierarchical, or phylogenetic structure". en. In: *Ecography* 40.8 (2017), pp. 913–929. ISSN: 1600-0587.

[191]   Duccio Rocchini et al. "Rasterdiv - an information theory tailored R package for measuring ecosystem heterogeneity from space: to the origin and back". In: *Methods in Ecology and Evolution* 12.6 (2021), p. 2195.

[192]   Tanya L. Rogers, Bethany J. Johnson, and Stephan B. Munch. "Chaos is not rare in natural ecosystems". en. In: *Nature Ecology & Evolution* 6.8 (Aug. 2022). Number: 8 Publisher: Nature Publishing Group, pp. 1105–1111. ISSN: 2397-334X.

[193]   Tanya L. Rogers and Stephan B. Munch. "Hidden similarities in the dynamics of a weakly synchronous marine metapopulation". en. In: *Proceedings of the National Academy of Sciences* 117.1 (Jan. 2020), pp. 479–485. ISSN: 0027-8424, 1091-6490.

[194]   Fredrik Ronquist and Isabel Sanmartín. "Phylogenetic methods in biogeography". In: *Annual Review of Ecology, Evolution, and Systematics* 42.1 (2011), pp. 441–464.

[195]   Christopher T Rota et al. "Occupancy estimation and the closure assumption". en. In: *Journal of Applied Ecology* 46.6 (Dec. 2009), pp. 1173–1181. ISSN: 1365-2664.

[196]   Jonathan Roughgarden. "Production functions from ecological populations: a survey with emphasis on spatially implicit models". In: *Spatial ecology: the role of space in population dynamics and interspecific interactions* 30 (1997), pp. 296–317.

[197]   Francisco Rowe and Dani Arribas-Bel. *Spatial modelling for data scientists*. 2022.

[198]   J. Andrew Royle. "N-mixture models for estimating population size from spatially replicated counts". en. In: *Biometrics* 60.1 (2004), pp. 108–115. ISSN: 1541-0420.

[199]   Håvard Rue and Leonhard Held. *Gaussian Markov random fields: theory and applications*. English. 1st edition. CRC Press, 2005. ISBN: 978-1-03-247790-9.

[200]   Håvard Rue, Sara Martino, and Nicolas Chopin. "Approximate Bayesian inference for latent Gaussian models by using integrated nested Laplace approximations". In: *Journal of the Royal Statistical Society: Series B (Statistical Methodology)* 71.2 (2009), pp. 319–392.

[201] Howard L. Sanders. "Marine benthic diversity: a comparative study". In: *The American Naturalist* 102.925 (May 1968). Publisher: The University of Chicago Press, pp. 243–282. ISSN: 0003-0147.

[202] J. R. Sauer et al. *The North American breeding bird survey results and analysis.* Tech. rep. Laurel, MD: Eastern Ecological Science Center, 1997.

[203] G. A.F Seber. "A note on the multiple-recapture census". In: *Biometrika* 52.1-2 (1965), p. 249. ISSN: 0006-3444.

[204] Catherine Sheard et al. "Ecological drivers of global gradients in avian dispersal inferred from wing morphology". en. In: *Nature Communications* 11.1 (May 2020), p. 2463. ISSN: 2041-1723.

[205] Richard B. Sherley et al. "Estimating IUCN Red List population reduction: JARA—A decision-support tool applied to pelagic sharks". en. In: *Conservation Letters* 13.2 (2020), e12688. ISSN: 1755-263X.

[206] Chris Sherlock. "Direct statistical inference for finite Markov jump processes via the matrix exponential". en. In: *Computational Statistics* 36.4 (Dec. 2021), pp. 2863–2887. ISSN: 1613-9658.

[207] Daniel Simpson, Finn Lindgren, and Håvard Rue. "In order to make spatial statistics computationally feasible, we need to forget about the covariance function". In: *Environmetrics* 23.1 (2012), pp. 65–74.

[208] Hans Skaug and Dave Fournier. "Automatic approximation of the marginal likelihood in non-Gaussian hierarchical models". In: *Computational Statistics & Data Analysis* 51.2 (2006), pp. 699–709.

[209] John Gordon Skellam. "Random dispersal in theoretical populations". In: *Biometrika* 38.1/2 (1951), pp. 196–218.

[210] Stephen A. Smith and Joseph W. Brown. "Constructing a broadly inclusive seed plant phylogeny". en. In: *American Journal of Botany* 105.3 (2018), pp. 302–314. ISSN: 1537-2197.

[211] Alan M. Springer et al. "Transhemispheric ecosystem disservices of pink salmon in a Pacific Ocean macrosystem". en. In: *Proceedings of the National Academy of Sciences* 115.22 (May 2018), E5038–E5045. ISSN: 0027-8424, 1091-6490.

[212] Christine C. Stawitz et al. "A state-space approach for detecting growth variation and application to North Pacific groundfish". In: *Canadian Journal of Fisheries and Aquatic Sciences* 72.9 (Apr. 2015), pp. 1316–1328. ISSN: 0706-652X.

[213] Charles Stein. "Inadmissibility of the usual estimator for the mean of a multivariate normal distribution". In: *Proceedings of the Third Berkeley Symposium on Mathematical Statistics and Probability, Volume 1: Contributions to the Theory of Statistics* 3.1 (Jan. 1956), pp. 197–207.

[214] R. William Stein et al. "Global priorities for conserving the evolutionary history of sharks, rays and chimaeras". en. In: *Nature Ecology & Evolution* 2.2 (Feb. 2018), pp. 288–298. ISSN: 2397-334X.

[215] Craig L. Stevens and Gregory A. Lawrence. "Estimation of wind-forced internal seiche amplitudes in lakes and reservoirs, with data from British Columbia, Canada". en. In: *Aquatic Sciences* 59.2 (June 1997), pp. 115–134. ISSN: 1420-9055.

[216] Henry Stommel. "Varieties of oceanographic experience". In: *Science* 139.3555 (1963), pp. 572–576. ISSN: 0036-8075.

[217] Floris Takens. "Detecting strange attractors in turbulence". In: *Dynamical systems and turbulence.* Ed. by D. Rand and Young. Springer, 1981, pp. 366–381.

[218] J.T. Thorson et al. *Comparison of near-bottom fish densities show rapid community and population shifts in Bering and Barents Seas.* Tech. rep. NOAA, 2019.

[219] J.T. Thorson et al. "Spatial delay-difference models for estimating spatiotemporal variation in juvenile production and population abundance". In: *Canadian Journal of Fisheries and Aquatic Sciences* 72.12 (June 2015), pp. 1897–1915. ISSN: 0706-652X.

[220] James T Thorson. "Development and simulation testing for a new approach to density dependence in species distribution models". In: *ICES Journal of Marine Science* 79.1 (Jan. 2022), pp. 117–128. ISSN: 1054-3139.

[221] James T. Thorson. "Guidance for decisions using the Vector Autoregressive Spatio-Temporal (VAST) package in stock, ecosystem, habitat and climate assessments". en. In: *Fisheries Research* 210 (Feb. 2019), pp. 143–161. ISSN: 0165-7836.

[222] James T. Thorson. "Measuring the impact of oceanographic indices on species distribution shifts: The spatially varying effect of cold-pool extent in the eastern Bering Sea". en. In: *Limnology and Oceanography* 64.6 (2019), pp. 2632–2645. ISSN: 1939-5590.

[223] James T. Thorson. "Standardizing compositional data for stock assessment". en. In: *ICES Journal of Marine Science: Journal du Conseil* 71.5 (Aug. 2014), pp. 1117–1128. ISSN: 1054-3139, 1095-9289.

[224] James T. Thorson, Lorenzo Ciannelli, and Michael A. Litzow. "Defining indices of ecosystem variability using biological samples of fish communities: A generalization of empirical orthogonal functions". en. In: *Progress in Oceanography* 181 (Feb. 2020), p. 102244. ISSN: 0079-6611.

[225] James T. Thorson and Melissa A. Haltuch. "Spatiotemporal analysis of compositional data: increased precision and improved workflow using model-based inputs to stock assessment". In: *Canadian Journal of Fisheries and Aquatic Sciences* 76.3 (May 2018), pp. 401–414. ISSN: 0706-652X.

[226] James T. Thorson, Jason Jannot, and Kayleigh Somers. "Using spatio-temporal models of population growth and movement to monitor overlap between human impacts and fish populations". en. In: *Journal of Applied Ecology* 54.2 (Apr. 2017), pp. 577–587. ISSN: 1365-2664.

[227] James T. Thorson and Kasper Kristensen. "Implementing a generic method for bias correction in statistical models using random effects, with spatial and population dynamics examples". In: *Fisheries Research* 175 (Mar. 2016), pp. 66–74. ISSN: 0165-7836.

[228] James T. Thorson, Stephan B. Munch, and Douglas P. Swain. "Estimating partial regulation in spatiotemporal models of community dynamics". en. In: *Ecology* 98.5 (May 2017), pp. 1277–1289. ISSN: 1939-9170.

[229] James T. Thorson, Kotaro Ono, and Stephan B. Munch. "A Bayesian approach to identifying and compensating for model misspecification in population models". In: *Ecology* 95.2 (2014), pp. 329–341. ISSN: 0012-9658.

[230] James T. Thorson, Malin L. Pinsky, and Eric J. Ward. "Model-based inference for estimating shifts in species distribution, area occupied and centre of gravity". en. In: *Methods in Ecology and Evolution* 7.8 (Aug. 2016), pp. 990–1002. ISSN: 2041-210X.

[231] James T. Thorson et al. "Demographic modeling of citizen science data informs habitat preferences and population dynamics of recovering fishes". In: *Ecology* 95.12 (Dec. 2014), pp. 3251–3258. ISSN: 0012-9658.

[232] James T. Thorson et al. "Density-dependent changes in effective area occupied for sea-bottom-associated marine fishes". en. In: *Proc. R. Soc. B* 283.1840 (Oct. 2016), p. 20161853. ISSN: 0962-8452, 1471-2954.

[233] James T. Thorson et al. "Diet analysis using generalized linear models derived from foraging processes using R package mvtweedie". en. In: *Ecology* 103.5 (2022), e3637. ISSN: 1939-9170.

[234] James T. Thorson et al. "Estimating fine-scale movement rates and habitat preferences using multiple data sources". en. In: *Fish and Fisheries* 22.6 (2021), pp. 1359–1376. ISSN: 1467-2979.

[235] James T. Thorson et al. "Geostatistical delta-generalized linear mixed models improve precision for estimated abundance indices for West Coast groundfishes". en. In: *ICES Journal of Marine Science: Journal du Conseil* 72.5 (June 2015), pp. 1297–1310. ISSN: 1054-3139, 1095-9289.

[236] James T. Thorson et al. "Identifying direct and indirect associations among traits by merging phylogenetic comparative methods and structural equation models". en. In: *Methods in Ecology and Evolution* 14.5 (2023), pp. 1259–1275. ISSN: 2041-210X.

[237] James T. Thorson et al. "Joint dynamic species distribution models: a tool for community ordination and spatio-temporal monitoring". en. In: *Global Ecology and Biogeography* 25.9 (Sept. 2016), pp. 1144–1158. ISSN: 1466-8238.

[238] James T. Thorson et al. "Predicting life history parameters for all fishes worldwide". en. In: *Ecological Applications* 27.8 (Dec. 2017), pp. 2262–2276. ISSN: 1939-5582.

[239] James T. Thorson et al. "Spatially varying coefficients can improve parsimony and descriptive power for species distribution models". en. In: *Ecography* 2023.5 (2023), e06510. ISSN: 1600-0587.

[240] James T. Thorson et al. "The importance of spatial models for estimating the strength of density dependence". In: *Ecology* 96.5 (Oct. 2014), pp. 1202–1212. ISSN: 0012-9658.

[241] James T Thorson et al. "Grand challenge for habitat science: stage-structured responses, nonlocal drivers, and mechanistic associations among habitat variables affecting fishery productivity". In: *ICES Journal of Marine Science* fsaa236 (June 2021). ISSN: 1054-3139.

[242] Uffe Høgsbro Thygesen et al. "Validation of ecological state space models using the Laplace approximation". en. In: *Environmental and Ecological Statistics* 24.2 (June 2017), pp. 317–339. ISSN: 1573-3009.

[243] Luke Tierney, Robert E. Kass, and Joseph B. Kadane. "Fully exponential Laplace approximations to expectations and variances of nonpositive functions". In: *Journal of the American Statistical Association* 84.407 (1989), pp. 710–716.

[244] S. Timothy and C. B. Anderson. "From raw air quality data to the nightly news: an overview of how EPA's AIRNow program operates". In: *Sixth Conference on Atmospheric Chemistry*. 2004.

[245] Joseph A. Tobias et al. "AVONET: morphological, ecological and geographical data for all birds". en. In: *Ecology Letters* 25.3 (2022), pp. 581–597. ISSN: 1461-0248.

[246] Vanessa Trijoulet, Gavin Fay, and Timothy J. Miller. "Performance of a state-space multispecies model: What are the consequences of ignoring predation and process errors in stock assessments?" en. In: *Journal of Applied Ecology* 57.1 (2020), pp. 121–135. ISSN: 1365-2664.

[247] Vivitskaia J. D. Tulloch et al. "Future recovery of baleen whales is imperiled by climate change". en. In: *Global Change Biology* 25.4 (2019), pp. 1263–1281. ISSN: 1365-2486.

[248] Florin Vaida and Suzette Blanchard. "Conditional Akaike information for mixed-effects models". In: *Biometrika* 92.2 (2005), pp. 351–370.

[249] Kristin Vanderbilt and Evelyn Gaiser. "The International Long Term Ecological Research Network: a platform for collaboration". en. In: *Ecosphere* 8.2 (2017), e01697. ISSN: 2150-8925.

[250] A. V. Vecchia. "Estimation and model identification for continuous spatial processes". en. In: *Journal of the Royal Statistical Society: Series B (Methodological)* 50.2 (1988), pp. 297–312. ISSN: 2517-6161.

[251] Jay M. Ver Hoef, Ephraim M. Hanks, and Mevin B. Hooten. "On the relationship between conditional (CAR) and simultaneous (SAR) autoregressive models". en. In: *Spatial Statistics* 25 (June 2018), pp. 68–85. ISSN: 2211-6753.

[252] Alexandre M. J.-C. Wadoux and Gerard B. M. Heuvelink. "Uncertainty of spatial averages and totals of natural resource maps". en. In: *Methods in Ecology and Evolution* 14.5 (2023), pp. 1320–1332. ISSN: 2041-210X.

[253] Wenjie Wang and Jun Yan. "Shape-restricted regression splines with R package splines2". en. In: *Journal of Data Science* 19.3 (Aug. 2021). Publisher: School of Statistics, Renmin University of China, pp. 498–517. ISSN: 1680-743X, 1683-8602.

[254] Joe H. Ward Jr. "Hierarchical grouping to optimize an objective function". In: *Journal of the American Statistical Association* 58.301 (1963). Publisher: Taylor & Francis, pp. 236–244.

[255] David I. Warton et al. "Bivariate line-fitting methods for allometry". en. In: *Biological Reviews* 81.2 (May 2006), pp. 259–291. ISSN: 1469-185X.

[256] David I. Warton et al. "So many variables: joint modeling in community ecology". In: *Trends in Ecology & Evolution* 30.12 (Dec. 2015), pp. 766–779. ISSN: 0169-5347.

[257] Taiyun Wei and Viliam Simko. *R package 'corrplot': visualization of a correlation matrix.* 2021.

[258] Robert Edwin Wengert. "A simple automatic derivative evaluation program". In: *Communications of the ACM* 7.8 (1964). Publisher: ACM New York, NY, USA, pp. 463–464.

[259] R. H. Whittaker. "Dominance and diversity in land plant communities numerical relations of species express the importance of competition in community function and evolution". en. In: *Science* 147.3655 (Jan. 1965), pp. 250–260. ISSN: 0036-8075, 1095-9203.

[260] Hadley Wickham. *ggplot2: elegant graphics for data analysis.* en. 2nd ed. Use R! Springer International Publishing, 2016. ISBN: 978-3-319-24275-0.

[261] C. K. Wikle and N. Cressie. "A dimension-reduced approach to space-time Kalman filtering". en. In: *Biometrika* 86.4 (Dec. 1999), pp. 815–829. ISSN: 0006-3444.

[262] Christopher K. Wikle, Andrew Zammit-Mangion, and Noel Cressie. *Spatio-temporal statistics with R*. English. 1 edition. Boca Raton: Chapman and Hall/CRC, Feb. 2019. ISBN: 978-1-138-71113-6.

[263] David L. Wilson. "The analysis of survival (mortality) data: Fitting Gompertz, Weibull, and logistic functions". en. In: *Mechanisms of Ageing and Development* 74.1 (May 1994), pp. 15–33. ISSN: 0047-6374.

[264] Mathieu Woillez, Jacques Rivoirard, and Pierre Petitgas. "Notes on survey-based spatial indicators for monitoring fish populations". en. In: *Aquatic Living Resources* 22.2 (Apr. 2009). Publisher: EDP Sciences, pp. 155–164. ISSN: 0990-7440, 1765-2952.

[265] Joanna M. Wolfe et al. "Convergent adaptation of true crabs (Decapoda: Brachyura) to a gradient of terrestrial environments". In: *bioRxiv* (2022).

[266] S. N Wood. *Generalized additive models: an introduction with R*. 1st ed. Boca Raton, FL: Chapman and Hall/CRC Press, 2006.

[267] Simon N. Wood. "Fast stable restricted maximum likelihood and marginal likelihood estimation of semiparametric generalized linear models". In: *Journal of the Royal Statistical Society Series B: Statistical Methodology* 73.1 (2011). Publisher: Oxford University Press, pp. 3–36.

[268] Simon N. Wood. "Thin plate regression splines". In: *Journal of the Royal Statistical Society Series B: Statistical Methodology* 65.1 (Aug. 2003), pp. 95–114. ISSN: 1369-7412.

[269] Simon N. Wood, Natalya Pya, and Benjamin Säfken. "Smoothing parameter and model selection for general smooth models". In: *Journal of the American Statistical Association* 111.516 (Oct. 2016), pp. 1548–1563. ISSN: 0162-1459.

[270] Sewall Wright. "An analysis of variability in number of digits in an inbred strain of guinea pigs". In: *Genetics* 19.6 (1934), p. 506.

[271] Sewall Wright. "The method of path coefficients". In: *The Annals of Mathematical Statistics* 5.3 (1934). Publisher: Institute of Mathematical Statistics, pp. 161–215. ISSN: 0003-4851.

[272] Yan Zhou et al. "Long-term insect censuses capture progressive loss of ecosystem functioning in East Asia". In: *Science Advances* 9.5 (Feb. 2023), eade9341.

[273] A F Zuur, I D Tuck, and N Bailey. "Dynamic factor analysis to estimate common trends in fisheries time series". In: *Canadian Journal of Fisheries and Aquatic Sciences* 60.5 (May 2003), pp. 542–552. ISSN: 0706-652X, 1205-7533.

[274] Alain F. Zuur et al. "Estimating common trends in multivariate time series using dynamic factor analysis". In: *Environmetrics* 14.7 (2003), pp. 665–685.

[275] Alain F. Zuur et al. *Mixed effects models and extensions in ecology with R*. 1st ed. New York: Springer, Mar. 2009. ISBN: 0-387-87457-7.

Index